Metamaterial Technology and Intelligent Metasurfaces for Wireless Communication Systems

Shilpa Mehta
Auckland University of Technology, New Zealand

Arij Naser Abougreen
University of Tripoli, Libya

A volume in the Advances in Wireless Technologies and Telecommunication (AWTT) Book Series

IGI Global
PUBLISHER of TIMELY KNOWLEDGE

Published in the United States of America by
IGI Global
Information Science Reference (an imprint of IGI Global)
701 E. Chocolate Avenue
Hershey PA, USA 17033
Tel: 717-533-8845
Fax: 717-533-8661
E-mail: cust@igi-global.com
Web site: http://www.igi-global.com

Library of Congress Cataloging-in-Publication Data

Names: Mehta, Shilpa, 1987- editor. | Abougreen, Arij, 1989- editor.
Title: Metamaterial technology and intelligent metasurfaces for wireless communication systems / edited by Shilpa Mehta, Arij Naser Abougreen.
Description: Hershey, PA : Information Science Reference, [2023] | Includes bibliographical references and index. | Summary: "With this book, our aim is to promote metamaterials research and shed light on how metamaterials and metasurfaces will be used in the 5G era and beyond. It will assist the community by contributing to various solutions for 5G and beyond communication systems"-- Provided by publisher.
Identifiers: LCCN 2023011138 (print) | LCCN 2023011139 (ebook) | ISBN 9781668482872 (h/c) | ISBN 9781668482919 (s/c) | ISBN 9781668482889 (eISBN)
Subjects: LCSH: Metamaterials. | Wireless communication systems--Materials.
Classification: LCC TK7871.15.M48 M35 2023 (print) | LCC TK7871.15.M48 (ebook) | DDC 621.382--dc23/eng/20230413
LC record available at https://lccn.loc.gov/2023011138
LC ebook record available at https://lccn.loc.gov/2023011139

This book is published in the IGI Global book series Advances in Wireless Technologies and Telecommunication (AWTT) (ISSN: 2327-3305; eISSN: 2327-3313)

British Cataloguing in Publication Data
A Cataloguing in Publication record for this book is available from the British Library.

All work contributed to this book is new, previously-unpublished material.
The views expressed in this book are those of the authors, but not necessarily of the publisher.

For electronic access to this publication, please contact: eresources@igi-global.com.

Advances in Wireless Technologies and Telecommunication (AWTT) Book Series

ISSN:2327-3305
EISSN:2327-3313

Editor-in-Chief: Xiaoge Xu, University of Nottingham Ningbo China, China

MISSION

The wireless computing industry is constantly evolving, redesigning the ways in which individuals share information. Wireless technology and telecommunication remain one of the most important technologies in business organizations. The utilization of these technologies has enhanced business efficiency by enabling dynamic resources in all aspects of society.

The **Advances in Wireless Technologies and Telecommunication Book Series** aims to provide researchers and academic communities with quality research on the concepts and developments in the wireless technology fields. Developers, engineers, students, research strategists, and IT managers will find this series useful to gain insight into next generation wireless technologies and telecommunication.

COVERAGE

- Mobile Technology
- Digital Communication
- Radio Communication
- Mobile Communications
- Grid Communications
- Wireless Broadband
- Mobile Web Services
- Global Telecommunications
- Network Management
- Telecommunications

IGI Global is currently accepting manuscripts for publication within this series. To submit a proposal for a volume in this series, please contact our Acquisition Editors at Acquisitions@igi-global.com or visit: http://www.igi-global.com/publish/.

The Advances in Wireless Technologies and Telecommunication (AWTT) Book Series (ISSN 2327-3305) is published by IGI Global, 701 E. Chocolate Avenue, Hershey, PA 17033-1240, USA, www.igi-global.com. This series is composed of titles available for purchase individually; each title is edited to be contextually exclusive from any other title within the series. For pricing and ordering information please visit http://www.igi-global.com/book-series/advances-wireless-technologies-telecommunication/73684. Postmaster: Send all address changes to above address.

Titles in this Series

For a list of additional titles in this series, please visit:
http://www.igi-global.com/book-series/advances-wireless-technologies-telecommunication/73684

Opportunities and Challenges of Industrial IoT in 5G and 6G Networks
Poshan Yu (Soochow University, China) Xiaohan Hu (Shanghai University, China) Ajai Prakash (University of Lucknow, India) Nyaribo Wycliffe Misuko (KCA University, Kenya) and Gu Haiyue (Shanghai University, China)
Information Science Reference • © 2023 • 304pp • H/C (ISBN: 9781799892663) • US $250.00

Applications of Artificial Intelligence in Wireless Communication Systems
Karan Kumar (Maharishi Markandeshwar University (Deemed), Mullana, India)
Information Science Reference • © 2023 • 284pp • H/C (ISBN: 9781668473481) • US $250.00

Designing and Developing Innovative Mobile Applications
Debabrata Samanta (Rochester Institute of Technology, Kosovo)
Information Science Reference • © 2023 • 426pp • H/C (ISBN: 9781668485828) • US $225.00

Multidisciplinary Applications of Computer-Mediated Communication
Hung Phu Bui (University of Economics Ho Chi Minh City, Vietnam) and Raghvendra Kumar (GIET University, India)
Information Science Reference • © 2023 • 322pp • H/C (ISBN: 9781668470343) • US $250.00

Role of 6G Wireless Networks in AI and Blockchain-Based Applications
Malaya Dutta Borah (National Institute of Technology, Silchar, India) Steven A. Wright (Georgia State University, USA) Pushpa Singh (GL Bajaj Institute of Technology and Management, Greater Noida, India) and Ganesh Chandra Deka (Ministry of Skill Development and Entrepreneurship, Government of India, India)
Information Science Reference • © 2023 • 285pp • H/C (ISBN: 9781668453766) • US $250.00

IGI Global
PUBLISHER of TIMELY KNOWLEDGE

701 East Chocolate Avenue, Hershey, PA 17033, USA
Tel: 717-533-8845 x100 • Fax: 717-533-8661
E-Mail: cust@igi-global.com • www.igi-global.com

Table of Contents

Detailed Table of Contents

As per demand of high speed and wide bandwidth, there is a wide need to improve the antenna performances and take advantage of a new technology; nevertheless, passive and active devices operating at high frequencies are a big challenge. In the present era of the digital world, demand for IoT-based smart devices has seen having tremendous growth. Basically, these devices involve real-time human-to-machine communication and interaction. Communication of uninterrupted quality depends on the high bandwidth and speed of the internet. The development of 5G wireless network technology is the response to the crucial factors that lead to this demand because of its ability to provide extremely fast internet speed, high bandwidth, high performance, reduced latency, and high reliability and better gain. Metamaterials (LHM) hold great promises for such devices and for many applications like medical, technologies, communications, defense, etc. Metamaterials have interest properties, making them more amenable to transmit and receive in small quantities.

The study of metamaterials has experienced rapid growth, particularly with the discovery of left-handed metamaterials. One promising technology for controlling electromagnetic waves in small structures is transmission-line metamaterial (TL-MTM), which can accurately determine various parameters, such as working frequency, bandwidth, and phase offsets. Antennas are essential components for receiving and transmitting signals in RF systems. Although the microstrip antenna is a popular choice due to its low cost and ease of manufacturing, it is limited in terms of size and frequency range. In this chapter, more than 100 papers are reviewed based on their methodology and performance to discuss various techniques for improving antenna performance, such as the use of artificial neural networks (ANN) and metamaterials. The insights presented in this chapter will aid in identifying, analyzing, and synthesizing the latest developments in the field of metamaterials.

This chapter presents the machine learning (ML) concept for standard RF component design in microwave frequency. It will explain the use of the deep machine learning concept for antenna and other RF components, such as RF filter, and all relevant analysis will be based on the CST simulations. The comparative study of the ML approach and the antenna design tool (such as CST) will be presented in the form of their performance. This chapter will explain the perspectives of ML in RF system design and analysis. This chapter will present the design of the antenna and filter as examples. The simulated results will be obtained by the CST MW Studio and the ML will be implemented in MATLAB.

The chapter explores the development and manufacturing of tailored nonlinear metamaterials and metasurfaces for wireless communication. It covers design and fabrication techniques, material selection and characterization, and design considerations. The chapter also examines simulation and modelling techniques, advanced fabrication methods, integration into wireless communication systems, and characterization and testing processes. It addresses reliability and scalability challenges, the role of nonlinear effects, and the significance of reliability testing and performance characterization. Economic, scalability, and sustainability considerations are discussed, along with case studies and future trends in design and fabrication techniques. The chapter strongly emphasizes the potential of nonlinear metamaterials for enhancing wireless communication and aids as a valuable source for experts in the field.

In this chapter, the authors discuss metamaterial absorbers. The need for active absorbers and also different types of active absorbers have been listed in detail such as switchable absorber/reflector, switchable absorber, and tunable absorbers. The recent trends and the work investigated by researchers around the globe have also been mentioned in brief, which can set a benchmark for further research in the field of metamaterial absorbers. The proposed MTM structure is made up of four-square patches. The structure was designed on an FR4 substrate measuring 11 mm by 11 mm with a thickness of 1.60 mm. Other properties of the substrate include relative permittivity of 4.4 and loss tangent of 0.019. The structure has a full ground plane to minimize transmission. The structure's primary resonance for absorber mode configuration is at 8 GHz. The simulations of the proposed structure were carried out using CST.

The artificially engineered materials termed as metamaterials (MMs) possess unnatural electromagnetic (EM) properties with μ < 0 and ε < 0 due to their geometrical and structural uniqueness. This resulted in diversified study of MMs from the GHz to THz range with one of the potential application as microwave absorber. In this chapter, a case study has been taken up with a unique design of a MM microwave absorber that exhibits multi-band as well as broad-band absorptance within 5GHz - 40GHz. The proposed structure consists of a square ring and a "+" shaped patch at its centre and the scaled "+" shaped patches at four corners. The proposed MM absorber exhibited absorptance peaks of 99.74%, 98.58%, 90.00%, 90.04%, 84.74%, and 94.94% at 6.33 GHz, 14.08 GHz, 28.92 GHz, 32.48 GHz, 33.84 GHz, and 37.69 GHz, respectively, with a broad-band lying within frequency range of 32 GHz to 37 GHz. The MM behavior is studied in terms of normalized matched impedance, effective permeability, and surface current distribution with potential application in stealth technology.

In this chapter, the authors introduce the concept of MMA through reviewing the development of conventional absorbers. To illustrate the important operating mechanism, design process, and functionalization of the metamaterial absorber, they show the absorption relation of the absorber, impedance matching condition, equivalent circuit model, the evaluation of thickness, and oblique incidence condition in the first part. Then, they show the basic design method of the absorber, including parameter sweep method, equivalent circuit method, and deep learning method. Next, the planar integration strategy, vertical stacking strategy, and dispersive material strategy for broadband absorption bandwidth are discussed. In addition, they illustrate the tunable design of absorber, involving lumped element design method, thermal tunable design method, electrical tunable design method, mechanical tunable method, and reconfigurable tunable method. Finally, they talk about the basic engineering application of the metamaterial absorber in practical engineering application.

Elizabeth Caroline Britto, IFET College of Engineering, India
Sathish Kumar Danasegaran, Centre for Advances in Signals Processing and Sensor Networks, Department of ECE, Sri Eshwar College of Engineering, India
K. Sagadevan, IFET College of Engineering, India

Breast cancer is the most prevalent type of cancer causing 1 of 8 cancer diagnoses globally. In the year 2020, there were approximately 2.3 million newly diagnosed cases and 0.685 million deaths from this disease. In order to avoid these, it is mandatory to detect breast cancer in the early stage. In breast cancer detection, terahertz (THz) technology is a promising technology that will not affect the skin due to heavy radiation and offers enhanced breast detection capability. The proposed THz antenna is based on a square-shaped split ring resonator (SRR) which helps to detect the tumor in the breast of the woman. The proposed square SRR antenna resonates at 2 THz with a return loss of -66.55 dB as the maximum and a VSWR of 1.0005. The square SRR structure produced the maximum gain of 8.721 dBi with 8.734 dB directivity. The designed antenna is suitable for biomedical fields such as detecting tumors, body scanner, and various THz applications.

N. Yogeesh, Government First Grade College, Tumkur, India

Metamaterials are a class of artificially engineered materials that exhibit unique physical and electromagnetic properties not found in natural materials. Fuzzy clustering is a machine learning technique that can be used to classify metamaterials based on their physical and electromagnetic characteristics. In this chapter, the authors provide an overview of metamaterial properties and classification challenges and introduce the basics of fuzzy clustering and its application in material classification. They then present a proposed approach for metamaterial classification using fuzzy clustering, along with case studies demonstrating the effectiveness of this approach.

They discuss the potential applications of metamaterial classification and how it can support the development of new metamaterial applications. This chapter provides a valuable resource for researchers and students interested in using fuzzy clustering to classify metamaterials based on their physical and electromagnetic properties.

Nonlinear metamaterials have interesting properties that could change the way technology works. Modelling the complex behaviour of metamaterials is needed to improve their performance and make new gadgets. This chapter looks at fuzzy logic models for complex effects in metamaterials. This part talks about fuzzy sets, membership functions, fuzzy rules, and inference systems. Metamaterials don't behave in a straight way, so there are problems with describing them. Input factors, membership functions, along with fuzzy rules are talked about in the modelling method of fuzzy logic for nonlinear metamaterials. Acoustic and visual metamaterial case studies show how fuzzy logic models can be used. This part also talks about the latest developments, possible uses, and pros and cons of fuzzy logic models in nonlinear metamaterials. This chapter tells scientists and engineers how to use fuzzy logic to model and understand nonlinear metamaterials.

In ambient energy harvesting applications, metamaterials and metasurfaces are useful in localizing and absorbing energy from the surroundings. This chapter reviews past achievements, recent developments and future trends in metamaterial-based radio-frequency (RF) energy harvesters. The chapter begins by introducing metamaterial energy harvesters from a general perspective, covering mechanical, acoustic, and electromagnetic metamaterial energy harvesters. The subsequent section focuses specifically on the working principle of electromagnetic metamaterial energy harvesters in the RF regime. To enhance understanding of the absorption mechanism and performance evaluation, a simulation of an Omega-ring-shaped metamaterial harvester operating at a frequency of 5.8 GHz is demonstrated. The absorption efficiency, harvesting efficiency, and conversion efficiency of the metamaterial harvester are determined and discussed. Furthermore, the state-of-the-art design of metamaterial RF harvesters and their respective performance are reviewed.

The authors of this chapter begin by discussing the fundamentals of metamaterials and metasurfaces before moving on to active and passive metamaterials. This chapter is broken up into six different sections. In the first section, an overview of metamaterials and metasurfaces is provided, and in the second half, active and passive metamaterials and metasurfaces are discussed. This is followed by an explanation of active metamaterials and metasurfaces in part three, fabrication methods for metamaterials and metasurfaces in part four, future directions and challenges in the field of metamaterials in part five, and a conclusion in part six.

Preface

A wireless communication system uses electromagnetic signals to transmit voice and data rather than cables or wires between sending and receiving facilities, thereby bypassing physical cables (Mehta & Li, 2021).

Wireless transmitters first appeared in the early 20th century through radiotelegraphy, which uses Morse code or other coded signals to communicate. Eventually, modulation enabled wireless transmission of music and voice, leading to the creation of radio. Using electromagnetic waves, wireless transmitters can transmit voice, data, video, or signals (Moozakis, 2023).

The evolution of wireless communication can be attributed to Martin Cooper. A wireless data transmission device was invented by him, and is considered to be the first generation of mobile communications. First generation (1G) technology was introduced in the 1980s and represents the beginning of today's technology. This generation could only support voice calls and data transmission using analog signals (Mehta et al., 2021). Thus due to low voice quality and short battery life an upgrade is needed. This led to the emergence of second generation (2G). During this generation, Global System for Mobile Communication (GSM) was introduced. It was not only possible to make voice calls but also send text messages. Additionally, digital communication approach was followed instead of analog communication (Mehta, 2022). However, strong network coverage is required for data transmission and video transmission is not possible thorugh messaging. Thus, to improve data transmission rate upgraded technologies were introduced such as 2.5 and 2.75 generations which covers General Packet Radio Service (GPRS), Enhanced Data Rates for GSM(EDGE) and Code Division Multiple Access (CDMA) respectively. Email transmission and Internet browsing were also possible during this generation. Additionally, cameras were introduced in cellular phones. To overcome the problems faced during earlier generations a new generation was introduced which was referred to as third generation (3G). This generation made use of Universal Mobile Telecommunication Systems (UMTS) technology which is one of the core network technolgoy for 3G. This network was capable of handling 2G and some advanced features leading to faster data rate delivery. During this generation it was possible to perform web browsing, emails, video downloading, image sharing etc. The main disadvantage of 3G phones were not backward generation incompatible. Later 3G evolved into upgraded versions such as 3.5G and 3.7G. Later another advanced generation evolved which was termed as fourth generation (4G). With IP-based voice, data, multimedia, and internet services delivered over IP, its main objective was to provide high-speed, robust, and comprehensive connectivity while increasing security measures and reducing telecommunications expenses. 4G could handle many applications, including better mobile web access, IP telephony, and immersive gaming experiences. Other applications include high-definition mobile television, seamless video conferencing, captivating 3D television, and efficient cloud computing. Among the key technologies involved in 4G were MIMO (Multiple Input Multiple Output) and OFDM (Orthogonal Frequency Division Multiplexing). 4G networks offered significant speeds at the cost of latency. With the advent of 4G, features like game downloads and seamless HD television streaming became possible. 4.5 G and 4.9G were the upgraded versions of 4G offering some advanced features and later paved the way for next advanced generation i.e. fifth generation (5G). 4G was upgraded to 4.5G and 4.9G which offered some advanced features and eventually paved the way for 5G (Net-Informations, 2023).

The field of mobile communications experiences a significant transformation approximately every decade. However, the time span between these transformations, known as "G's," is gradually decreasing. As 5G networks are becoming commercially available, there is already substantial anticipation for the development of systems beyond 5G leading to sixth generation (6G) (Tataria et al., 2021). Research on 6G of wireless technology has already commenced in both academic and industrial sectors since 2019, marking it as the starting point for 6G research. It is expected that this technology will be commercially available for deployment around 2030, following the established pattern of introducing a new generation of wireless technology every decade. At present, starting from 2020, telecommunication operators worldwide have begun implementing 5G technology in their networks. Several research groups are actively engaged in studying and exploring the concepts and technologies associated with 6G, resulting in numerous conceptual papers. The potential benefits that 6G will bring to society are yet to be fully understood. However, it is certain that machine learning and artificial intelligence will play pivotal roles in enabling smart cities within the 6G technology landscape. These innovative applications have the potential to enhance the quality of life for citizens by monitoring, analyzing, planning, and executing various integrated components (Kaur & Khan, 2022).

Preface

Intelligent or innovative materials are required for modern wireless communication systems. Because of their exceptional material characteristics, metamaterials can be integrated into high-quality measurements, biomedical applications, remote and near-field sensing, electronic appliances, to meet the demand of the electromagnetic behavior of smart wireless devices. Additionally, the application range of wireless communication systems is expanded by the selectivity of scattering parameters and adaptability of wavelength-based dimensions in metamaterial designs. To resolve crucial operations in cellular, aerospace, and space communication systems, electromagnetic-wave-compatible and high-frequency-interfacing devices can be used. Along with having a high Q factor, tunability based on scattering parameters is appropriate for electromagnetic radiation mechanisms because it allows for some flexibility and interference-free signal transmission and reception for communication modules. As a result of synchronizing with the sustainable development of upcoming smart systems, traditional communication methods are currently becoming less popular. The progress made in metamaterials has led to the emergence of a promising technology called Reconfigurable Intelligent Surface (RIS) for future 6G wireless communications. RISs offer significant advantages such as high array gain, cost-effectiveness, and low power consumption. As a result, they are anticipated to significantly expand signal coverage, enhance system capacity, and improve energy efficiency (Zhang & Dai, 2023).

Metamaterials are a rapidly developing and groundbreaking field of research. They are artificial structures made up of sub-wavelength elements that possess unique and beneficial properties not found in natural materials. Electromagnetic metamaterials are capable of manipulating electromagnetic waves in unconventional ways, resulting in a range of exotic phenomena and devices. Metamaterials have gained attention in the acoustic and visible regions. Recent research has demonstrated that metamaterials can improve various antenna properties such as bandwidth, gain, and size. Metamaterials have also shown significant potential in holography, anomalous reflection, subwavelength focusing, perfect absorber, cloaking, 5G communication, Internet of things (IoT), and more (Tian et al., 2022).

Metasurfaces are gaining a lot of attention because they are highly effective at modifying electromagnetic waves across a broad range of frequencies, from microwave to visible light. In recent times, programmable metasurfaces have been able to manipulate the amplitude or phase of electromagnetic waves in real-time, making them particularly attractive for wireless communication applications (Tang et al., 2019).

Metasurfaces have been utilized to create innovative and highly efficient wireless transceiver systems, leading to a significant change in how transceiver designs are approached and a decrease in hardware costs for wireless communication systems. Among these systems, those based on microwave metasurfaces are particularly intriguing. This is because microwave frequencies are the most ideal for wireless communication, owing to their relatively low path loss, ability to operate without a direct line of sight, and the small size of the transceiver components (such as antennas and transmission lines). Electromagnetic metasurfaces are arrays consisting of unit cells that are carefully designed to provide properties not found in natural materials. Reconfigurable metasurfaces are extremely thin in thickness and relatively large in their transverse size, and they are composed of scattering unit cells. The specific arrangement and bias of the unit cells determine how the metasurface transforms the incident wave into desired reflected and transmitted electromagnetic waves that can be reciprocal or nonreciprocal. Metasurfaces are also capable of creating new frequencies and can be loaded with active components such as diodes or transistors that can be biased through direct current (DC) and radio-frequency (RF) modulation signal waves. The DC biasing signal is used to change the diode's operating point, while the RF biasing signal provides a time modulation signal for temporal or spatiotemporal modulation of the effective permittivity or conductivity. Both DC and RF biasing signals can be used to control the unit cells' characteristics and response and the metasurface's operation and functionality. A computer processor, possibly running an artificial intelligence application or a field-programmable gate array, can be used for real-time full control of the metasurface. To enable effective communication between different parts of a network (including small-cell base stations, macro base stations, and satellites), transmissive and reflective metasurfaces can be installed on or inside buildings. These metasurfaces can provide a range of functionalities, such as active beamsteering, signal amplification, frequency conversion, channel allocation, and coding. The reflective and transmissive metasurfaces used in communication links may have different characteristics and functionalities, such as beamsteering, signal boosting, frequency conversion, real-time signal coding, multiple access communication, and nonreciprocal transmission and reflection. The specific functionality required for each metasurface is determined by the system's core and can be controlled and programmed using software (Ataloglou et al., 2023).

This book's objective is to discuss recent developments in metamaterials and metasurfaces for wireless applications. Numerous experts have contributed research and review chapters to this book. This book will serve as a valuable resource that provides information about the current state of the field and inspires ideas for future challenges by offering the most recent insights and thoughts of top researchers on significant topics in this emerging discipline.

The book provides balanced research and practical works that will interest readers.

There are 12 chapters in the book. Readers are free to read chapters that particularly interest them without following the order in which they appear, as all chapters are independent.

Brief summaries of each chapter of this book are provided below.

Chapter 1

This chapter presents the concept of metamaterials and discusses in details the design of fifth generation antenna.

Chapter 2

This chapter provides insight into recent advancements in miniaturized patch antennas and the application of Tx-line metamaterials to tackle complex design challenges in compact antenna development.

Chapter 3

This chapter explains the prospect of dynamic metasurface antenna and its complexity in envisioning potential applications under 5G and 6G technologies for various modalities, including wireless sensing and communication.

Chapter 4

This chapter explores the methods and processes involved in creating and constructing nonlinear metamaterials and metasurfaces. It specifically emphasizes their usage in the field of wireless communication.

Chapter 5

The chapter has been devoted to an in-depth exploration of metamaterial antennas and their various applications. The primary focal point of this chapter resides in the comprehensive analysis of the metamaterial absorber. Additionally, a thorough examination of the meta-surface absorber, accompanied by the inclusion of pertinent mathematical equations, has been undertaken.

Chapter 6

This chapter delves into the investigation of a distinctive structure of a metamaterial absorber through a detailed case study.

Chapter 7

The primary objective of this chapter is to concentrate on the cutting-edge development of metamaterial antennas (MMAs), meticulously examining their operational principles, design approaches, broadband characteristics, tunable configurations, and engineering applications. The comprehensive analysis and insights presented in this chapter serve as a valuable point of reference for meeting the practical requirements of engineering design.

Chapter 8

In this chapter, the THz antenna was simulated for various widths and thicknesses of the square ring. The suggested metamaterial-based square SRR antenna is utilized to detect cancers earlier and more accurately.

Chapter 9

In this chapter, it has been demonstrated that fuzzy clustering emerges as a powerful technique for the classification of metamaterials, with particular emphasis placed on their physical and electromagnetic attributes. The effectiveness of the proposed approach, encompassing metamaterial classification through fuzzy clustering, has been showcased in terms of the identification of diverse metamaterial types and the revelation of their functional characteristics, thereby shedding light on their prospective applications. The challenges associated with metamaterial classification have been underscored in this chapter, along with the demonstration of how fuzzy clustering can serve as a means to surmount these challenges.

Chapter 10

This chapter provides guidance to scientists and engineers on the application of fuzzy logic for the modeling and comprehension of nonlinear metamaterials.

Chapter 11

This chapter presents a comprehensive review of the utilization of metamaterials and metasurfaces in energy harvesting, with a particular emphasis on the realm of RF energy harvesting.

Chapter 12

This chapter provides an overview of active and passive metamaterials and metasurfaces, followed by fabrication approaches. It also focuses on challenges and future directions while working with metamaterials.

Shilpa Mehta
Auckland University of Technology, New Zealand

Arij Naser Abougreen
University of Tripoli, Libya

REFERENCES

Ataloglou, V.G., Taravati, S., & Eleftheriades, G.V. (2023). Metasurfaces: physics and applications in wireless communications. *National Science Review,* 10(8). 10.1093/nsr/nwad164

Kaur, J., & Khan, M. A. (2022). Sixth Generation (6G) Wireless Technology: An Overview, Vision, Challenges and Use Cases. *2022 IEEE Region 10 Symposium (TENSYMP),* 1-6. doi: 10.1109/TENSYMP54529.2022.9864388

Mehta, S. (2022). *Reconfigurable Mixer Design for Software-Defined Radios* [Doctoral dissertation]. Auckland University of Technology.

Mehta, S., & Li, X. J. (2021). Design of an Active Inductor based Reconfigurable Mixer for Software-Defined Radios. *2021 International Symposium on Networks, Computers and Communications (ISNCC),* 1-6. 10.1109/ISNCC52172.2021.9615861

Mehta, S, Li, XJ, Donelli, M. (2021). Recent advancement in the design of mixers for software-defined radios. *Int J RF Microw Comput Aided Eng., 32*(2), e22963.

Moozakis, C. (2023). *Mobile Computing*. Retrieved from https://www.techtarget.com/searchmobilecomputing/definition/wireless

Net-Informations. (2023). *1G Vs. 2G Vs. 3G Vs. 4G Vs. 5G*. Retrieved from https://net-informations.com/q/diff/generations.html

Pranavnath. (2023). *Generations of Wireless Communication*. Retrieved from https://www.tutorialspoint.com/generations-of-wireless-communication

Tang, W., Li, X., Dai, J. Y., Jin, S., Zeng, Y., Cheng, Q., & Cui, T. J. (2019). Wireless Communications with Programmable Metasurface: Transceiver Design and Experimental Results. *China Communications*, *16*(5), 46–61. doi:10.23919/j.cc.2019.05.004

Tataria, H., Shafi, M., Molisch, A. F., Dohler, M., Sjöland, H., & Tufvesson, F. (2021). 6G Wireless Systems: Vision, Requirements, Challenges, Insights, and Opportunities. *Proceedings of the IEEE, 109*(7), 1166–1199. doi:10.1109/JPROC.2021.3061701

Tian, S., Zhang, X., Wang, X., Han, J., & Li, L. (2022). Recent advances in metamaterials for simultaneous wireless information and power transmission. *Nanophotonics*, *11*(9), 1697–1723. doi:10.1515/nanoph-2021-0657

Zhang, Z., & Dai, L. (2023). Reconfigurable Intelligent Surfaces for 6G: Nine Fundamental Issues and One Critical Problem. *Tsinghua Science and Technology, 28*(5), 929-939.

Chapter 1
Metamaterials-Based Antenna for 5G and Beyond

Smrity Dwivedi
IIT BHU, India

ABSTRACT

As per demand of high speed and wide bandwidth, there is a wide need to improve the antenna performances and take advantage of a new technology; nevertheless, passive and active devices operating at high frequencies are a big challenge. In the present era of the digital world, demand for IoT-based smart devices has seen having tremendous growth. Basically, these devices involve real-time human-to-machine communication and interaction. Communication of uninterrupted quality depends on the high bandwidth and speed of the internet. The development of 5G wireless network technology is the response to the crucial factors that lead to this demand because of its ability to provide extremely fast internet speed, high bandwidth, high performance, reduced latency, and high reliability and better gain. Metamaterials (LHM) hold great promises for such devices and for many applications like medical, technologies, communications, defense, etc. Metamaterials have interest properties, making them more amenable to transmit and receive in small quantities.

INTRODUCTION

In todays era, the innovation to 5G wireless technology enabled IoT devices at a millimetre-wave band will revolutionize the electronics and telecommunication field (John Colaco et al. 2021). This and technologies of wireless communications is one of the biggest human achievements in terms of 5G and IoT technology. Now, 5G technology and IoT technology thinking will become reality for smart devices

DOI: 10.4018/978-1-6684-8287-2.ch001

(A. Helena et al. 2020). It could be a big dream come true and a game-changer for humans particularly in the field of education, health, agriculture, autonomous vehicles, industry and many other related applications. Hence, in this proposed chapter, author has elaborated a design and analysis with specifications of metamaterials based antenna for 5G and beyond technologies for various 5G enabled advanced IoT Devices. IoT-based smart devices are usually expanding their support for Internet access (broadband) beyond devices, such as smart/mobile phones, laptops, and so on. The IoT based smart devices have been amalgamated with cutting-edge technology to manage and communicate smoothly using 5G and beyond wireless technologies. 5G wireless technologies grasps an important role in setting up the platform for real-time communications of IoT devices due to its potential to offer greater bandwidth with millimetre bands compared with 4G or 3G wireless technology previously. 5G wireless network technology combines the spectrum and access networks to meet the customer's capacity and coverage needs to fulfill the demand (G. GSMA 2019). The internet of everything specifies the synchronisation of various smart electronic devices such as tablets, smartphones, laptops, multiple machines such as smart vehicles equipped with sensors with IoT communication, and wireless or wired connection of consumer appliances connected through the internet (G. GSMA 2019) in present era. 5G technology will provide massive support for IoT devices as a useful part of the human digital world as the digital population is growing more than the human population as per need. The latest 5G enabled IoT smart device's trends are extending to sensor-based IoT competencies to actuators, robots, and drones for distributed synchronization system. The biggest and crucial challenge facing 5G enabled IoT devices is cyber security as hackers proliferate and infiltrate and target the server's proxy. Hence secured communication of a machine to a machine has a dynamic role in emerging IoT smart devices (J W. Ejaz et al. 2016). During the COVID-19 epidemic, this advancement would also allow real-time video and audio quality data for patient data analysis and wireless uninterrupted communication of detected health parameters. In this chapter, few designs have been given for antennas for effective multi-band operation useful for 5G based IoT devices and applications. In this chapter, antenna loaded the metamaterial SRR structure on the microstrip patch to boost the performance has discussed. The performance is then analysed in comparison with and without loading metamaterial SRR structure.

CONCEPT OF METAMATERIAL

There has been a great deal of attention given in the field of metamaterials over a past few years (C. L. Holloway et al. 2009; V. G. Veselago 1968). Metamaterials are fully synthetic materials developed to get unique properties not normally found in nature. It was artificial dielectric in terms of electromagnetics. Appropriately designed metamaterials can affect electromagnetic radiation or sound, which is not found in bulk materials. Basically, metamaterials are divided into two major categories, resonant and non resonant types, depending upon the oscillation of waves as well as periodicity of structures. Resonant types of metamaterials have well specific permittivity and permeability, which further divide into sub categories. Double negative (DNG), negative permittivity (ENG), negative permeability (MNG) and negative refractive index types are widely used resonant type metamterials, whereas anisotropic and hyperbolic are the non resonant type metamaterials, which have certain specified bands for electromagnetic waves. Today's, metamaterials based structures are more commonly used in waveguides, antennas design, filters design and lots of other applications. Left handed metmaterials (LHM), which has negative permittivity and negative permeability is also known as DNG materials, are used for design of coplanar waveguides, filters (low pass, high pass) and microstrip lines Antennas (D. R. Smith et al. 2000; C. L. Holloway et al. 2003). LHM based microstrip antennas have high efficiency, low profile and high bandwidth, which is broadly usable in wide band applications. Simple microstrip antennas have radiation not only from patch, but also from substrate material that supports patch and ground plane, which is basically a surface waves, further creates radiation losses. So, the antenna has low bandwidth, low efficiency, poor directivity as well as higher losses (E. Shamonina et al. 2007; D. V. Sivukhin 1957). These losses can only be reduced by keeping the values of permittivity and permeability as low as possible. On the other hand, non resonant type metamaterials such as photonic band gap (PBG) or electromagnetic band gap structures (EBG), plays an important role to achieve higher bandwidth with greatest potential applicability in broadband communication systems. These PBG structures have smaller size, which varies in sub wavelengths as well as have a greatest tolerance value to resist the structural deformations. PBG structures have magical properties than resonant metamaterials with periodicity ~λ/4, whereas resonant metamaterials have periodicity ~λ/10. For advance communication, PBG structure is more valid for high bandwidth and can be applicable with antenna design in mm-wave technology. Plane waves with photons enter into the band gap (stop band), coming after finding the dispersion diagram for getting the cut off frequency with phase constant. Metamaterials is also used for artificial magnetic conductors (R. A. Silin et al. 2001). Negative refractive index metamaterials is used for incident waves which have propagation faster than light propagation and for angle insensitive devices. According to national research in metamaterials, lots of works have been going on over a decades, in which the most the works are dedicated to microwave components and antennas using metamaterials for frequency less than 10 GHz. Although metamaterials are not a new topic, but for the research and development point of view, works are still left.

A Metamaterial is a material that is artificially engineered to acquire unconventional electromagnetic, thermal, acoustic, and mechanical properties that are not found in other naturally occurring materials because of its extraordinary unusual and unique kind of properties such as negative permeability and negative permittivity. When the values of permeability and negative permittivity both are simultaneously negative, then the electric field, magnetic field, and the propagation vector form left-handed medium materials in a double-negative region and wave propagation are moving curiously in reverse signifying material has negative refractive index (L. H. Wen et al. 2018). Different studies and researches are being carried out by scientists and researchers all over the world that will benefit humanity through the use of this metamaterial. The metamaterial electromagnetic properties are the best defined by famous maxwells equations given as (D. Yanget al. 2017). The preferences of using metamaterials in the design of the microstrip antenna are to minimize the size and increase other parameters such as bandwidth, gain, and return loss as well as efficiency (C. Wang et al. 2018). This will enable to apply and adapt concepts to innovations in microstrip antenna designs from laboratories for practical applications of engineering with a good contract (C. Wang et al. 2018). Since the permeability negative material, the unit cell split-ring resonators are the most used structure used as metamaterials. The metamaterial has various applications such as smart antenna, medical devices, optical devices, smart sensors, smart IoT devices, smart solar power, radomes, optical lenses, invisible submarines, radar, etc. The metamaterial structures have significant roles in the applications given above for boosting their performance in terms of power or energy harvesting, bandwidth, and gain enhancement, squeezing the size of the devices for optimum performance and other benefits. The design for microstrip patch antenna in the various field of electronics, communication, information technology, and electrical has enormously increased owing to its high characteristics such as bandwidth, return loss, etc. This chapter aims at providing a multi-frequency band and improvement in the performance of the previously proposed antenna applicable for various 5G enabled IoT smart Devices for interrupted quality services. This chapter, therefore, provides all the necessities required to fulfil the demand for next-generation IoT devices.

METAMATERIAL ANTENNAS

Antennas are one of the important components now a days that makes wireless communications possible in todays era. The work of antennas is to interface the radio system with the external environment (such as LNA, oscillators, receiver etc.). Wireless communication systems require antennas at the both sides like transmitter and receiver to operate properly without causing disturbance. The concept of metamaterials has been widely applied in the design of microwave, millimetre-wave and terahertz devices and antennas now a days to make communications and other important task wisely. With the rapid development of flexible portable devices such as mobile phones, laptops, note pads, watches, wearable devices, etc., antennas with different changing functions, based on variable structures, are in demand for next generation of wireless communication systems such as 6G and beyond. The application of metamaterials related components over the past decade has achieved great successes in the fields of both science and engineering. Variable metamaterials have been designed from radio frequencies up to optical frequencies (terahertz and infrared etc.), and different functions have been realized, e.g., negative refractive index (NRI), anisotropy and bianisotropy (Lapine et al. 2007). As an interdisciplinary topic that is being used everywhere, metamaterials can be classified into different categories based on different criteria as given different sections. From an operating frequency point of view, they can be classified as microwave metamaterials, terahertz metamaterials, and photonic metamaterials for antenna, polarizers, absorber etc. From a spatial arrangement point of view, there are 1D metamaterials, 2D metamaterials, and 3D metamaterials present. From a material point of view, there are metallic and dielectric metamaterials that is being utilized. One of the most important applications of metamaterials is antenna design that is being discussed here. Due to the unusual properties of metamaterials, we can achieve antennas with novel characteristics which cannot be realized with traditional materials like normal material with positive refractive index. Usually, antennas consist of a combination of conductors, dielectrics and other conventional materials that have certain geometry during design. That design follows either traditional analytical methods and rules that depends on experience. In between, artificial intelligence and powerful optimization techniques used by full-wave simulations (Simulation software) are also utilized to obtain the highest possible performance by fine tuning of the structure's parameters (M. O. Akinsolu et al. 2020). It is fact that the antenna design is primarily focused on determining the optimized geometrical shape of conventional materials and then it is bound by the material characteristics for every material. In an effort to overcome all limitation in normal or conventional material, metamaterials with their unique properties have gained focus in antenna design. Irrespective of the fact that metamaterials are microscopically composed of conventional materials such as conductors and dielectrics, their macroscopic characteristics are completely different due to their smart shapes like SRR etc. The realisation of negative constitutive parameters in the microwave regime is applicable with metamaterials, while their subwavelength size (very small dimension) is another useful feature. As a result, the construction of metamaterials into antennas can offer advanced flexibility and enable novel design strategies that is followed. Therefore, the investigation of the benefits of metamaterial inspired antennas and their capability of having an active role in modern wireless communications is of great significance in advance communication.

State of Art (Theoretical background):- As already discussed above about the metamaterial, these are engineered, man made materials that are able of changing the electromagnetic waves in a different way than conventional, natural materials occurring in nature. For instance, structures that exhibit negative permittivity and permeability are most common examples of metamaterials (A. Alu et al. 2007). Three main metamaterial categories can be identified as per finding: i) SNG (Single-Negative) metamaterials that possess either negative permittivity or negative permeability, ii) DNG (Double-Negative) metamaterials that simultaneously have negative permittivity and permeability both and iii) ZIM (zero-index materials) that have either zero permittivity or zero permeability any one. Also, EBG (Electromagnetic Bandgaps) that prohibits the propagation of EM waves and AMC (Artificial Magnetic Conductors) that have zero magnetic field are also usually regarded as metamaterials. In 1968 as discussed, Veselago was the first one to theoretically study the electrodynamics of DNG media and given their interesting properties such as reversal of the Doppler Effect, negative index of refraction and left-handed propagation. While the conventional materials support only the typical forward-wave propagation. The wave in SNG media is evanescent since the propagation constant is real ($\gamma \in R$). Left-handed or DNG media support backward wave propagation, where the phase velocity and group velocity are opposite to each other and the wave travels anti parallel to the power flux. Some important structures are discussed below using diagram.

The demonstration and effects of negative permittivity in the microwave regime is possible with a periodic arrangement of thin metallic wires of diameter α and periodicity p as shown in Figure 1 and was initially proposed by Sir John Pendry and his group.

Figure 1. Periodic arrangement of thin conductive wires for the realisation of negative permittivity
Source: Milius et al. (2021)

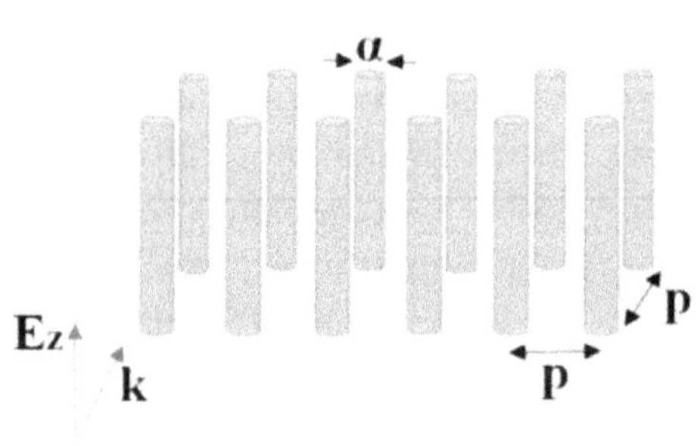

As shown in Figure 2, the real part of the effective permittivity is negative below the plasma frequency and the medium can be characterized as ENG.

Figure 2. Real part of the effective permittivity of periodically arranged infinitely long thin wires
Source: Milius et al. (2021).

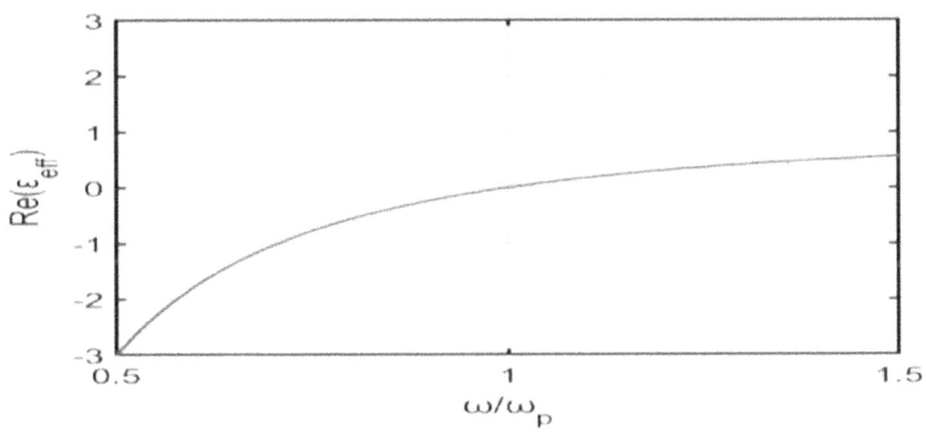

It is the fact that there are no materials in nature with negative permeability. Subsequently, it was Pendry again who proposed a Split Ring Resonator (SRR) as depicted in Figure 3, which shows a negative permeability behaviour around its

Figure 3. Geometry of a square split ring resonator unit cell. Blue parts depict metallization.
Source: Milius et al. (2021)

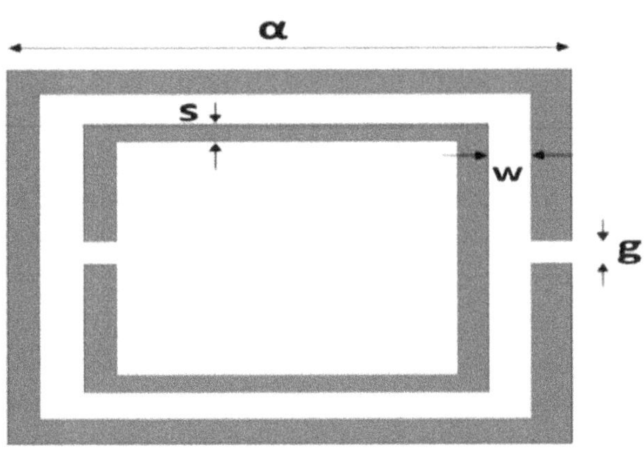

resonance frequency. Under the presence of an external time-varying magnetic field perpendicular to the SRR's surface, currents are induced on both the conductive inner and outer rings and charges are accumulated across the gaps between them when it is excited.

As Figure 4 shows, the SRR exhibits a narrow-band resonance where its real permeability is negative. It should be mentioned that a single SRR or thin wire cell is equivalent to an atom in conventional materials. Hence, in order to create an effective macroscopic ENG or MNG behaviour, a number of periodically arranged thin wires or SRRs is required.

Figure 4. Real part of the SRR's effective permeability
Source: Milius et al. (2021)

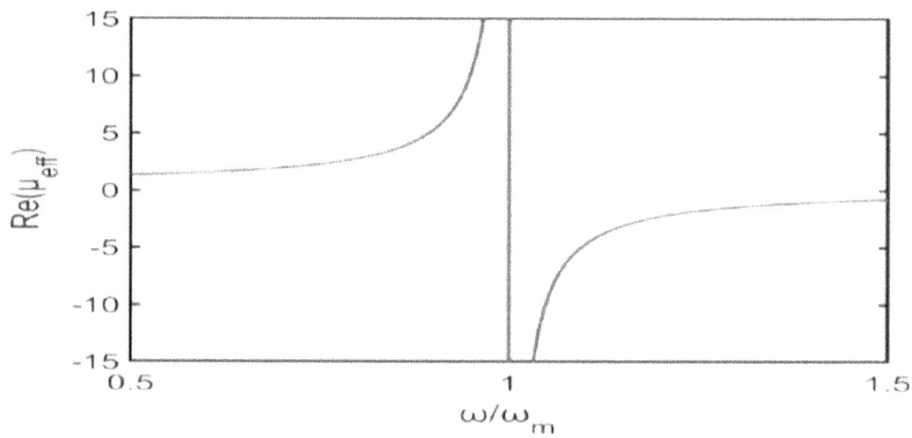

DNG metamaterials can be synthesized by combinations of ENG and MNG layers. The experimental demonstration of such a unique material was first accomplished by Smith et al. (D. R. Smith et al. 2000) in 2000 using a composite structure consisting of wires and SRRs, which is given in Figure 5.

The unit cell of a conventional, right-handed transmission line consists of a series inductor L_R and a shunt capacitor C_R, when the loss of dielectrics and conductors (equivalent to shunt admittance and series resistance) is neglected in the circuit. By introducing an additional shunt inductance L_L and series capacitance C_L, as shown in Figure 6, the dispersion characteristics are alternated and this type of transmission lines is referred as composite right/left handed (CRLH) TL (C. Caloz et al. 2006).

The progress of the communication system has used voluminous challenges to wireless systems, particularly in the field of antennas. The crucial role of design and development of antenna must be considered cautiously and comprehensively

Figure 5. DNG medium consisting of unit cells that are synthesized by wires (ENG) placed in front of SRRs (MNG). Both blue and grey parts depict metallization.
Source: Milius et al. (2021)

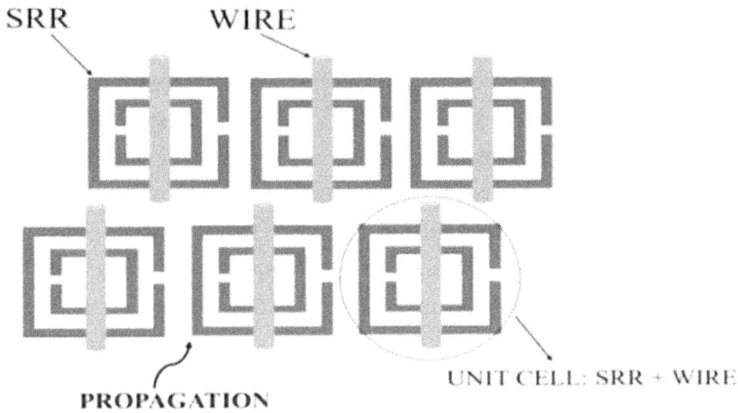

Figure 6. Unit cell of a composite right/left-handed transmission line consisting of series inductor and shunt capacitor (RH) and series capacitor and shunt inductor (LH)
Source: Milius et al. (2021)

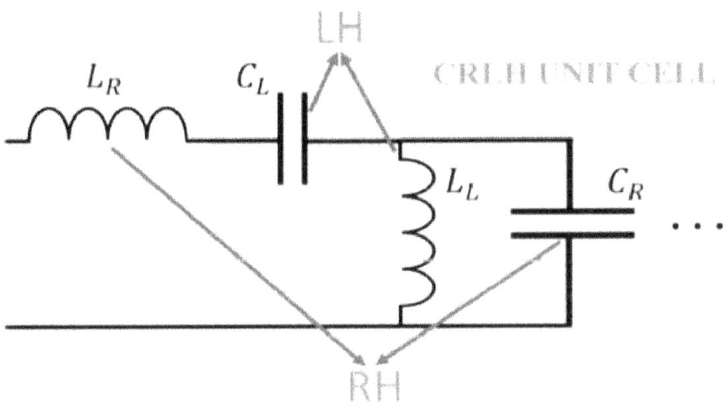

as the antenna is a key element in the wireless communication system and also any application associated with radio engineering and technology. The present day telecommunication system introduces various operating frequency ranges and numerous wireless standards on a single stage upto terahertz range. Therefore, multiband antennas are fascinating in mobile technologies stretching from 2G to 5G and beyond, comprising with IEEE 802.11 standards. On the other hand, advancement in long term evolution (LTE) imposes lower frequency operating ranges upto 4G.

Hence it stimulates a novel antenna design operating over a large bandwidth as well as frequency and it can be integrated to any handheld devices within the assigned dimensions. The development of these antennas utilizes different strategies and procedures to achieve novel structures, some of which are concisely given here. Generally, microstrip antennas are used because their TM01 and TM10 orthogonal modes offer port isolation and reduce cross-polarization factor (Z. Tang et al. 2018). To enhance bandwidth, high gain, better radiation pattern, dipole antenna comprised of electric and magnetic poles are used previously (H. Huang et al. 2018; C. Ding et al. 2018). Multiple PCB layers are used to extend bandwidth for multiband operation by exciting resonant modes on the plates for range of frequency (E. Tatartschuk et al. 2012). Inserting parasitic elements and various coupling techniques to the antenna improves antenna parameters such as radiation pattern, bandwidth, and isolation and gain (N. Gneiding et al. 2014). Resonant cavity antennas are used for enhancing antenna directivity, front to back ratio, and gain for better bandwidth (C. Saha et al. 2012). Recently, the metamaterial (MTM) inspired antenna has gained substantial attention from antenna researchers. The initiation of MTM has significantly shown new design methodologies of materials, particularly materials having unusual properties when compared to available materials in nature as shown in Figure 8 given below (R. K. Saraswat et al. 2020). One of the significant classes of MTMs is electromagnetic MTMs which are engineered sub-wavelength structures. Mostly, MTMs are defined as engineered artificial materials comprising of periodic or non-periodic structures of unit cells with its diameter less than the wavelength of light being propagated through it (Z. Troudi et al. 2019). Such an artificially implemented array of unit cells can create an electromagnetic output that can be customized evenhandedly by magnetic and electric wave components. Hence, a periodic array of unit cells can be considered as an efficient material described by its permittivity and permeability parameters. This arrangement agrees to constitute electromagnetic responses at the desired frequency; such response will not exhibit in material presents in nature. For instance, magnetic materials are much bounded at microwave frequencies. Hence, array periodic structures of metallic rings demonstrate such magnetic response at high frequencies. Rotman demonstrated negative effective permittivity by assembling an array of metal wires (Almpanis E et al. 2017). In Table 1, special characteristics of materials have been given. Also materials classification has mentioned in Figure 9.

Negative Permeability Material: The negative permeability materials are the materials having μ negative and ε positive and represent magnetic plasma as well. The absence of magnetic response at the optical range is due to the weak coupling of the magnetic field component of light with an atom which is equal to the Bohr magneton. Some of the arrangements which exhibit magnetic response at microwave and higher frequencies with negative permeability are coupled like nanorods, nanoplates, and strips, an array of metallic staple, and split ring resonator (SRR).

Figure 7. Classification engineered material based on electromagnetic properties
Source: Caloz et al. (2004)

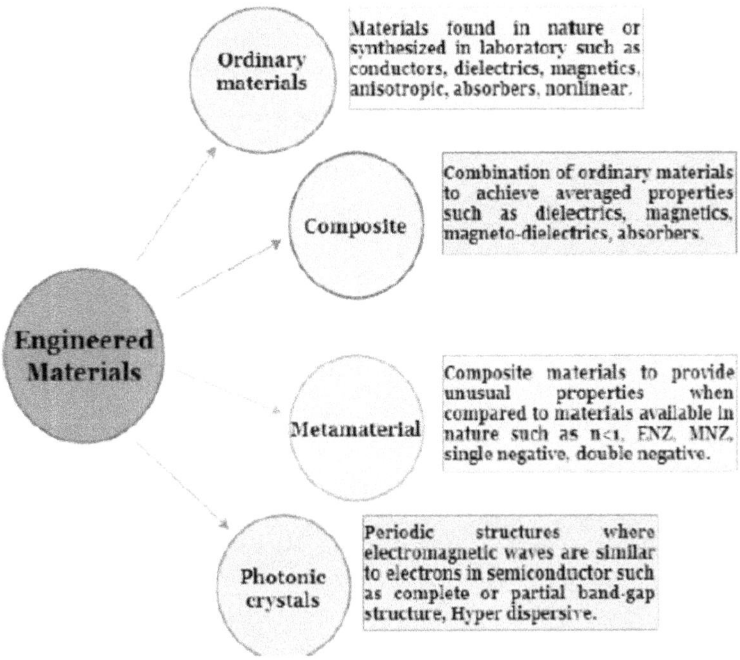

Table 1. Special material properties in ε-μ domain

Permittivity (ε)	Permeability (μ)	Description
$\varepsilon = -\varepsilon_0$	$\mu = -\mu_0$	Characterizes anti air in left hand medium yields to the perfect lens
$\varepsilon = 0$	$\mu = 0$	Characterizes anihility yields to the perfect tunneling effects
$\varepsilon > \varepsilon_0$	$\mu = \mu_0$	Characterizes materials available in nature
$\varepsilon = \mu$		Characterizes perfect impedance matching in right hand medium and left hand medium ensuing no reflections

The most general structure used for achieving negative permeability is an SRR (Sihvola, A 2007). Pendry and his team established a structure comprising of two opposite facing concentric split-ring resonators at sub-wavelength dimensions as shown in Figure 9. The gap in each metallic ring circularly prevents the development of current inside it. The charges gather at the edge of the gap in the metal ring and exhibit capacitance there. Therefore, the presence of an inductor and capacitor in a metal ring makes an SRR makes it resonant element. The analytical expressions for determining effective inductance and capacitance for this structure are given in

Figure 8. Material classification is based on permittivity and permeability
Source: Caloz et al. (2004)

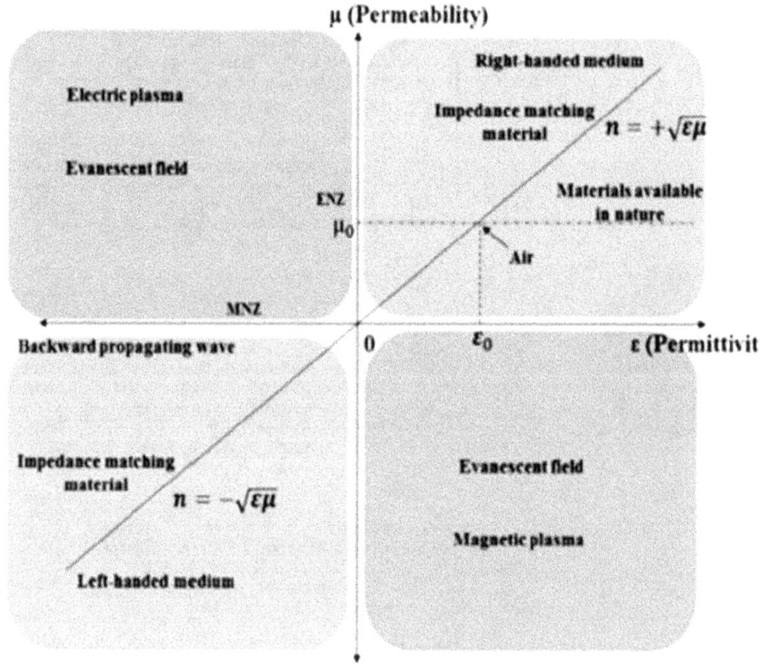

Figure 9. Magnetic material resonator: Single ring split (left side), double ring split (right side)
Source: Caloz et al. (2004)

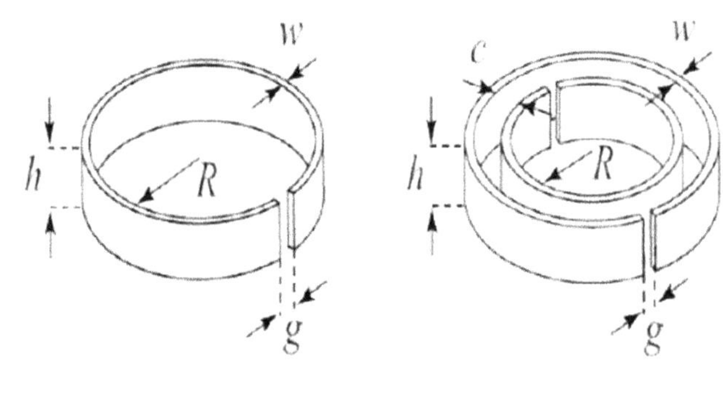

(Sihvola, A 2007). Several other analytical expressions are available in many literatures for appropriately determining inductance, capacitance, and the resonant frequency of different resonator structure such as SRR, loop gap, and CSRR (Tamayama, Y et al. 2014). The negative permeability material structures are not just limited to circle/ring structures, but various other geometrical structures can also be used to create negative permeability for example spiral, swiss roll, square, hexagonal (Thippeswamy et al. 2021), (Zhang, S et al. 2019). The transmission and reflection characteristics of the SRR depend on shape and structure. For single SRR with two varying split gaps along the ring gives rise to a change in the resonant frequency of the SRR over a wide range of frequency available. For double split-ring resonators, the influence of the geometrical parameters on SRR is listed in Table 2 by varying one parameter at a time and keeping other parameters constant (Zhang, Y et al. 2008). So far numerous design structures have been available in literature in the field of MTMs technology. The most commonly used are rectangular and circular SRR. The rectangular SRR dominates circular SRR in terms of miniaturization and dense packing.

Negative Permittivity Material: The negative permittivity materials are the materials having ε negative and μ positive and represent electric plasma. To stimulate and design the surface Plasmons, it is necessary to have roughness in the surface, a dielectric coupler to vacuum like hemisphere or prism, and a metallic strip structure. An array of thin metallic wires are organized periodically in a dielectric medium or in vacuum to show the low-frequency stop band characteristics from zero to a certain cut-off frequency. Rotman (W. Rotman 1962) systematically investigated various arrangements of 1D, 2D, and a 3D array of wire network structures to induce resonance in the polarized electric field at lower than the plasma frequency in design.

Table 2. Effect of geometrical properties on SRR

Geometrical Parameters	Effect on SRR
Increasing the split gap of SRR	As the split gap increases, the value of the capacitance decreases leads to a decrease in total capacitance, and a decrease in capacitance value increase the resonant frequency.
Increasing in spacing between two adjacent rings of SRR	The increase in spacing between the rings reduces the mutual inductance and capacitance, respectively. Due to a decrease in inductance and capacitance resonant frequency increase as analytically expressed.
Increasing the side length of SRR	Increase the side length of SRR increases the effective value of inductance and thereby shifts cutoff frequency towards the lower frequency.
Increasing in metal width	The increase in metal width reduces mutual inductance and mutual capacitor and thereby increase the resonant frequency
Multiple split gaps in SRR	Multiple gaps in SRR intercept magnetic resonance because of the induced current by an electric field and shift resonant frequency towards the higher frequency

Source: Sihvola (2007)

Such an artificial composite wire structure creates a negative electric permittivity with manageable strength.

Now, one of the most important applications of metamaterials is antenna design. Due to the unusual properties of metamaterials, antennas with novel characteristics can be achieved which cannot be realized with traditional materials. Effect has been discussed below of metamaterials over antenna design.

Electrically small antennas: Now a day's electrically small antennas (ESA) are highly desired for wireless applications as they can be easily integrated with devices without compromising the system form factor. The performance of antennas in terms of bandwidth, gain and directivity etc. is extremely critical and is governed by fundamental limitations in size, i.e., Chu's limit. Although modern integrated circuit technology can miniaturize electronic circuits to a very small size however, in traditional designs, the performance of the antenna is related to its physical size always comparable with wavelength. The antenna usually has dimensions in the order of the operating wavelength. Consequently, this sets boundaries for the size of the whole wireless system is useful of this issue makes metamaterials a hot topic for research in the development of electrically small antennas. This is an opportunity for using a zero refractive index metamaterial (ZIM) medium as its operating wavelength is infinite at the given design frequency. Since the wavenumber in ZIM antennas is zero, but in theory, the physical size of such antennas can be made independent of its operating frequency. Multiband antennas:- Mobile communication systems need to support multiple communication standards, such as 5G, Bluetooth, WiFi, NFC etc. To save space it is desirable to have a single antenna that can simultaneously accommodate these standards just like reconfigurable property (Shivani Chandra et al. 2021). Multiband antennas are normally designed with different resonant structures using different mateials. The disadvantage of this technique is the large size of the resulting antenna, which is dictated by the lowest operating frequency. Because of this, metamaterial structures can support negative and positive modes along with a zeroth-order mode as well, this property can be exploited to realize a singular antenna that operates over multiple bands and is miniature to boot the antenna.

Antenna lenses and polarizers: Directivity and gain of an antenna can be improved with the use of dielectric lenses. The only drawback of this technique is the high cost of the 3D lens. Moreover, the location of the lens needs to be carefully selected in relation with the phase center of the antenna in advance. The high cost of the lens can be offset by using a 2D lens made from metamaterial, which can be easily integrated with the planar antenna structure to reduce the profile and size of the antenna system which is the best feature. Polarizers are universal components employed in diverse application including antennas, imaging, display and microscopy etc. A polarizer based on a chiral medium are used to transform a linearly polarized wave into a circularly polarized wave, and to date various polarization convertors have been proposed as given (H. L. Zhu et al. 2013). Non-alignment in polarization will result in signal fading. This issue can be eliminated by utilizing circularly polarized electromagnetic waves as given (J. D. Kraus 2003). Conventional methods to fabricate polarization convertors are to use optical gratings, anisotropic media, the Brewester as well as Faraday effects.

Other types of antennas based on metamaterials: As other examples of radiating structures based on metamaterials include leaky wave antennas, magnetodielectric microstrip antennas, ultra-wideband (UWB) antennas with notched, reconfigurable antennas etc. All these designs exhibit better performance than the corresponding conventional designs.

Discussion of already published paper using metamaterials for 5G antenna:

(a) In this paper, author has discussed and used metamaterial surface as a reflector due to its in-phase reflection characteristic and high-impedance nature to improve the gain of an antenna. The radiating element of the proposed metamaterial based antenna is made up of copper material which is backed by the dielectric as shown in figure given below, i.e., Rogers-4003 with a standard thickness,

Figure 10. Geometry of antenna: (a) Front view without metamaterial, (b) back view, (c) 4X4 array design using metamaterial

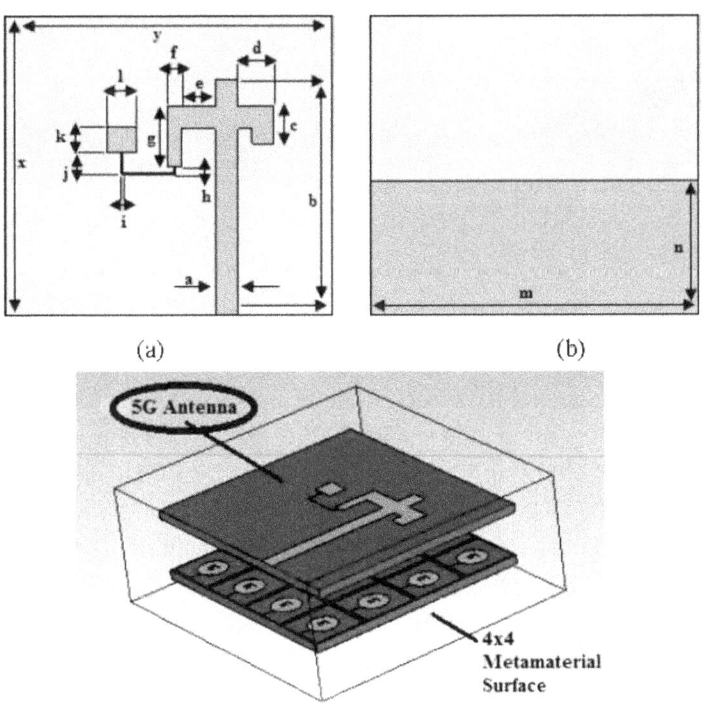

loss tangent and a relative permittivity of 1.524 mm, 0.0027 and 3.55, correspondingly. The proposed antenna with and without metamaterial surface operates at the central frequency of 3.32 GHz and 3.60 GHz, correspondingly. The traditional antenna yields a boresight gain of 2.76 dB which is further improved to 6.26 dB, using the metamaterial surface (Fazal Muhammad et. al. 2020). So, this antenna is well used for 5G applications and to use it for beyond 5G, few modifications will be done.

(b) In this paper, author has used low dielectric constant of 2.2, which is of Rogers RT/Duroid substrate, and a dielectric loss tangent of 0.0010 in the proposed design as given in figure below. Author has developed a metamaterial-based multi-

Figure 11. (a) 3D view of split-ring resonator, (b) unit cell of proposed split-ring resonator

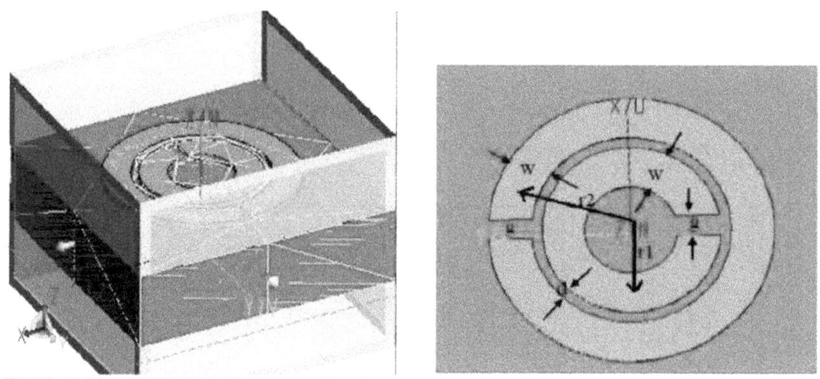

(a) (b)

band microstrip rectangular shape patch antenna with a wide high-performance bandwidth because of the demand. The design has a resonant frequency of 26 GHz. The simulations have been done using FEKO software to analyzed its performance. The simulation and analysis reveal a good return loss of -34.4 dB at 26 GHz, -13.49 dB at 40 GHz, -13.63 dB at 53.5 GHz, high bandwidth of 5.368 GHz at 26 GHz, 3.76 GHz at 40 GHz, 2.88 GHz at 53.5 GHz, desirable voltage standing wave ratio, $1 \leqslant VSWR \leqslant 2$, high gain of 10 dBi at 26 GHz, 5 dBi at 40 GHz, and high antenna radiation efficiency of 99.7% at 26 GHz, and 61% at 40 GHz, 50% at 53.5 GHz. The bandwidth, return loss, antenna radiation efficiency and power density indicate an improvement of 5.368 GHz to 5.630 GHz, -34.82 dB to -57.10 dB, 99.7% to 99.8% and 2208 kW/m2 to 2800 kW/

m2 respectively after loading and incorporating artificial magnetic split-ring resonator-based metamaterial on the patch (Calico et al. 2021).

(c) In this paper, Author has used a symmetrical monolayer metamaterial superstrate which is placed on a microstrip patch antenna resonant at 3.5 GHz. MTM metamaterial based CSRR Complementary Split Ring Resonators, constituting two rectangular rings is opposite splits on both ends, which are printed on Rogers RT 5880 type substrate with relative permittivity of 2.2, tangential loss of 0.0009, and a thickness of 0.508mm. This resonator is capable of exhibiting negative permittivity as shown in figure given below. Moving an

Figure 12. (a) CSRR, (b) the geometry of the antenna with MTM

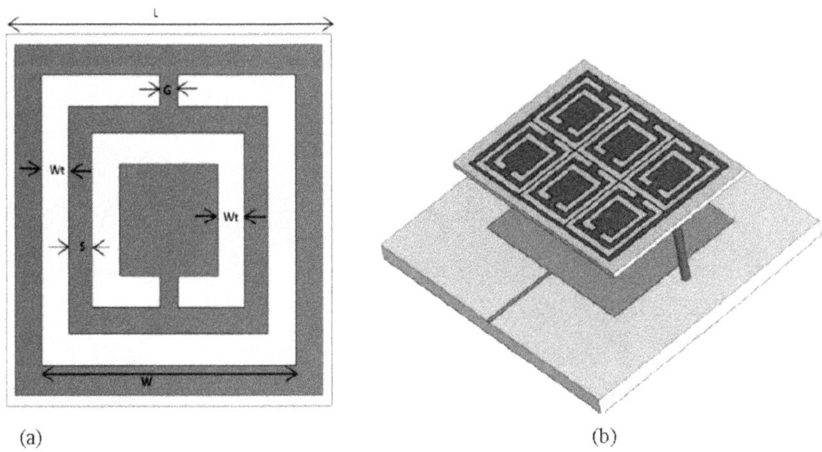

MTM superstrate loaded with a 2 × 3 grating on a single radiating element, provides a gain increase from 4.66 dBi to 5.67 dBi was found at its operating frequency, and the radiation pattern was directional and stable over the entire frequency band. For a 4×1 antenna array loaded with MTM, the gain reaches about 9.84 dBi. the proposed antenna is suitable for 5G mobile applications (Abderrahim Bellekhiri et al. 2022)

(d) In this paper, author has presented a high-isolation metamaterial based dual band multiple-input multiple-output (MIMO) antenna for 5G millimeter-wave communication networks. The proposed antenna is a pentagon-shaped monopole that provides a dual-band response with a wide operating bandwidth at 5G. The antenna is designed on 0.508-mm-thick Rogers RT5880 substrate of relative permittivity $\varepsilon r = 2.2$. It exhibits a small physical size of $5.5 \times 5.4 \times 0.508$ mm^3, excluding the feeding line as shown in figure given below. The MIMO system is constructed of two symmetric radiating elements arranged adjacently with the mutual coupling of – 18.5 dB at both resonant frequencies. The dual band metamaterial is designed and placed between the two radiators to reduce the mutual coupling. Here, embedding a 3 × 1 metamaterial array enhances the isolation to – 39 dB and – 38 dB at 28 GHz and 38 GHz, respectively. The proposed design is capable of covering both 28/28 5G bands and has the merits of broad bandwidth, low profile, high gain (>5 dB), improved isolation (– 38 dB), low envelope correlation coefficient (ECC) (9.99 dB). The system performance is verified experimentally with good agreement between the simulated and measured data. These properties demonstrate the system applicability for 5G millimeter-wave communication networks (Bashar Ali Esmail, 2023).

Figure 13. (a) The configuration of the proposed metamaterial unit cell, (b) the configuration of the proposed two-port MIMO antenna with metamaterials
Source: Bashar Ali Esmail (2023)

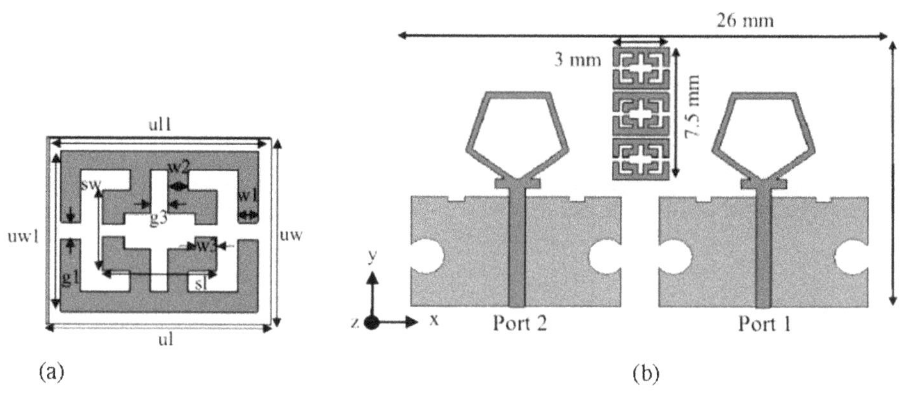

FUTURE CHALLENGES

Dimensions of electrically small antennas are considerably small in relation to the required operating wavelength. However, the design of ESA is challenging due to their limitations, which requires compromises between antenna size and performance in terms of bandwidth and radiation efficiency as well as gain. Although metamaterials have proved to be effective in reducing the size of the radiating elements by enabling zeroth-order modes, the main challenges remain unchanged which can improved further. For instance, the narrowband nature of metamaterials directly affects the bandwidth of the small resonator antennas. To date most of the designs are focused exclusively on reducing the size of the antenna by leveraging the sub wavelength resonance of metamaterial cells, however the overall antenna performance is suboptimum and can be improved. In particular, the interaction between the metamaterial cell and the rest of the antenna is not considered even though this interaction is vital for determining the antenna performance as well. Although ESA designs based on metamaterials are relatively straightforward, but accomplishing an overall excellent performance from them will require considerations of other factors such as feeding mechanism, metamaterial structure, position of the unit cells, antenna housing, farfield region etc. Moreover, theoretical formulations for such antennas are important to provide an insight of the key to characterize the parameters as well as for providing design guidelines. To enhance the design of ESA for desirable performance characteristics what is needed is formulation of accurate theory behind metamaterial inspired antennas. More specifically, technique for increasing the bandwidth and improving the radiation efficiency of ESAs has to be improved. Small, energy efficient and wideband antennas will be vital components for next generation 5G and beyond of wireless communication systems.

Also, the gain of an antenna is affected by its size and aperture efficiency. Recent techniques proposed to improve the gain include near-zero refractive index (NZRI) superstrates and artificial magnetic conductor (AMC) surfaces. This is because NZRI superstrates can focus the electromagnetic energy as well as AMC acts like as reflecting surface. Metamaterial inspired gradient index (GRIN) lenses have also been given to enhance gain and this is over a wide bandwidth. The improvement in gain of these techniques is mainly due to the increase in the aperture size of the antenna. Moreover, NZRI and AMC surfaces are easier to fabricate and cheaper than metallic reflectors that require extreme surface smoothness at higher frequencies. The viability of concurrent gain enhancement and beam steering will truly be possible when metamaterials can be dynamically configured so that they adapt in real time to the changes in the transmitter-receiver environment. GRIN lens integrated beam steerable antennas achieve beam steering by either mechanical means or by switching between radiating elements positioned at different orientation in plane. This is a cumbersome approach and hence there is a need for a dynamic solution which can be electronically controlled and agile as well. But, GRIN lenses are incompatible for various low profile various applications. Beamforming is a key technology in 5G and beyond systems that reduces carbon emissions by focusing the wireless signal towards a specific receiving device, rather than transmitting the signal in all directions in system. The direct connection results in a faster, high quality, and more reliable communications connection than it would be without beamforming as reported. To achieve further improvements in the system performance and thereby reduce the greenhouse gases effect requires increase in the beam gain and radiation efficiency of beam steerable arrays. This entails the further research on adaptive NRZI and AMC metasurfaces and low-profile metamaterial GRIN lenses.

In 5G wireless systems the use of massive multiple-input multiple-out (MIMO) is essential to increase the cell capacity and downloading data rates of user equipment as per demand. This is achieved with multiple antennas that co-exist on the same platform and operate on the same frequency bands even at higher frequency. Hence, the performance of massive MIMO can be undermined by the mutual coupling interaction between the radiating elements constituting the antenna array. The suppression of unwanted coupling in arrays which are based on metamaterial decoupling techniques has been demonstrated to be quite effective. This is achieved by inserting the metamaterial decoupler between the adjacent radiating elements in design. The benefit of using metamaterial is that it provides a low-profile solution and, in most cases, has no impact on the size of the array during the design. The space permitting it can in some cases be retrofitted to existing antenna arrays. The main disadvantage of metamaterial inspired decoupling techniques reported to date in literature is their narrowband limitation and low bandwidth. For existing and next generation and beyond of wireless communication systems to operate across multiple bands makes the narrowband metamaterial decoupling an ineffective solution. Mutual coupling isolation of -20 dB is acceptable in MIMO systems. Similarly in the case of CW and FMCW radar systems the performance is dependent on the isolation between the transmitter and receiver. In the case of radars, isolation of better than 100 dB is required for interference immunity. Hence, clearly decoupling over a wideband is a challenge and a low-profile metamaterial decoupling is highly preferred in place of using a conductive shield.

A critical parameter which can affect an antenna's performance is frequency. An antenna is designed at a specific frequency and therefore its performance is effective over a range of frequencies centered on its resonant frequency during the design. However, because the antenna's other properties (e.g., radiation pattern, gain and impedance) change with frequency, its resonant frequency may just be close to the center frequency of these properties. Therefore, a design technique is based on metamaterial, which needs to be investigated that minimizes such frequency variations.

Also, the need for wireless power transfer (WPT) for electric vehicles (EV) is growing in demand. However, currently there are technical bottlenecks problem for current WPT technologies, such as the low-energy efficiency, short transmission range and electromagnetic exposure, etc. The metamaterials exhibit great potentials to concentrate and enhance the magnetic flux density of the electromagnetic field by means of its left-handed characteristics.

MTM-INSPIRED ANTENNA

Antenna miniaturization is due to the placing of metaresonators into the antenna structures using SRR design. In this approach, the properties of an antenna are due to single or more than two unit cells interact with the antenna structures and can work for high frequency. MTM inspired antenna can be achieved by engraving the unit cell structures into the radiating patch or ground plane or between the ground and radiating plane. This approach is very similar to introducing a slit/slot into the antenna structure, but this unit cell arrangement acts as a resonant structure due to its periodic behavior also.

Design of Frequency Reconfigurable Metamaterial Antenna (Shivani Chandra et al., 2020; Smrity Dwivedi 2022)

In order to use an antenna at different frequencies in a particular frequency band, a Frequency Reconfigurable Antenna design is proposed using Split Ring Resonators (SRR). It uses a Rogers RT duroid 5880 (30 × 28.4 mm^2) as a substrate with a thickness of 0.508 mm. It uses a CPW feed and a metal surface at the back of substrate of same size which act as a reflector. This reflector is used to increase the gain and other characteristics of the antenna. Two PIN diodes of same size are used as switches for connecting the Split Ring Resonators. By ON/OFF the switches we get the four different modes and different resonating frequencies at those modes. It exhibits the VSWR < 2 and simulated on CST®MWS®. It resonates at four different frequencies – 5.32 GHz, 5.662 GHz, 6.469 GHz and 6.547 GHz with a maximum gain of 6.44 dBi, 7.75 dBi, 8.54 dBi and 8.39 dBi respectively.

Now a day's the antenna with same structure for different purpose resonating at different frequencies is mostly used and achieved an attention in every field. Moreover we can not only achieve reconfiguration in frequency, it can be achieved with respect to polarization i.e. by using same structure we can achieve linear, circular or elliptical polarizations. One can get different gain and stable radiation pattern with reduced or negligible side lobes at those specific frequencies. The Frequency Reconfigurable Antenna design is proposed using a coplanar waveguide (CPW) feed. It also makes the use of Split Ring Resonators (SRR). With the help of these resonators, reconfiguration is achieved in terms of frequency. Gain achieved is also high due to the use of metal plate at the back of substrate at a certain distance, which acts as a reflector. Because of the reflector at the back, major problem occurs is the existence of side lobes but they are reduced by optimizing the feed length and length of the ground surface. Circular Split Ring Resonators used in the design act as a Meta-material. They are two rings structure having a cut perpendicular to each other.

They are depicting the properties of negative permeability and negative permittivity. These properties are achieved due to their structure which helps in enhancing the gain and radiation pattern of the antenna. They are periodic structures having a special feature of unique band gap. It makes the antenna different from normal antennas as mostly other antennas reflects the signal back to the source but it has the property to store and re-radiate the signal energy. They can also be used in array structure.

The CPW feeding used in the designed structure enables the antenna to be used for having a wide bandwidth. It has ground plane on the side of the feeding line. It helps in providing the series and parallel connection between active and passive elements which helps in matching and improvement in gain and radiation pattern. Two PIN diodes are used which are acting as a switches. They are meant for connecting the circular SRR to main ring structure of the feeding line. This diode can be ON/OFF as a switch therefore it has four different modes 00, 01, 10 and 11. Four different resonating frequencies are obtained at these modes. In a frequency band of 4 GHz to 7 GHz, we are getting frequencies 5.32 GHz, 5.662 GHz, 6.469 GHz and 6.547 GHz. These are having the frequency range of 5G spectrum. Therefore, it can be used as a 5G antenna and for other communication purposes.

Initial Design: The first step in designing the structure is the designing of circular ring structure extending the feeding line. It also acts as an antenna of length (L) is 30 mm and width (W) is 28 mm. It is resonating at a frequency of 6.484 GHz with a gain of 8.45 dBi. Other parameters for ground plane are Lg and Wg whose values are 14.642 and 13.45 respectively. The above mentioned design without use of circular SRR is shown below in Figure 14.

Figure 14. CPW feed antenna without split ring resonators

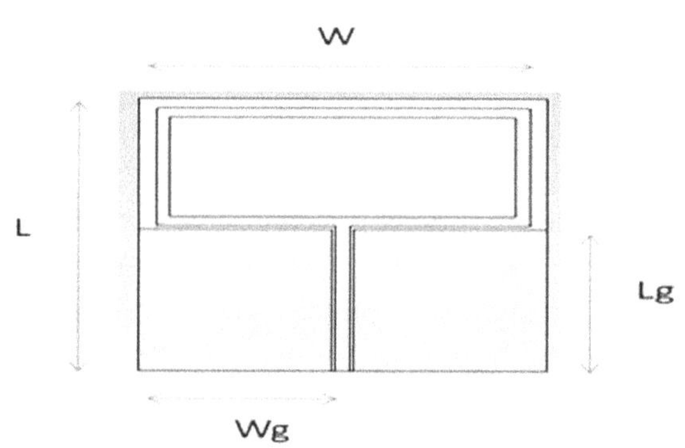

Final Design: The basic design of CPW feed antenna is shown in Figure 15 does not make the use of split ring resonators. The final design structure uses split ring resonators connected to the ring like structure through PIN diodes which acts

Figure 15. The proposed design structure of the antenna

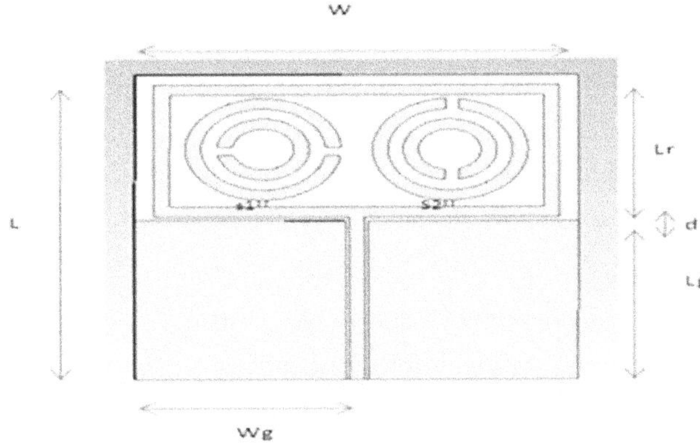

as switches. In this we use a Rogers RT duroid 5880 as a substrate material of size 30 mm x 28.4 mm. The spacing gap between the ground plane and the feeding line is 0.75 mm. The ring like structure is at height g from the ground whose value is 0.358 mm. The diodes S1 and S2 used as a switch are of size 0.65 mm x 0.5 mm. The other parameters of circular SRR are as bigger ring is of outer radius equal to 5 mm and inner radius is 4 mm. And smaller ring is having outer radius equal to 3 mm and inner radius of 2 mm. Both the ring structures have a gap or cut perpendicular to each other so to make it behave as a Meta –material structure.

RT/duroid 5880 covers a low dielectric steady and low dielectric misfortune, making them appropriate for high recurrence/broadband applications. Assisting with keeping up the consistency are the haphazardly arranged microfibers fortifying the PTFE composites. It has uniform electrical properties over wide recurrence run. It can handily cut, shared and machined to shape. It is impervious to solvents and reagents utilized in carving or plating edges and gaps.

Moreover, perfect for high dampness conditions and entrenched material. It is most reduced electrical misfortune for strengthened PTFE material. Bandwidth must be high for wider applications. CPW feed antenna has comparatively higher bandwidth than any other feeding techniques antenna. Bandwidth is high in that case due to extra lobes generated. CPW is lossy and less coupling compared to micro strip line. Yes, losses in CPW are more. Only at higher characteristic impedances the loss can be made lesser than micro strip. But at 50 ohms, CPW is more lossy than micro strip. Hence less Q means therefore more bandwidth.

The basic and proposed design is simulated on CST Microwave. And the results obtained through this are studied to examine the performance of the antenna. As there are two PIN diodes which act as a switch, there are four different modes 00, 01, 10 and 11. The frequency and radiation pattern obtained in these modes are studied.

First the initial design of the antenna is simulated. That is the basic structure consists of substrate and ground surface. It is the coplanar waveguide (CPW) feed antenna with a ring shaped antenna. This structure does not contain split ring resonators and pin diodes. The feeding parameters are selected such that it gives good results including S11 characteristics, VSWR and other far field and 3D plots.

Figure 16. (a) S1, 1 parameter; (b) VSWR; (c) polar plot; (d) 3D plot of basic design structure using CPW feed antenna without split ring resonators

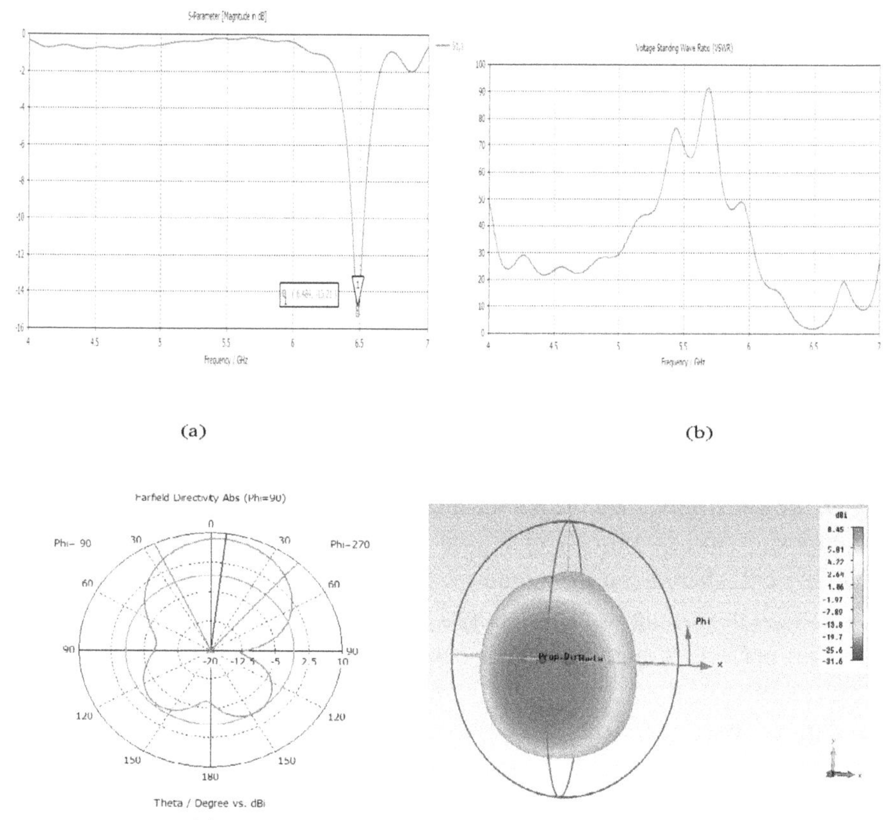

(a) (b) (c) (d)

From the above results, the reflection parameter S11 is achieved and it is about -15 dB which is less than the required normal condition of -10 dB. It is achieved at the frequency of 6.49 GHz. The other characteristics such as VSWR ratio plot and far field plot is shown. The maximum gain achieved in this configuration is around

8.5 dBi. And the polar plot of this is shown above which has very less or reduced side lobes. Major part of it consist a main lobe which is helpful for the radiation of antenna at the particular direction. Now, the final design is simulated which make the use of split ring resonators and pin diodes as switches. After using the Switches S1 and S2, different modes with their characteristics are studied. They are classified as different states – State 1 (00), State 2 (01), State 3 (10), and State 4 (11).

State 1: When both the switches S1 and S2 are in OFF position, i.e. 00 mode. At this mode, the antenna is resonating at a frequency of 6.469 GHz. The maximum gain obtained is 8.54 dBi as shown in the 3D plot below. The S11 parameter obtained is around -16 dB.

Figure 17. (a) S1, 1 parameter at 6.469 GHz; (b) VSWR at 6.469 GHz; (c) polar plot at 6.469 GHz; (d) 3D plot at 6.469 GHz

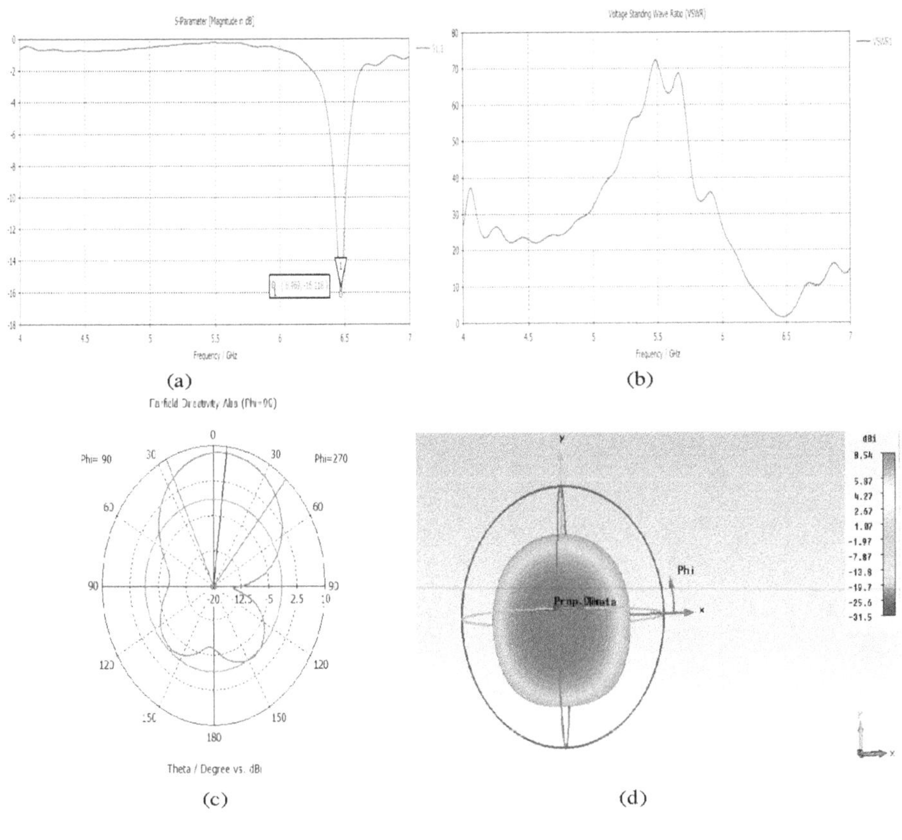

State 2: When the switch S1 is OFF and S2 is in ON position, i.e. 01 mode. At this mode, the antenna is resonating at a frequency of 6.547 GHz. The maximum gain obtained is 8.39 dBi. The S11 parameter obtained is around -11 dB. The S11 parameter and gain obtained in this state is less than the same results and plot obtained in previous state. It makes the use of switch 2 which is in ON condition.

Figure 18. (a) S1, 1 parameter at 6.547 GHz; (b) VSWR at 6.547 GHz; (c) polar plot at 6.547 GHz; (d) 3D plot at 6.547 GHz

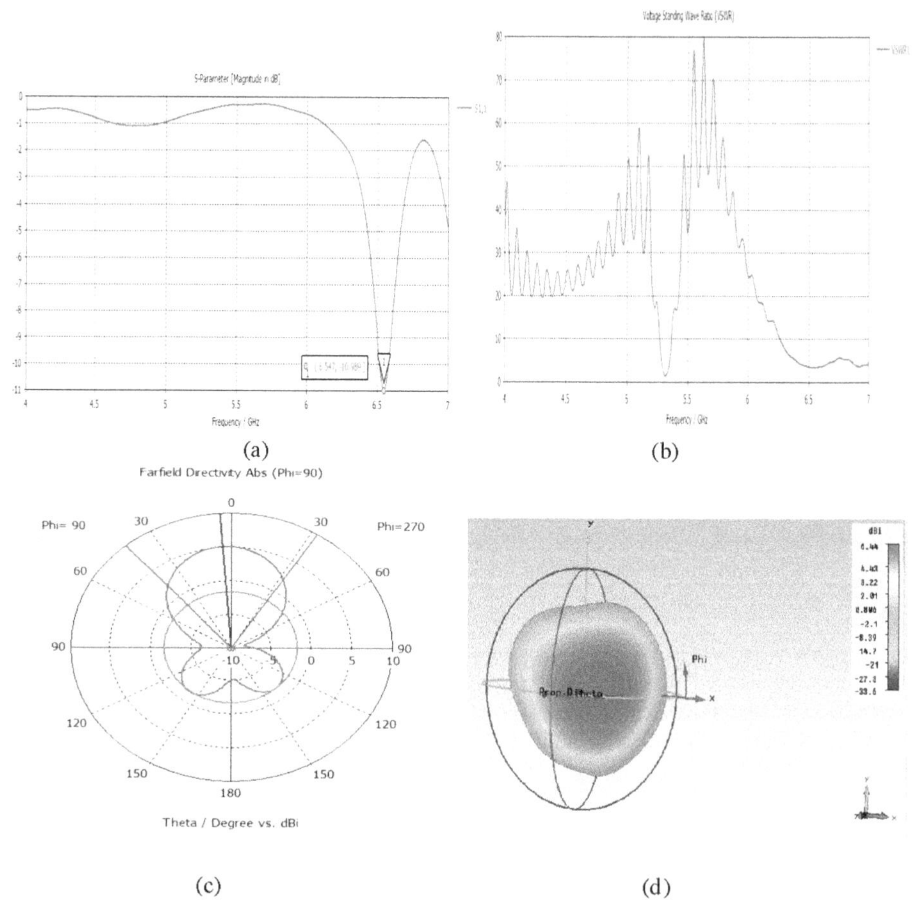

State 3: When the switch S1 is ON and S2 is in OFF position, i.e. 10 mode. At this mode, the antenna is resonating at a frequency of 5.32 GHz. The maximum gain obtained is 6.44 dBi. And the S11 parameter obtained in this mode is around -15 dB.

Figure 19. (a) S1, 1 parameter at 5.32 GHz; (b) VSWR at 5.32 GHz; (c) polar plot at 5.32 GHz; (d) 3D plot at 5.32 GHz

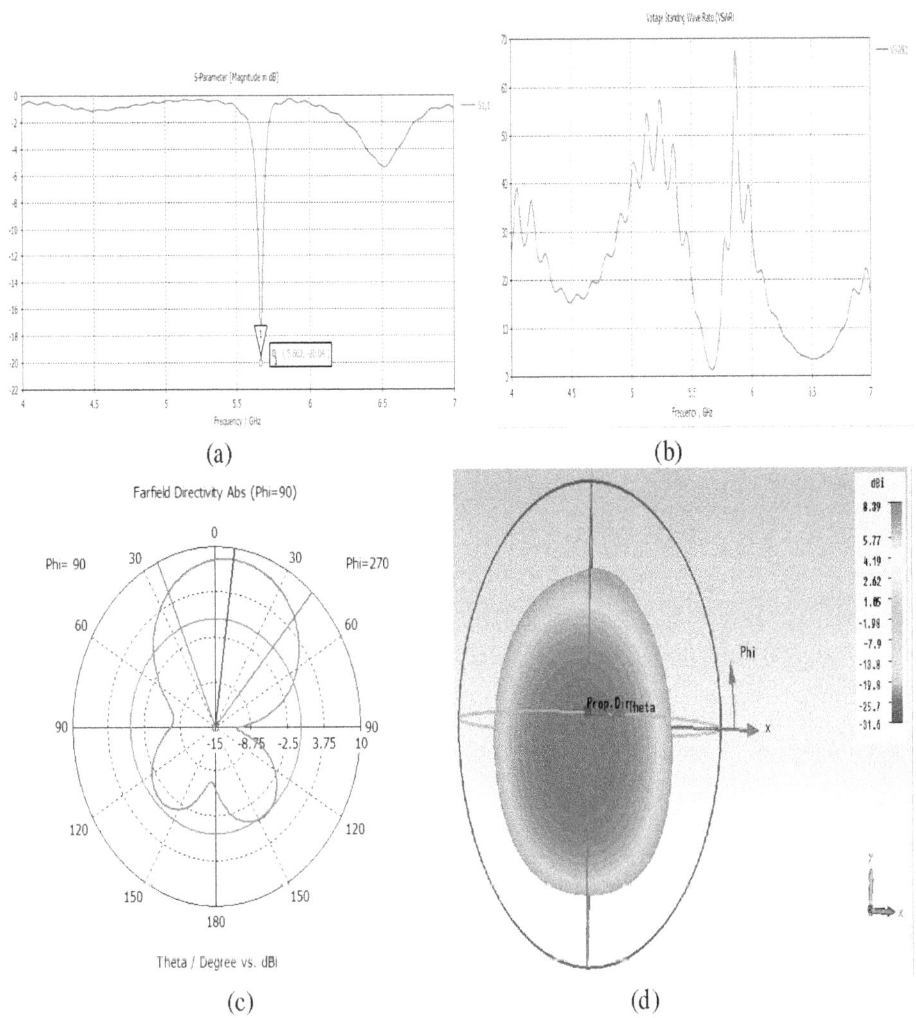

State 4: When the switch S1 is ON and S2 is in ON position, i.e., 11 mode. At this mode, the antenna is resonating at a frequency of 5.662 GHz. The maximum gain obtained is 7.75 dBi. The S11 parameter obtained in this state is -20 dB.

The reconfiguration in terms of frequency is achieved in the proposed design. It varies from 5.32 GHz to 6.5 GHz with a difference of about 1200 MHz's. It is obtained through the use of diode connected in the design structure. The designed

Figure 20. (a) S1, 1 parameter at 5.66 GHz; (b) VSWR at 5.66 GHz; (c) polar plot at 5.66 GHz; (d) 3D plot at 5.66 GHz

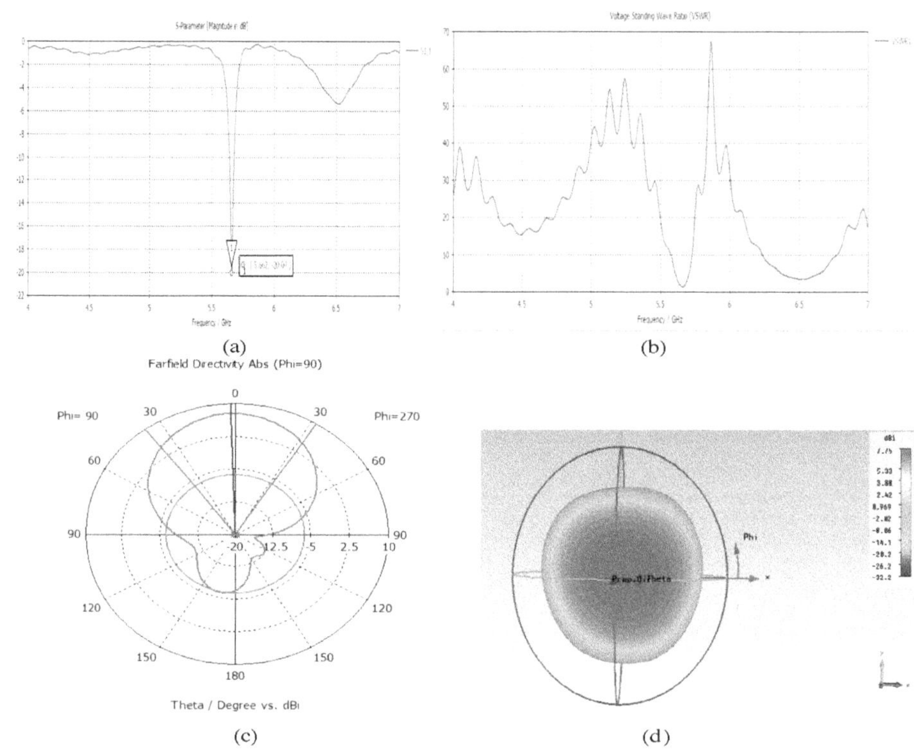

structure makes the use of SRR to increase the gain and stability. It can be used for 5G applications for the resulted frequencies as shown in previous graphs. Moreover, the design is flexible and proper matching is achieved. On removing the metal plate at the back of substrate, gain of the antenna is reduced to around 4 dBi, but on using the plate it increases to around 8.5 dBi.

Table 3. Configuration of switches

States	Mode	S1	S2	Resonant Frequency (GHz)	Gain (dBi)
State 1	00	OFF	OFF	6.469	8.54
State 2	01	OFF	ON	6.547	8.39
State 3	10	ON	OFF	5.32	6.44
State 4	11	ON	ON	5.66	7.75

Metamaterial Waveguide Antenna (Smrity Dwivedi 2022)

Using CST microwave studio, design has been drawn for the structure mentioned. Figure 21 (a) represents the front view and (b) represents the back view of structure where BC-SRR is clearly visible. In this design, the C band waveguide is coupled as transition structure and other side is air box to excite the structure.

Figure 21. Schematic view of metamaterial waveguide antenna: (a) Front view, (b) back view

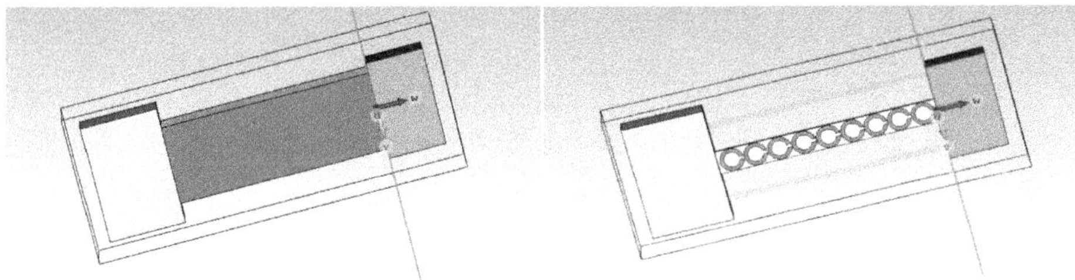

(a) (b)

This is very unique featured material to design SRR due to its negative permittivity and negative permeability. Metamaterial is an engineered material having with periodic structure, gives low loss and high gain as well as light weight material. It also gives the high bandwidth and good for getting optical behaviour. Figure 22 shows the negative permittivity and negative permeability of metamaterial to design the SRR.

From Figure 23 (a) and (b), it is clear that the maximum gain from the given antenna ~ 10 dBi. From the plot, the side lobes are observed. This is happened because of coupling issue between two waveguides and impedance mismatching.

CONCLUSION

Study of MTM opened a new dimension in antenna design because of its unique electromagnetic characteristics such as negative permittivity and permeability, negative refractive index, and its manipulation. On the other hand, the evolution of electronic gadgets reserves a compact area for antennas and looking for better performance. Several different approaches are examined for MTM-based antenna, in specific i) MTM structure loading having an ε/μ negative material, high permittivity/permeability cell reconfigurable antenna ii) Metamaterial

Figure 22. Epsilon and mue plots for metamaterial structure

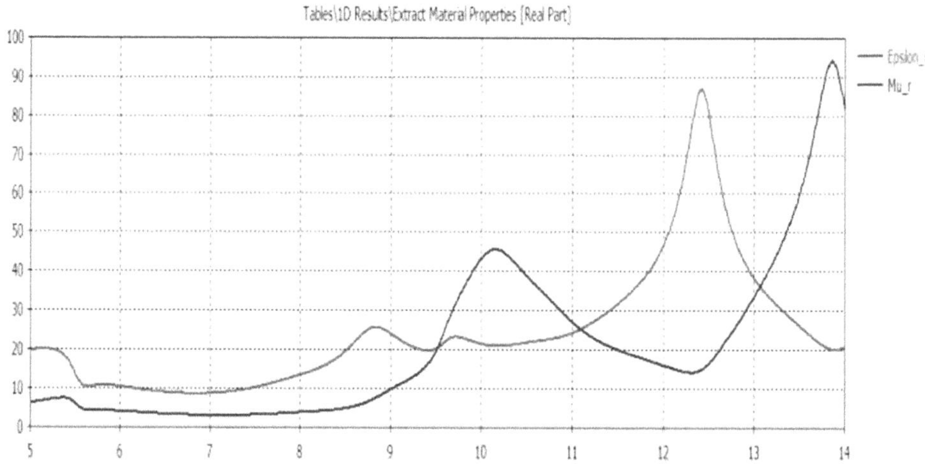

Figure 23. (a) Radiation pattern 3D plot, (b) radiation pattern in 2D (polar) plot

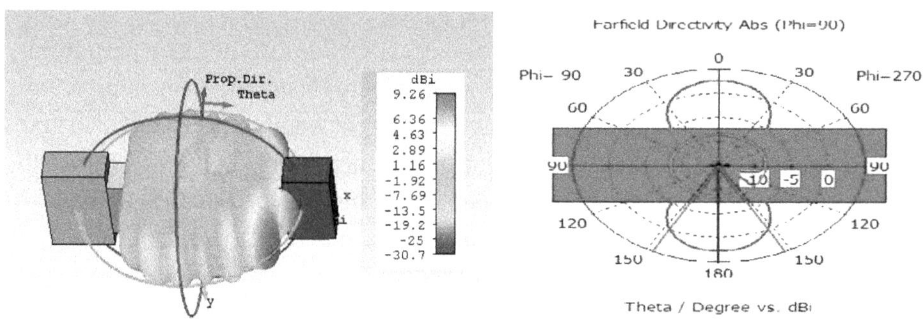

(a) (b)

waveguide antenna. Although, there are numerous applications of MTM in antenna design, however, here we have discussed explicitly its recent progress in antenna miniaturization, enhancement of gain and bandwidth, and suppression of mutual coupling in different antenna systems. Antenna miniaturization can be achieved by inserting MTM unit cell structures on top of the patch and/or bottom of the ground plane. An array of periodic sub wavelength unit cells commonly termed as metasurface placing onto the antenna will change effective inductance and capacitance of the antenna system resulting in miniaturization again. Apart from antenna miniaturization, MTM based antenna is a promising technology for improving the gain and bandwidth of an antenna. Also, for 5G and beyond, it is the best suited material for antenna design for several applications in which IoT is one the main important application.

REFERENCES

Akinsolu, M. O., Mistry, K. K., Liu, B., Lazaridis, P. I., & Excell, P. (2020). Machine learning-assisted antenna design optimization: A review and the state-of-the-art. *Proc. 14th Eur. Conf. Antennas Propag. (EuCAP),* 1–5. 10.23919/EuCAP48036.2020.9135936

Almpanis, E., Pantazopoulos, P. A., Papanikolaou, N., Yannopapas, V., & Stefanou, N. (2017). A birefringent etalon enhances the Faraday rotation of thin magneto-optical films. *Journal of Optics*, *19*(7), 075102. doi:10.1088/2040-8986/aa7420

Alu, A., Engheta, N., Erentok, A., & Ziolkowski, R. W. (2007, February). Single-negative, double-negative, and low-index metamaterials and their electromagnetic applications. *IEEE Antennas & Propagation Magazine*, *49*(1), 23–36. doi:10.1109/MAP.2007.370979

Balanis, C. A. (2012). *Antenna theory: Analysis and design.* John Wiley & Sons. doi:10.1109/LMWC.2004.828009

Bashar & Koziel. (2023). High isolation metamaterial-based dual-band MIMO antenna for 5G millimeter wave applications. *International Journal of Electronics and Communication, 158*, 154470.

Bellekhiri, A., Chahboun, N., Laaziz, Y., & El, A. (2022). A New Design of 5G Planar Antenna based on metamaterials with a high gain using array antenna. *ITM Web of Conferences, 48.*

Caloz, C., & Itoh, T. (2004). Array factor approach of leaky-wave antennas and application to 1-D 2-D composite right left-handed (CRLH) structures. *IEEE Microwave and Wireless Components Letters*, *14*(6), 274–276. doi:10.1109/LMWC.2004.828009

Caloz, C., & Itoh, T. (2006). *Electromagnetic Metamaterials Transmission Line Theory and Microwave Applications*. Wiley.

Chandra & Dwivedi. (2020). Simulation analysis of High Gain and Low Loss Antenna with Metallic Ring and spoof SPP Transmission Line. In *7th International Conference on Signal Processing & Integrated Networks*. IEEE Explore.

Chandra & Dwivedi. (2021). Graphene based Radiation Pattern reconfigurable antenna. *International Conference for convergence in Technology*.

Colaco & Lohan. (2021). Metamaterial Based Multiband Microstrip Patch Antenna for 5G Wireless Technology-enabled IoT Devices and its applications. *J. Phys.: Conf. Ser.*

Ding, C., Sun, H.-H., Ziolkowski, R. W., & Jay Guo, Y. (2018, December). A dual layered loop array antenna for base stations with enhanced crosspolarization discrimination. *IEEE Transactions on Antennas and Propagation*, *66*(12), 6975–6985. doi:10.1109/TAP.2018.2869216

Dwivedi. (2022). *Design and Analysis of Metamaterial Waveguide Antenna for Broadband Applications*. doi:10.1007/978-981-19-5224-1_42

Dwivedi. (2023). *Antenna Array for reconfiguration*. IGI Global. doi:10.4018/978-1-6684-5955-3

Ejaz, J. W., Anpalagan, A., Imran, M. A., Jo, M., Naeem, M., Bin Qaisar, S., & Wang, W. (2016). *2016 Internet of Things (IoT) in 5G Wireless Communications*. IEEE Access.

Gneiding, N., Zhuromskyy, O., Shamonina, E., & Peschel, U. (2014, October). Circuit model optimization of a nano split ring resonator dimer antenna operating in infrared spectral range. *Journal of Applied Physics*, *116*(16), 164311. doi:10.1063/1.4900479

GSMA. (2019). *5G the internet of things and wearable devices: Radiofrequency exposure*. GSMA Public Policy.

Helena, Ramos, Varum, & Matos. (2020). Antenna Design Using Modern Additive Manufacturing Technology: A Review. *IEEE Access, 8*, 177064-177083.

Holloway, Dienstfrey, Kuester, O'Hara, Azad, & Taylor. (2009). A Discussion on the Interpretation and Characterization of Metafilms/Metasurfaces: The Two Dimensional Equivalent of Metamaterials. *Metamaterials, 3*(2), 100-112.

Holloway, Kuester, Baker-Jarvis, & Kabos. (2003). A Double Negative (DNG) Composite Medium Composed of Magneto-Dielectric Spherical Particles Embedded in a Matrix. *IEEE Transactions on Antennas and Propagation, 51*(10), 2596-2603.

Huang, H., Li, X., & Liu, Y. (2018, June). A novel vector synthetic dipole antenna and its common aperture array. *IEEE Transactions on Antennas and Propagation*, *66*(6), 3183–3188. doi:10.1109/TAP.2018.2819894

Kraus, J. D., & Marhefka, R. J. (2003). *Antennas for all applications* (3rd ed.). McGraw-Hill.

Lapine, M., & Tretyakov, S. (2007). Contemporary notes on metamaterials. *IET Microwaves, Antennas & Propagation*, *1*(1), 3–11. doi:10.1049/iet-map:20050307

Milius, C., Andersen, R. B., Lazaridis, P. I., Zaharis, Z. D., Muhammad, B., & Jes, T. B. (2021). Metamaterial-Inspired Antennas: A Review of the State of the Art and Future Design Challenges. *IEEE Access : Practical Innovations, Open Solutions*, *9*, 89846–89865. doi:10.1109/ACCESS.2021.3091479

Rotman, W. (1962, January). Plasma simulation by artificial dielectrics and parallelplate media. *IRE Transactions on Antennas and Propagation*, *10*(1), 82–95. doi:10.1109/TAP.1962.1137809

Saha, C., & Siddiqui, J. Y. (2012, July). Theoretical model for estimation of resonance frequency of rotational circular split-ring resonators. *Electromagnetics*, *32*(6), 345–355. doi:10.1080/02726343.2012.701540

Saraswat, R. K., & Kumar, M. (2020, October). A quad band metamaterial miniaturized antenna for wireless applications with gain enhancement. *Wireless Personal Communications*, *114*(4), 3595–3612. doi:10.100711277-020-07548-z

Sehrai, D. A., Muhammad, F., Kiani, S. H., & Kim, S. (2020). Gain-Enhanced Metamaterial Based Antenna for 5G Communication Standards. *Computers, Materials & Continua*, *64*(3), 1587–1599. doi:10.32604/cmc.2020.011057

Shamonina & Solymar. (2007). Metamaterials: How the Subject Started. *Metamaterials*, *1*(1), 12-18.

Sihvola, A. (2007). Metamaterials in electromagnetics. *Metamaterials (Amsterdam)*, *1*(1), 2–11. doi:10.1016/j.metmat.2007.02.003

Silin, R. A., & Chepurnykh, I. P. (2001). On Media with Negative Dispersion. *Commun. Technol. Electron.*, *46*, 1121–1125.

Sivukhin, D. V. (1957). The Encrgy of Elcctromagnetic Fields in Dispersive Media. *Opt. Spektrosk.*, *3*, 308–312.

Smith, D. R., Padilla, W. J., Vier, D. C., Nemat-Nasser, S. C., & Schultz, S. (2000). Composite Medium with Simultaneously Negative Permeability and Permittivity. *Physical Review Letters*, *84*(18), 4184–4186. doi:10.1103/PhysRevLett.84.4184 PMID:10990641

Tamayama, Y., Yasui, K., Nakanishi, T., & Kitano, M. (2014). A linear-to-circular polarization converter with half transmission and half reflection using a single-layered metamaterial. *Applied Physics Letters*, *105*(2), 021110. doi:10.1063/1.4890623

Tang, Z., Liu, J., Cai, Y.-M., Wang, J., & Yin, Y. (2018, April). A wideband differentially fed dual-polarized stacked patch antenna with tuned slot excitations. *IEEE Transactions on Antennas and Propagation*, *66*(4), 2055–2060. doi:10.1109/TAP.2018.2800764

Tatartschuk, E., Gneiding, N., Hesmer, F., Radkovskaya, A., & Shamonina, E. (2012, May). Mapping inter-element coupling in metamaterials: Scaling down to infrared. *Journal of Applied Physics*, *111*(9), 094904. doi:10.1063/1.4711092

Thippeswamy, M. C., Kuchibhatla, S. A. R., & Rajagopal, P. (2021). Concentric shell gradient index metamaterials for focusing ultrasound in bulk media. *Ultrasonics*, *114*, 106424. doi:10.1016/j.ultras.2021.106424 PMID:33819870

Troudi, Z., Machac, J., & Osman, L. (2019, October). Miniaturised planar band-pass filter based on interdigital arm SRR. *IET Microwaves, Antennas & Propagation*, *13*(12), 2081–2086. doi:10.1049/iet-map.2018.5708

Veselago, V. G. (1968). The Electrodynamics of Substances with Simultaneously Negative Values of ε and μ. *Physics Uspekhi*, *10*, 509–514. doi:10.1070/PU1968v010n04ABEH003699

Wang, C., Chen, Y., & Yang, S. (2018, November). Dual-band dual-polarized antenna array with flat-top and sharp cutoff radiation patterns for 2G/3G/LTE cellular bands. *IEEE Transactions on Antennas and Propagation*, *66*(11), 5907–5917. doi:10.1109/TAP.2018.2866596

Wen, L.-H., Gao, S., Mao, C.-X., Luo, Q., Hu, W., Yin, Y., & Yang, X. (2018). A wideband dual-polarized antenna using shorted dipoles. *IEEE Access : Practical Innovations, Open Solutions*, *6*, 39725–39733. doi:10.1109/ACCESS.2018.2855425

Yang, D., Liu, S., & Zhao, Z. (2017, May). A broadband dual-polarized printed dipole antenna with low cross-polarization and high isolation for base station applications. *Microwave and Optical Technology Letters*, *59*(5), 1107–1111. doi:10.1002/mop.30467

Zhang, S., Chen, X., & Pedersen, G. F. (2019). Mutual coupling suppression with decoupling ground for massive MIMO antenna arrays. *IEEE Transactions on Vehicular Technology*, *68*(8), 7273–7282. doi:10.1109/TVT.2019.2923338

Zhang, Y., Hong, W., Yu, C., Kuai, Z. Q., Don, Y. D., & Zhou, J. Y. (2008). Planar ultrawideband antennas with multiple notched bands based on etched slots on the patch and or split ring resonators on the feed line. *IEEE Trans. Antennas Propag.*, *56*(9), 3063–3068. doi:.928815 doi:10.1109/tap.2008

Zhu, H. L., Cheung, S. W., Chung, K. L., & Yuk, T. I. (2013). Linear-to-circular polarization conversion using metasurface. *IEEE Transactions on Antennas and Propagation*, *61*(9), 4615–4623. doi:10.1109/TAP.2013.2267712

Chapter 2
Review of Antenna Performance Based on Metamaterials

Devendra Kumar Somwanshi
https://orcid.org/0000-0003-4331-0917
Poornima College of Engineering, Jaipur, India

ABSTRACT

The study of metamaterials has experienced rapid growth, particularly with the discovery of left-handed metamaterials. One promising technology for controlling electromagnetic waves in small structures is transmission-line metamaterial (TL-MTM), which can accurately determine various parameters, such as working frequency, bandwidth, and phase offsets. Antennas are essential components for receiving and transmitting signals in RF systems. Although the microstrip antenna is a popular choice due to its low cost and ease of manufacturing, it is limited in terms of size and frequency range. In this chapter, more than 100 papers are reviewed based on their methodology and performance to discuss various techniques for improving antenna performance, such as the use of artificial neural networks (ANN) and metamaterials. The insights presented in this chapter will aid in identifying, analyzing, and synthesizing the latest developments in the field of metamaterials.

INTRODUCTION

Antennas play a crucial role in wireless communication systems by coupling radio waves to electrical currents used in receivers and transmitters. They intercept and convert electromagnetic waves into voltage for reception or generate radio frequency currents to convert them into radiated electromagnetic waves. Antennas are essential

DOI: 10.4018/978-1-6684-8287-2.ch002

components in various applications, including radio and television broadcasting, wireless LAN, mobile telephony, radar, and satellite communications. Composed of metallic conductors, antennas create alternating magnetic fields when current passes through them, or induce voltages when exposed to distant radio transmitters. These magnetic and electric fields combine to form electromagnetic waves capable of propagating over long distances. Antenna designs can be omnidirectional, transmitting and receiving signals from all directions, or directional, focusing on a specific direction. Ongoing research explores metamaterials and miniaturization of antennas, while numerical modeling aids in predicting their properties prior to construction and testing.

Fabricated structures and composite materials capable of replicating known material responses or demonstrating unique response functions not naturally occurring have garnered significant interest as metamaterials. The remarkable characteristics of these materials arise from the incorporation of intentionally fabricated inclusions or inhomogeneities that are embedded within or connected to a host medium or surface. Artificial electromagnetic materials have a rich history, with notable contributions from J. C. Bose in 1898, who conducted experiments on twisted elements with chiral characteristics, and Karl Ferdinand Lindman in the early twentieth century, who explored metallic helices as artificial chiral media. Following these early explorations, researchers delved into the study of artificial dielectrics, particularly for the development of lightweight microwave antenna lenses. Subsequent investigations expanded to encompass artificial chiral materials, with a focus on their applications in areas like microwave radar absorbers. Advancements in electromagnetic bandgap (EBG) structured materials, single-negative (SNG) materials, and double-negative (DNG) materials, along with their captivating properties, have been the driving force behind the recent surge of interest in metamaterials.

Significant emphasis is placed on the study of complex materials that demonstrate negative permittivity and permeability at specific frequencies. In 1967, Veselago conducted theoretical research exploring the behavior of plane waves in materials exhibiting simultaneous negative permittivity and permeability. This investigation has attracted considerable interest in understanding the unique properties and potential applications of such materials. Metamaterials, a classification of materials with distinctive electromagnetic properties, have been assigned various names to describe their characteristics. These terms include designations like "left-handed" media, "backward-wave media" (BW media), and "double-negative (DNG)" metamaterials. These labels help identify and differentiate the unique features exhibited by these materials. A medium that possesses positive permittivity and permeability ($\varepsilon > 0$, $\mu > 0$) is referred to as a double positive (DPS) medium, which includes naturally occurring dielectrics. On the other hand, epsilon-negative (ENG) media are characterized by negative permittivity and positive permeability ($\varepsilon < 0$, μ

> 0) and are commonly found in plasmas and noble metals such as silver and gold within specific frequency ranges, notably in the infrared (IR) and visible domains. In specific frequency regimes of gyrotropic materials, munegative (MNG) media are characterized by positive permittivity and negative permeability ($\varepsilon > 0, \mu < 0$). Through research efforts, scientists have successfully created artificial materials that display properties of double positive (DPS), epsilon-negative (ENG), and munegative (MNG) media. However, the realization of double-negative (DNG) media, featuring negative permittivity and permeability ($\varepsilon < 0, \mu < 0$), has only been achieved through the use of artificial constructs. Material properties, typically characterized by unchanging values for permittivity and permeability, exhibit a more nuanced nature in reality as they fluctuate with varying frequencies. To capture this frequency-dependent behavior, several material models have been devised, enabling a comprehensive description of how materials interact with electromagnetic waves across different frequency ranges.

Understanding the interaction between electric and magnetic fields is crucial in studying metamaterials. The electric field influences electron motion and dipole moments, while the magnetic field, constrained by the medium's wave impedance, is comparatively smaller. Comprehending this behavior allows for modeling the medium's electric susceptibility and permittivity. In contrast, certain media prioritize magnetic field responses, described by magnetic susceptibility and permeability, mirroring the dual nature of the electric field. Magnetic dipoles, arising from current loop moments, can be mathematically represented using magnetic charge and current analogs. Ongoing metamaterial research focuses on miniaturizing antennas, particularly for personal wireless devices like cell phones. Numerical modeling is instrumental in predicting antenna properties prior to construction and testing.

LITERATURE REVIEW ON PERFORMANCE ENHANCEMENT OF ANTENNA USING METAMATERIALS

In this section, an extensive review of over 100 research papers from the 2005 literature has been conducted, providing a concise summary of their findings and contributions.

Fahad Qureshi et al. (2005) developed a 1.77 GHZ frequency operated low-profile metamaterial ring antenna. The antenna design included two metamaterial unit cells with inductive posts and lumped-element inductors. It demonstrated a well-matched return loss bandwidth of 120 MHz and exhibited a favorable radiation pattern.

S. Wang et al. (2005),developed a thin resonant cavity antenna using metamaterial ground planes with negative reflection phase values. The antenna consisted of a ground plane, a partially reflective surface (PRS) array, and a dipole excitation source.

Ray tracing analysis and simulations were performed to optimize the resonance condition for maximum directivity at broadside.

Laure Mercier et al. (2006) demonstrated smart metamaterials for EBG resonator antennas, enabling beam steering and frequency tunability. Integration of the smart metamaterial enhanced a 4x4 patch antenna array's gain by 6dB and provided tunability across three frequencies (2-3GHz)..

Chun-Yih Wu et al. (2007) designed a metamaterial structure based high-gain patch antenna which consist of metallic rings and Rogers 5880 substrate layers. By incorporating a radome with layered arrays of metallic rings, the antenna improved front-to-back ratio and gained 1.8-2.3 dB in the 5.8-GHz band. It provided a gain equivalent to a 1×2 antenna array.

T. C. Han et al. (2007) aimed to enhance a microstrip antenna's gain and directivity at 3.9 GHz using Left-Handed Metamaterial (LHMTM). CST Microwave Studio simulations showed improved return loss, gain, directivity, and reduced Half Power Bandwidth (HPBW). Antenna fabrication on an FR-4 board increased directivity indicated by the reduced HPBW with return loss improvement of 12 dB.

George V. Eleftheriades et al. (2007) proposed a compact NRI-TL miniaturization techniques. The approach included a 2.6% total area NRI-TL power divider, along with area-reduced metamaterial ring antenna and printed NRI-TL branch-line coupler, maintaining performance. Slightly higher insertion loss was registered with proposed method.

Majid et al. (2007) designed a microstrip patch antenna with left-handed metamaterial structure operating at 2.08-2.20 GHz with low losses at 2.1 GHz. Through this gain and directivity was improved. The design yielded a 1 dB gain and directivity increase, suggesting potential performance enhancement without an antenna array using the modified NRW approach.

A. Ourir et al. (2007) developed a compact resonant cavity for a steerable directive beam antenna. This was achieved through phase-varying metamaterial. Significant size reduction compared to traditional antennas was achieved through the cavity. The cavity consisted a metallic ground plane and PRS reflector. The antenna demonstrated enhanced directivity and achieved beam deflection of approximately $\pm 20°$ through varying the gaps of the cavity. Numerical simulations and experiments supported these findings.

Le-Wei Li et al. (2008) used integrating planner metamaterial patterned structures to design a patch antenna by which bandwidth was increased. LHMs and SRRs were utilized to reduce size and improve radiation. Experimental tests on a rectangular patch antenna showed a wider bandwidth, surpassing the limited bandwidth of conventional antennas. Results demonstrated high gain and effective suppression of vertical radiation, consistent with numerical predictions. Optimization was performed using a full-wave FEM-based simulator to refine parameters.

Shing-Lung et al. (2008) presented a wideband planar EBG structure. The wideband EBG structure achieved a band-gap bandwidth of 58.8% (2.87 GHz to 5.26 GHz), outperforming the Mushroom-type EBG structure with a bandwidth of only 10.8%. Comparative studies showed improved impedance matching and a wider bandwidth (9.6%) for the microstrip patch antenna with resulting in an impressive bandwidth of 12.5%. At last results were compared with Mushroom-type EBG structure.

Ho-Yong Kim et al. (2008) presented an electrically small antenna with a negative permittivity metamaterial structure, aiming for wide Q performance. The antenna underwent optimization and 3D field simulation. The measured Q value of 7.21 was lower than the calculated QL of 108 based on Chu limits, indicating the impact of the negative permittivity metamaterial structure. Future improvements can be explored by combining an effective metamaterial structure with the small antenna design.

Kim et al. (2008) used lumped capacitors and inductors and transformed a T-shaped monopole into a metamaterial antenna. The addition of these elements enabled the maintenance of the desired half-wave resonance within an oversized wire antenna cavity. Through the periodic loading of inductors and capacitors, the resonance frequencies were effectively shifted to higher values. The resulting metamaterial antenna showcased narrower radiation beams at different frequencies and lengths, showcasing the advantages of well-designed metamaterial wires.

Pyo et al. (2009) presented a metamaterial-based antenna design with EBG cells and triangular slots on upper and lower planes. The antenna, fabricated on a RO6010 substrate, exhibited resonant frequency shift and approximately 14.7% enhanced efficiency due to the inclusion of ground plane triangular slots.

Marco et al. (2009) presented a printed monopole antenna with MTM loading, achieving a broad dual-mode operation with a 3.84 GHz bandwidth (3.15-6.99 GHz) on FR4 substrate. The MTM loading enabled exhibiting high efficiencies of 89.2%-92.6% at 5.5 GHz.

J.G. Joshi et al. (2009) proposed an offset-cut DSSR metamaterial antenna resonating at 8.27 GHz with a 490 MHz bandwidth, offering 46 MHz enhancement and 1.5 dBi gain improvement compared to the normal-cut DSSR antenna resonating at 8.11 GHz with a 444 MHz bandwidth and -28dB return loss.

C. Huang et al. (n.d.) developed a compact microstrip balun that incorporated metamaterial transmission lines and classical Wilkinson power dividers. Fabricated on a glass substrate, the balun underwent thorough testing using an RF network analyzer. The results showed promising performance in terms of power splitting and phase characteristics, surpassing the previous design in both size reduction and functionality.

Prathaban Mookiah et al. (2009) studied a metamaterial-based substrate for miniaturized rectangular patch antenna array in MIMO communications. The metamaterial enabled size reduction and improved bandwidth. The antenna array demonstrated similar capacity gains to FR4 array but maintained consistent isolation for different inter-element spacing configurations.

Hafid Griguer et al. (2009) used a metamaterial cell called SHR-W to enhance antenna bandwidth. By incorporating the cell into a square patch antenna on an FR4 substrate, they achieved an 800 MHz bandwidth. When a 3D metamaterial substrate with adjusted negative effective parameters was used, the bandwidth increased to 7 GHz. The metamaterial cell allowed for frequency shift and increased bandwidth.

M. T. Xie et al. (2010) introduced a grounded dielectric substrate AMC unit cell. The unit cell exhibited a reflection phase close to zero at 5.8 GHz and -135° at 6.3 GHz. By incorporating the unit cell into a dipole antenna, lower return loss (-29 dB) and improved gain (8 dBi) were achieved compared to a conventional PEC ground plane. The proposed AMC design offers potential for low-profile and low-backlobe antenna applications.

M.Z.M. Zani et al. (2010) proposed a compact circular antenna on a metamaterial substrate for C-band applications, offering advantages over FR-4. The metamaterial antenna reduced size by 41.667% and achieved a return loss of -24.2 dB. It exhibited higher directivity gain (5.66 dBi) and focused radiation compared to the FR-4 antenna. While the FR-4 antenna had a higher overall gain (8.06 dB), the metamaterial substrate showed potential for further gain enhancements.

Xue-Hua Song et al. (2010) studied an I-shaped left-handed patch antenna using the FDTD method and compared it to a conventional patch antenna. The left-handed design showed improved return loss and maximum gain. The calculated permittivity and permeability values indicated enhanced resonance and concentrated electromagnetic energy. The photonic crystal patch antenna demonstrated lower return loss, higher gain, and enhanced energy coupling and radiation.

J.F. Pintos et al. (2010) introduced a compact UHF band antenna integrated into a portable terminal using meta-material shape, magnetodielectric material, variable capacitance diode, and matching network. HFSS simulations confirmed adherence to constraints and minimal insertion losses.

Atul Thakur et al. (2010) investigated the application of magneto dielectric materials with balanced permeability and permittivity for RF antennas. By utilizing porous and spinel nanoferrites of specific composition, they observed low magnetic and dielectric losses within the 10-200 MHz frequency range.

Loïc Markley et al. (2011) discussed recent progress in near-field focusing and imaging using meta-screens and near-field antenna arrays. Experimental verification of super-resolution was achieved in both planar and free-space setups. Subwavelength imaging was demonstrated in one and two dimensions using antenna arrays. Buried object detection was also explored. These findings highlight the potential of these technologies for subwavelength imaging across different frequencies, from microwaves to optics.

J.G. Joshi et al. (2011) improved the bandwidth and gain of a dual-band microstrip patch antenna using a split ring resonator (SRR) array and metallic via hole loading. Two configurations of metamaterial-loaded antennas were tested, resulting in enhanced performance with increased bandwidth and gain at multiple resonant frequencies. The inclusion of metallic via holes further improved the antenna's performance, validating the proposed design.

G. M. F. Freitas et al. (2011) investigated a composite medium of split ring resonators (SRR) and thin wires (TW) with negative effective permittivity and permeability in the frequency range of 2.1-2.5 GHz. Artificial neural networks were used to optimize the metamaterials' parameters, demonstrating the potential for creating left-handed materials with negative refraction.

Youngki Lee et al. (2011) proposed a WiMax MIMO antenna using an SRR array structure for enhanced isolation. The antenna achieved a wide bandwidth (3.38 GHz to 3.7 GHz) with a 10 dB return loss. At 3.5 GHz, the antenna exhibited negative permeability and demonstrated high isolation across the frequency range. HFSS V11.1 was employed for design and analysis.

Mondher Labidi et al. (2011) developed a circularly polarized antenna array for X-band frequency using circular Split Ring Resonators (SRRs). The array comprised eight individual SRRs arranged on an FR4-epoxy substrate with a permittivity of 4.4 and $\tan\delta=0.0197$.

Ying Peng et al. (2011) introduced a compact, low-profile dipole antenna with a wide bandwidth of 7 GHz for high-speed data transmission at 60 GHz. The antenna employed a metamaterial resonant pit as a reflector and 100 Ω coplanar strip lines for feeding. It achieved a gain of 4.6 dB and demonstrated improved performance in terms of bandwidth and gain compared to conventional designs. Simulation using Ansoft HFSS v11 validated its effectiveness in achieving multi-Gb/s data transmission at 60 GHz.

Hassan Mirzaei et al. (2011) proposed a compact printed monopole antenna with frequency reconfigurability using a negative-refractive-index transmission line and varactor diode. The antenna achieved a wide 10-dB return loss bandwidth of 1.6 to 2.23 GHz. It offered a smaller size and broader tuning range compared to other antennas, although the circuit model's accuracy was limited at higher frequencies.

Yuandan Dong et al. (2011) developed a low-profile circularly-polarized patch antenna. For this they used composite right/left-handed mushroom-like structures and a responsive impedance surface. The antenna achieved a 39% size reduction compared to a regular patch, but had lower radiation efficiency and bandwidth. Dual-frequency operation was also demonstrated.

Singh et al. (2011) studied the use of metamaterials in antenna design. They developed miniaturized antennas using CRLH metamaterial transmission lines and utilized composite metamaterials for a steerable printed antenna. They discussed phase reflection, EBG ground planes, and metamaterial lens for improved antenna performance.

M. Malathong et al. (2011) introduced a chiral metamaterial polarizer to enhance circular polarization in patch antennas. By incorporating a modified C8 bi-layer structure in front of the antenna, the polarizer increased the axial ratio bandwidth by 4 MHz. CST Microwave Studio was employed for simulation purposes.

Surabhi Dwivedi et al. (2012) introduced a triband open patch antenna inspired by metamaterials, operating at a center frequency of 2.5 GHz. The antenna utilized four slots to resonate at three frequencies and employed a Parted Ring Resonator for size reduction. By using an alternative dielectric material, the antenna achieved a 36.7% reduction in substrate size and a 12.43% increase in bandwidth. The antenna's performance was simulated using HFSS 11.v software.

Bashir D. Bala et al. (2012) proposed a metamaterial antenna on FR4 substrate using split ring resonators and thin wires. Operating at 2.42 GHz in left-handed metamaterial region, it achieved a 5.58% bandwidth, outperforming conventional patch antennas. Simulated using HFSS 11.v software with six unit cells.

V.V. Gozhenko et al. (2012) investigated the effect of a mesh metamaterial slab. For this they used the patch antenna with frequency range of 9.8-12.0 GHz and analyzed it's E-plane radiation pattern. The metamaterial slab, comprised of spaced plates, was introduced to assess its influence on the antenna's radiation characteristics.

Abdallah Dhouibi et al. (2012) designed a miniaturized Luneburg lens antenna with a compact planar feeding structure and substrate-integrated GRIN material. The lens, incorporating a complementary closed-ring resonator array, exhibited highly directive emission and estimated directivity of 13.5-16 dBi, making it suitable for advanced communication systems.

N. Fhafhiem et al. (2012) proposed a high gain circularly polarized antenna design using a double polarizing metallic electromagnetic bandgap (EBG) structure. The antenna achieved a gain of approximately 19.04 dB at 2.45 GHz, with a 3 dB axial ratio over the frequency range of 2.43-2.61 GHz. The antenna's vertical and horizontal half-power beamwidths were around 18° and 18.5°, respectively.

Sarawuth Chaimool et al. (2012) developed a miniaturized square patch antenna by incorporating an artificial magneto-dielectric metamaterial substrate. The antenna featured a winding circle array with varying periodicities on an FR4 dielectric substrate. By adjusting the twisting and substrate dimensions, the resonant frequency could be tuned. The proposed antenna achieved a significant 65% size reduction. It exhibited a high radiation efficiency of 72% and a respectable directivity of 6.24 dBi. The antenna design was verified through HFSS simulations.

X. Zhao et al. (2012) proposed a compact loop antenna design incorporating split ring resonators (SRR) and a modified rectangular loop element. The antenna achieved a wide 10 dB bandwidth of 1.6 GHz (53% fractional bandwidth) and demonstrated frequency scanning capabilities. Measured results showed a 10 dB bandwidth of 2.3 to 3.8 GHz (50% fractional bandwidth) with rotating main beams and an average gain above 2 dBi, surpassing traditional loop antennas.

Marco A. Antoniades et al. (2012) introduced a novel stacked dipole metamaterial antenna with multiple resonances at unrelated frequencies. Unlike existing designs, this antenna utilizes a biconical transmission line for improved performance. Simulation results demonstrate three distinct resonances at different frequencies, along with high radiation efficiency. This antenna shows promise for applications in multistandard wireless communication systems and compact devices like cell phones.

D.R. Luna et al. (2013) proposed a hybrid optimization approach using a genetic algorithm and MLP-ANN models to enhance the effective permittivity of a metamaterial. The method focused on a metal thin wire artificial medium and aimed to find an optimal solution. Although the hybrid approach provided valuable insights, it occasionally resulted in imbalanced outcomes with resonant frequencies deviating too closely or far from the desired frequency.

P. M. Paul et al. (2013) proposed a compact microstrip antenna design utilizing meta-material structures. The antenna featured a square patch on an FR4 substrate, with complementary split rings (CSRR) and a conducting disk for size reduction and performance optimization. By adjusting the CSRR dimensions and quantity, the resonant frequency could be controlled. Experimental results demonstrated resonant frequencies of 2.21 GHz and 2.57 GHz for different CSRR configurations.

Mahmoud A et al. (2013) proposed a compact two-element metamaterial MIMO antenna for wireless applications. The antenna, operating at 5.8 GHz, utilized a left-handed transmission line unit cell and a defected ground structure (DGS) filter to reduce coupling. With dimensions of 2.36 x 2.6 cm, the antenna achieved over 50% size reduction compared to conventional patch antennas, while maintaining a return loss better than 15 dB and a mutual coupling isolation of 45 dB. Experimental results validated the design's performance.

Clinton P. Scarborough et al. (2013) introduced a low-sidelobe horn antenna with a super-broadened C-band using metamaterial technology. The antenna employed a metallic wire structure as a unit cell on the inner walls, achieving a gain range of 16 to 24 dB. Simulations demonstrated reduced sidelobes and backlobes compared to other horns, maintaining performance over a wide bandwidth. To address the design challenge of a low-index liner, an unsupported wire matrix was utilized instead of a dielectric substrate.

E. S. Pires et al. (2013) introduced three compact metamaterial-inspired antennas: a dual-band antenna operating at 2.4 GHz and 5.5 GHz, a wire antenna with series capacitance, and a folded dipole antenna operating at 2.5 GHz. These antennas exhibited bandwidths from 39 MHz to 115 MHz and gains ranging from 0.5 dBi to 2 dBi, offering versatility for various applications.

Tao Liu et al. (2013) utilized a metamaterial absorber (MA) to decrease the radar cross section (RCS) of a waveguide slot antenna. The MA consisted of two metallic layers separated by a lossy dielectric spacer. Simulation results demonstrated a peak absorptivity of 99.8% at 5.57 GHz with a 220 MHz bandwidth. The integrated antenna with the MA exhibited an RCS reduction exceeding 3 dBsm from 5.48 GHz to 5.7 GHz, with a maximum reduction of 16.6 dBsm at 5.58 GHz. Experimental measurements confirmed the effectiveness of the MA, showcasing its potential for achieving broadband RCS reduction.

Hayrettin Odabasi et al. (2013) explored the radiation properties of electrically small resonator antennas, including ELC, CELC, and SRR configurations. Simulations indicated fractional bandwidths from 0.030 to 0.055 and -10dB bandwidths from 0.99% to 1.88%. The radiation patterns and directivities varied among the antennas, with CELC resonators exhibiting promising features for electrically small metaresonator-based designs.

Kumud Ranjan Jha et al. (2013) explored the effect of a low-relative dielectric permittivity substrate on a microstrip antenna at 600 GHz. By incorporating periodic air cylinders, the effective permittivity of the photonic bandgap material decreased. The microstrip antenna designed on the PBG crystal exhibited a downward shift in resonance frequency with increased patch length and a marginal increase in gain, suggesting potential benefits for enhancing directivity.

Yang Shen et al. (2013) studied a patch antenna with composite rectangular split ring resonators (SRRs) using FDTD and FEM methods. The antenna, operating in multiple frequency bands including GSM, DCS, PCS, and WLAN, achieved significant gain enhancement by incorporating SRRs, indicating its potential for improved performance.

Albert Sabban et al. (2013) designed high-efficiency antennas for communication and medical applications using metamaterials. They utilized a new class of metamaterials and designed microstrip loaded and printed slot antennas with optimized dimensions and polarization. The antennas demonstrated resonant frequencies around 370-400 MHz and gains of 4-5 dBi. The results showed improved efficiency and potential for further enhancements in antenna design.

Yi Zhang et al. (2014) presented a compact dual-port dual-mode metamaterial loop antenna for WLAN applications. The antenna incorporated capacitive arc strips and hybrid feed networks to achieve pattern diversity. It offered a 5% shared frequency band, >25 dB isolation, and compact size.

Wei Liu et al. (2014) optimized and fabricated an 8x8 thin planar mushroom antenna array on LTCC substrate for 60-GHz applications. The array achieved a maximum gain of 8.0 dBi and demonstrated excellent performance in millimeter-wave frequencies, as confirmed by experimental results.

A. Dadgarpour et al. (2014) introduced metamaterial stacking to redirect radiation from a planar antenna. The proposed approach utilized a modified H-shaped resonator unit-cell structure arranged in a 3x4 array within the antenna's substrate. Experimental validation confirmed improved performance and a 17-degree shift in the main radiation beam.

Yahong Liu et al. (2014) introduced a perfect absorber metamaterial (PAM) with 98% absorptivity at 10.52 GHz. When applied to a rectangular patch antenna, the PAM reduced the RCS peak by 15 dB in the forward direction, maintaining radiation pattern and gain, but with no improvement in the backward direction.

Xiaoliang Mama et al. (2014) proposed a horn antenna which was a dual circularly polarized. In the antenna a chiral metamaterial was used which was a two-layered metallic arc structure. The antenna operated at 12.0-12.5 GHz (LCP) and 14.0-14.5 GHz (RCP). The integrated metamaterial improved axial ratio and incurred minimal gain degradation of 0.6 dB. Experimental validation in an anechoic chamber confirmed the antenna's performance.

G.K. Pandey et al. (2014) proposed a low-cost, compact metamaterial-based patch antenna for UWB applications. It achieved a wide −10 dB impedance bandwidth of 3 GHz to over 14 GHz (129% bandwidth) on an FR4 substrate. The antenna demonstrated promising radiation efficiency and unique characteristics for UWB applications.

Li et al. (2014) proposed an electrically small metamaterial (MTM) antenna with ELC loading. The antenna achieved a dual-band triple-mode response by incorporating two strip lines and a CELC structure. The measured bandwidth, with merged modes, was 20.5% on a Taconic RF-35 substrate.

Ji-jun Wang et al. (2014) investigated a composite helices patch antenna with left-handed effect. They used FDTD methods to analyze its characteristics and calculate the equivalent permittivity and permeability. The composite helices structure improved the return loss and gain.

Mahdi Oliaei et al. (2014) developed a compact Substrate Integrated Waveguide (SIW) antenna with high gain and a wide bandwidth of 23.8 GHz to 25.7 GHz. The antenna incorporated loaded dielectrics for improved performance and narrow beam widths. With a size of 82 mm × 40 mm × 2.54 mm, the antenna achieved 12.5 ± 1 dB gain over the K band and approximately 95% radiation efficiency, making it suitable for various communication distances.

He Huang et al. (2015) proposed a compact monopole antenna which was loaded with multiband metamaterial and specially used for WLAN and WiMAX applications. The antenna achieved wideband coverage from 2.3 to 4.1 GHz and from 5 to 6.6 GHz, with measured impedance bandwidths of 1.7 GHz and 1.6 GHz. It exhibited omnidirectional H-plane patterns, bidirectional E-plane patterns, and gains ranging from 2.34 dBi to 3.2 dBi.

Ahmed Elmi and Ahmed Kishk (2019) proposes a reconfigurable metasurface antenna design based on graphene for wireless communication systems. The proposed design uses graphene-based switches to achieve reconfigurability, and the experimental results show that the proposed design achieves good performance.

Hassan B. Kazemian et al. (2020) proposes a compact reconfigurable metasurface antenna design for multiband wireless communication systems. The proposed design uses a combination of circular and square-shaped patches, and the experimental results show that the proposed design can achieve good performance across multiple frequency bands.

Qian Gao et al. (2021) proposes an intelligent reflecting surface (IRS) design with adjustable phase and amplitude for wireless communication systems. The proposed design uses a combination of unit cells and varactors to achieve tunability, and the experimental results show that the proposed design can significantly improve the signal-to-noise ratio of wireless communication systems.

COMMON FINDING OF PERFORMANCE ENHANCEMENT OF ANTENNA USING METAMATERIALS

- Various types of metamaterials like SRR, NRI-TL, ZORA, EBG, DNG etc are used for performance enhancement of antenna.
- In the field of antenna the most popular applications of metamaterial are mobile phones and radar cross section.
- Different type of unit cells like split ring, spiral shape, square ring, mashroom, H- shaped, Jshaped, CRR etc are used.
- Using metamaterials we can improve gain, bandwidth, insulation, radiation pattern and efficiency upto over 90%
- Metamaterials in association with microstrip patch antenna improve the gain by 19.47dB, directivity 2dbi, bandwidth of 189% and reduce the size by 61%.
- Most of the researchers prefer circular and rectangular split ring resonators.
- Artificial magnetic conductor is better than perfectly magnetic conductors.
- CRLH metamaterial Tx line reduce the size of half wave dipole antenna (upto lamda/6) as compared to half wave dipole antenna without CRLH metamaterial without affecting radiation pattern.
- Use of electronic band gap structure in microstrip antenna array improved the gain of the patch antenna array by 6dB
- Use of Meta material can also increases the directivity without using the antenna array by 2dBi.
- Optimization of antenna parameters is proposed by most researchers.

Table 1. Solution approach for performance enhancement of antenna using metamaterial

Type of Meta-Material Used	Antenna Used / Frequency Band	Software/ Hardware Used	Technique Adopted	Remarks (Data Used)
MEBG	MPA / 14 GHz	TLM-based software Microstripes	EBG superstrates	Distance between the PRS and ground plane reduced to 3.35mm at 14GHz
LHMTM (Split Ring Resonator and the Thin Wire Structure)	MPA / 2.4	Math CAD/ CST Microwave Studio	Nicolson-Ross-Weir (NRW) Approach	Gain Improved by 1 dB.
Phase-varying 1-D MTM	MPA/ 10 GHz	HFSS	MTM-based Fabry-Pe´rot cavity with a minimal reflection 10.5	A deflection of the antenna beam of +/-200 has been observed.
3D-SHR-W LH-MTM	MPA/10GHz	HFSS	MPA printed on a 3D SHR-W MTM substrate	–ve MTM
Artificial Magnetic Conductor AMC	MPA / 5.8GHz	Not mentioned	Two dipoles Used	8dBi Gain
Chiral metamaterial	CP MPA / 2.37-2.4GHz	CST	Chiral structures allowed to propagate only a dominant polarized wave through	AR bandwidth and radiating pattern of the proposed chiral MTM based CP antenna were improved.
Dual log-spiral	MPA / Xband	CST	10×10 DLSRs used	Variation of 'L' against N in the log-spiral conductor was used
Double polarizing metallic EBG	Circular polarize metallic antenna / 2.45GHz	Not mentioned	45°oriented curved strip dipole Used	The gain improved 19.04 dB
Split Ring Resonator (SRR) array	MIMO /3.4-3.6 GHz	HFSS	The SRRs printed on top of FR4 substrate	Higher Isolation of 15 dB was used
S-shaped metallic rings 10X10 array.	Planar MPA/ 5.8GHz	Not mentioned	4 layer + 3 layer of RT-duroid substrate	Antenna gain improved to 2.3 dB.
Spiral loop resonator	MPA / 1.8GHz	CST	AMD metamaterial used as a substrate	65% size reduction was obtained
NRI-TL	Coplanar waveguide \1.6 to 2.23 GHz.	Full wave EM simulator	varactor diode used	Return loss = 10 dB bandwidth 1.6 to 2.23 GHz.
Periodically loaded capacitive arc strips	Planar loop antenna /2.4-2.52 GHz	HFSS	Frequency band = 2.40 to 2.52 GHz	High Isolation achieved

continued on following page

Table 1. Continued

Type of Meta-Material Used	Antenna Used / Frequency Band	Software/ Hardware Used	Technique Adopted	Remarks (Data Used)
Inductive spiral loop	MPA / 2.48GHZ	HFSS/T-Tech QC5000 milling machine	Stacking unit based material used	Efficiency and gain was not significantly improved
Mushroom structure	MPA array / 59.2–61.8 GHz	CST	Substrate integrated waveguide	Simulated gain WAS higher than 22.8 dBi
Complementary non resonant metamaterial	Luneburg lens antenna / 8-12 GHz	CST	Substrate-Integrated Luneburg Lens Antenna	Directive emission was high
CRLH mushroom	MPA / 2.6GHz	HFSS	Combination of CRLH and RIS are used	Gain improved
Perfect absorber metamaterial	Low RCS patch antenna / 10.5 GHz	CST	The PAM patterns printed on the surface of a MPA to reduce the antenna RCS	The absorptivity at 10.52GHz is 93%, 95.4%, and 99.8%, respectively
H-shaped unit-cell	MPA / 9.25GHz	HFSS	MTM structure was configured for high refractive index to implement beam tilting in a bow-tie	E-plane main beam tilted with 17 degrees
Metahorn	Horn antenna / 3.4-7GHz	FEKO/HFSS	Wire Grid used	Inaccuracies in dimension observed
Inverted L	Monopole MPA / 2.4-GHz WLAN and 2.5- GHz WiMAX	HFSS	Design a rectangular patch then cut inverted L	Generate all WLAN and WiMax band
Diamond shaped split rings	MPA / 490 MHz	IE3D	MM characteristics exhibits	Bandwidth of 490 MHz
ELC and EBG Loading	MPA / 2.49, 2.55, 3.0, 3.68, and 5.03, 6.04 GHz	HFSS	Two novel tri band MTM antennas with one ELC and another EBG loaded	2.49, 2.55, 3.0, 3.68 and 5.03 6.04 GHz are achieved

- The experiment on various types of antennas is performed by researchers to choose the best one for a specific problem.
- The outcomes are obtained by varying the parameters of the antenna.
- Bulky size reduction of antennas is the main objective of most experiments conducted.

Figure 1 represents that HFSS based on finite element method and CST based on Methed of Moments anaysis are most popular to design and development of antenna

Figure 1. Classification of software used in solution approach by metamaterial

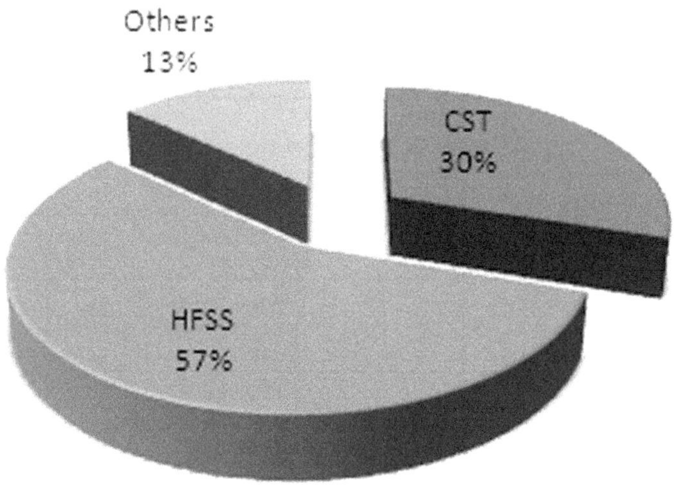

and other microwave devices. Simillarly by Figure 2 the most popular antenna and devices are of microstrip type when the design is metamaterial inspired.

Figure 3 represents that wireless mobile communication is most popular field for researchers. They are trying to do more size reduction of antenna and other high frequency devices used in mobile, Wi Fi, WiMAX, GPS etc. Figure 4 shows the

Figure 2. Classification of antennas used in solution approach by metamaterial

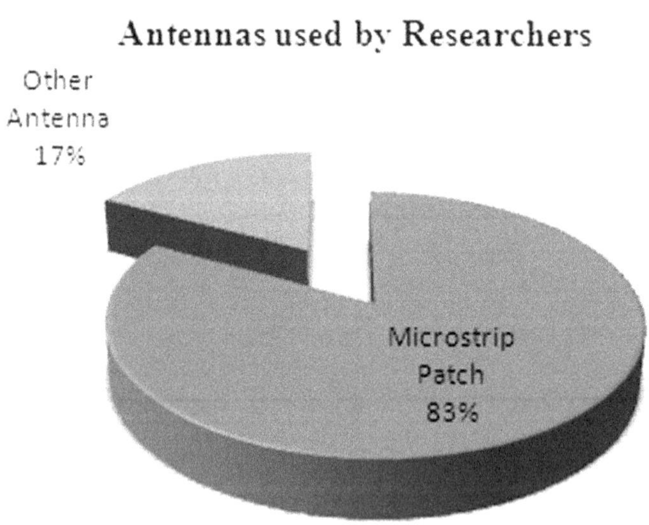

Figure 3. Classification of applications used in solution approach by metamaterial

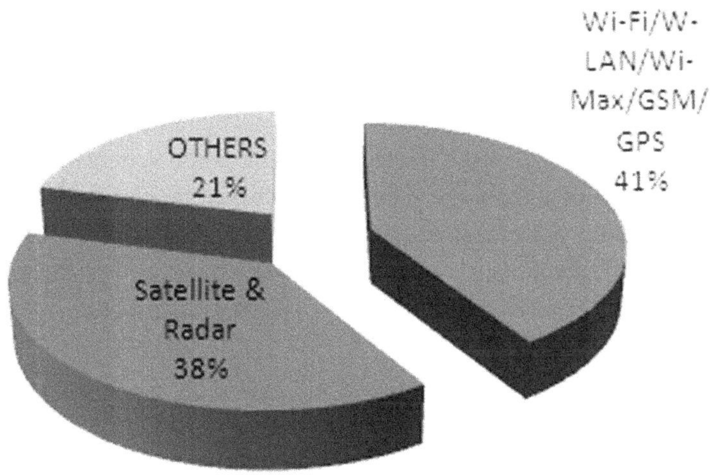

Figure 4. Classification of unit cells used in solution approach by metamaterial

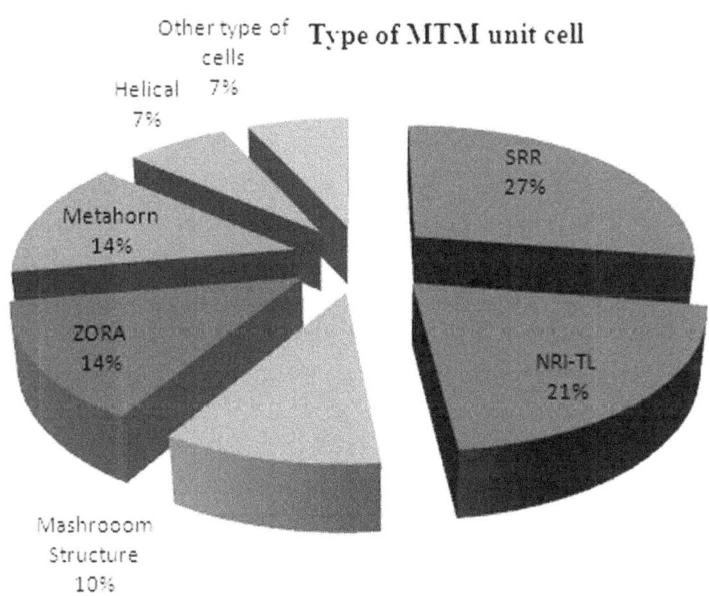

various types of metamaterial unit cells used by researchers for various applications. There is no certain region behind the choice of a particular type of unit cell. By the figure it is obvious that NRI-TL and SRR type of metamaterials are widely used in various applications.

Strengths

- Advances in nanotechnology have enabled the fabrication of metamaterials for a wide frequency range, including optical frequencies.
- Metamaterials can be used in antenna design to achieve size reduction, gain enhancement, and bandwidth improvement.
- Using metamaterials in microstrip antennas can suppress surface waves, resulting in gain enhancement and improved radiation pattern.
- Other types of metamaterials, such as chirality, perfectly absorbing, and low index metamaterials, have potential for beam-forming applications.
- Designing a metamaterial with desired properties is a challenging synthesis process that requires accurate field solvers and efficient numerical optimizers.
- Combining field solvers with numerical optimizers is the most promising approach to designing efficient metamaterials.
- Improvement of antenna gain can be achieved through conventional methods without the need for bulky structures.
- Greater bandwidth and return loss are exhibited by antennas when Uniplanar compact photonic Band Gap (PBG) structures are utilized.
- Utilization of a series of single layer SRR samples with different gap sizes enables the acquisition of successive narrowband THz resonances over a wide THz range.

Weaknesses

- Metamaterials have a limited range of wavelengths that they can operate effectively in, as their electromagnetic properties need specific values at each point in the grid. This leads to devices often operating at a single wavelength.
- Metamaterials work by creating local electromagnetic fields through the movement of electrons in the metal. They work passively, but this can lead to losses.
- Metamaterials require high precision fabrication to maintain their quality and minimize losses.
- The shape of a metamaterial cannot be changed during operation, as their design is strongly connected to their electromagnetic properties in a fixed position.
- The choice of a particular type of unit cell in metamaterial design is not clearly stated in the reviewed papers, but NRI-TL and SRR types are commonly used in various applications.
- The optimization methods used are not universally applicable but are specific to each individual problem they address.

CONCLUSION

The review of more than 100 papers highlighted the importance of antenna design in RF systems for transmitting and receiving signals. Microstrip antennas are widely used due to their low cost and ease of fabrication, but they have limitations in size and frequency range. The use of artificial neural networks and metamaterials, particularly transmission-line metamaterial (TL-MTM) technology, has shown promise in improving antenna performance. Metamaterials offer precise control over electromagnetic waves and can manipulate their propagation in confined spaces, enabling the design of electrically small antennas with sufficient wireless coverage. This review sheds light on the latest developments in miniaturized patch antennas and the use of Tx-line metamaterials to address challenging compact antenna design issues.

REFERENCES

Abdalla, M., Abdelnaby, U., & Mitkees, A. A. (n.d.). Compact and Triple Band Meta- material Antenna for All WiMAX Applications. *Proceedings of ISAP2012.*

Abraham, J. K., Vinoy, K. J., Tellakula, R. A., & Varadan, V. K. (2003). Experimental Results of Planar Meta Materials. *IEEE- APS International Symposium.*

Ali, M., Sayem, A., & Kunda, V. (2007, March). A reconfigurable stacked microstrip patch antenna for satellite and terrestrial links. *IEEE Transactions on Vehicular Technology*, *56*(2), 426–435. doi:10.1109/TVT.2007.891412

Antoniades, M. A., & Edward, S. (2012, December). Multiband Compact Printed Dipole Antennas UsingNRI-TL Metamaterial Loading. *IEEE Transactions on Antennas and Propagation*, *60*(12), 5613–5626. doi:10.1109/TAP.2012.2211324

Antoniades, M. A., Eleftheriades, G. V., & Rogers, E. S. (2009). A broadband dual-mode monopole antenna using NRI-TL metamaterial loading. *IEEE Antennas and Wireless Propagation Letters*, *8*, 258–261. doi:10.1109/LAWP.2009.2014402

Arun Kumar, K. (2010). Optimization of Multislotted Rectangular Microstrip Patch Antenna Using ANN Becterial Foraging Optimization. *Asia-Pacific Symposium on Electromagnetic Compatibility (APEMC)*, 449 – 452.

Bala, B. D., Rahim, M. K. A., & Murad, N. A. (2014, January 30). Small electrical metamaterial antenna based on coupled electric field resonator with enhanced bandwidth. *Electronics Letters*, *50*(3), 138–139. doi:10.1049/el.2013.3884

Bashir, D. (2012). Bandwidth Enhanced Microstrip Patch Antenna Using Metamaterials. *IEEE Asia-Pacific Conference on Applied Electromagnetics (APACE)*, 280 – 282.

Bhagat, P. P. (2012). Analysis and Synthesis of Microstrip Patch Antenna using Artificial Neural Networks. *IEEE Asia-Pacific Conference on Antennas and Propagation*. 10.1109/APCAP.2012.6333177

Campbell, S. D., & Ziolkowski, R. W. (2013). Lightweight, Flexible, Polarization-Insensitive, Highly Absorbing Meta-Films. *IEEE Transactions on Antennas and Propagation, 61*(March), 1191–1200. doi:10.1109/TAP.2012.2227658

Chaimool, S., Pinsakul, A., & Akkaraekthalin, P. (2012). Patch Antenna Miniaturization Using Artificial Magneto-Dielectric Metasubstrate. *Proceedings of ISAP2012*, 906-909.

Choi, D., Shin, C., Kim, N., & Shin, H. (2005). Design and SAR Analysis of Broadband PIFA with Triple Band. *Progress in Electromagnetics Research Symposium, 1*(3), 290–293. doi:10.2529/PIERS041212211650

Dadgarpour, A., Zarghooni, B., Virdee, B., & Denidni, T. A. (2014). Beam Tilting Antenna Using Integrated Metamaterial Loading. *IEEE Transactions on Antennas and Propagation, 62*(May), 2874–2879. doi:10.1109/TAP.2014.2308516

Delaune, D., Guan, N., & Ito, K. (2008). Simple Multiband Antenna for Mobile Phone Application Based on a Dual-arm Monopole Structure. PIERS Online, 4(1).

Delgado, H. J. (2005). A Novel Neural Network for the Synthesis of Antennas andMicrowave Devices. *IEEE Transactions on Antennas and Propagation, 16*, 1590–1600. PMID:16342499

Dhouibi, A., Burokur, S. N., Lustrac, A., & Priou, A. (2012). Compact Metamaterial-Based Substrate-Integrated Luneburg Lens Antenna. *IEEE Transactions On Antennas And Propagation, VOL., 11*, 1504–1507.

Dong, Y., Toyao, H., & Itoh, T. (2011). Compact Circularly-Polarized Patch Antenna Loaded With Metamaterial Structures. *IEEE Transactions on Antennas and Propagation, 59*(11), 4329–4333. doi:10.1109/TAP.2011.2164223

Dudczyk, J., & ... Applying the neural networks to formation of radiationpattern of microstrip antenna. *IEEE International Radar Symposium*, 1-4.

Eleftheriades, G. V., Antoniades, M. A., & Qureshi, F. (2007, February). Antenna applications of negativerefractive- index transmission-line (NRI-TL) structures. *IET Microwaves, Antennas & Propagation, 1*(1), 12–22. doi:10.1049/iet-map:20050345

Erentok, A., & Ziolkowski, R. W. (2007). Efficient MetamaterialInspired ElectricallySmall Electric-Based Antennas: Two- and ThreeDimensional Realizations. *Proc. of Antennas and Propag. Society International Symposium*, 1309-1312.

Fhafhiem, N., Krachodnok, P., & Wongsan, R. (2012). The Circularly Polarized Resonator Antenna using Double Polarizing Metallic EBG. *IEEE Asia-Pacific Conference on Antennas and Propagation*. 10.1109/APCAP.2012.6333133

Foroozesh, A., & Shafai, L. (2006). Size reduction of a microstrip antenna with dielectric superstrate using meta-materials: Artificial magnetic conductors versus magneto- dielectrics. *Proc. IEEE Antennas Propag. Soc. Int. Symp.*, 11–14. 10.1109/APS.2006.1710439

Freitas, G. M. F., Rêgo, S. L., & Vasconcelos, C. F. L. (2011). *Design of Metamaterials Using Artificial Neural Networks. In International Microwave & Optoelectronics Conference*. IMOC.

Gao, C., & Wang, Y. (2002). Analysis of EBG Structures Implemented on CPW Components by using EM-ANN Models. *Antennas and Propagation Society International Symposium,* 4, 368 – 371. 10.1109/APS.2002.1016999

Geyi, W. (2012). Optimum Design of Antennas. *Asia-Pacific Conference on Applied Electromagnetics*, 116-117.

Gianvittorio, J., & Yah, A. (2002, February). Fractal Antennas: A Novel Antenna Miniaturization Technique, and Applications. *IEEE Antennas & Propagation Magazine*, *44*(1), 20–36. doi:10.1109/74.997888

Gianvittorio, J. P., & Rahmat, S. Y. (2002, February). Fractal Antennas: A Novel Antenna Miniaturization Technique, and Applications. *IEEE Antennas & Propagation Magazine*, *44*(1), 20–36. doi:10.1109/74.997888

Gopalakrishnan, R., & Gunasekaran, N. (2005). Design of equilateral triangular microstrip anten using artificial neural networks. *IEEE International Workshop on Antenna Technology: Small Antennas and Novel Metamaterials*, 246 - 249.

Gozhenko, Wigle, Vedral, Hendrix, & Pinchuk. (2012). *Focusing effect of a metamaterial slab on the radiation pattern produced by a patch antenna*. Elsevier B.V.

Griguer, H. (2009). Patch Antenna Bandwidth Enhancement through the Use of Metamaterials. *International Conference on Telecommunications*, 323-327.

Guo, Y.-X. (2003). Compact internal multiband antennas for mobile handsets. *IEEE Antennas and Wireless Propagation Letters*, *2*, 143–146. doi:10.1109/LAWP.2003.815278

Guo, Y. X., Ang, I., & Chia, M. Y. W. (2003). Compact internal multiband antennas for mobile handsets. *IEEE Antennas and Wireless Propagation Letters*, *2*, 143–146. doi:10.1109/LAWP.2003.815278

Han, Rahim, Masri, & Karim. (2007). Left handed metamaterial design for microstrip antenna application 2007. *APMC 2007. Asia-Pacific, 1-4 Microwave Conference.*

Hassan Mirzaei & Eleftheriades. (2011). Frequency Reconfigurable Metamaterial Inspired Antenna. IEEE Antennas and Wireless Propagation Letters, 10.

Hee, T. W., Hall, P. S., & Song, C. T. (2003). Fractal PIFA, Dipole and Monopole Antennas. *IEEE Topical Conference on Wireless Communication Technology,* 275-276. 10.1109/WCT.2003.1321522

Ho-Yong. (2008). The Design of a Small Monopole Type Antenna using Negative Permittivity Metamaterial Structure. *Antenna Technology: Small Antennas and Novel Metamaterials, 2008. iWAT 2008. International Workshop on, 2008,* 275-278.

Huang, C., Chen, D., Leidich, S., & Gessner, T. (2010). Compact Meta-material Transmission Line Balun Based on Meander Lines Structure and MEMS Technology. *Microelectronic Engineering*, *87*(4), 584–587. doi:10.1016/j.mee.2009.08.018

Huang, Liu, & Gong. (2015). Multiband Metamaterial-Loaded Monopole Antenna for WLAN/WiMAX Applications. *IEEE Antennas and Wireless Propagation Letters, 14*, 662-665.

Joshi, J. G., Pattnaik, S. S., & Devi, S. (2011). *Metamaterial loaded dual band microstrip patch antenna. In Antenna Week.* IAW.

Joshi, J. G., Pattnaik, S. S., Devi, S., Lohokare, M. R., & Vidyasagar, C. (2009). Offset fed diamond shaped split ring (DSSR) planar metamaterial antenna. *IEEE Conference on Applied Electromagnetics*, 1-4. 10.1109/AEMC.2009.5430686

Jouvaud, C., de Rosny, J., & Ourir, A. (2013, April 11). Adaptive metamaterial antenna using coupled tunable split-ring resonators. *Electronics Letters*, *49*(8), 518–519. doi:10.1049/el.2013.0398

Kataja, J., & Nikoskinen, K. (2011). The Parametric Optimization of Wire Dipole Antenna. *IEEE Transactions on Antennas and Propagation*, *59*(2), 350–356. doi:10.1109/TAP.2010.2096181

Kern, D. J., Werner, D. H., Wilhelm, M. J., Church, K. H., & Mittra, R. (2002). Multi-band high impedance frequency selective surfaces. *Proc IEEE AP-S/URSI Symp Dig*, 264.

Kim, D., & Kim, M. (2008, January 31). Narrow-beamwidth T-shaped monopole antenna fabricated from metamaterial wires. *Electronics Letters*, *44*(3), 179–180. doi:10.1049/el:20082854

Labadi, M., Jamel, B. T., & Choubani, F. (2011). New design of antenna array using Left Handed Meta-Material "LHM" based on circular Split Ring Resonator "SRR". *Mediterranean Microwave Symposium (MMS),* 52-56.

Lee, C. H. L., & Itoh, T. (2006). Compact dual-band antenna using an anisotropic metamaterial. *Proceedings of the 36th European Microwave Conference*, 1044-1047. 10.1109/EUMC.2006.281112

Lee, Y., Ga, D., & Choi, J. (2011). Design of a MIMO Antenna with Improved Isolation Using MNG Metamaterial. *International workshop on Antenna Technology,* 231 – 234.

Li. (2013). Design of Electrically Small Metamaterial Antenna With ELC and EBG Loading. *IEEE Antennas and Wireless Propagation Letters, 12*, 678-681.

Li, L.-W. (2008). Design of a Novel Rectangular Patch Antenna with Planar Metamaterial Patterned Substrat. *International Workshop on Antenna Technology: Small Antennas and Novel Metamaterials*, 124-126.

Li, Z. (2000). PBG, PMC and PEC Ground Planes: A Case Study of Dipole Antennas. *IEEE Antennas and Propagation Society International Symposium (APSURSI),* 2, 674 - 677.

Liu, W., Chen, Z., & Qing, X. (2014, September). 60-GHz Thin Broadband High-Gain LTCC Metamaterial- Mushroom Antenna Array. *IEEE Transactions on Antennas and Propagation*, *62*(9), 4592–4601. doi:10.1109/TAP.2014.2333052

Liu, Y., & Zhao, X. (2014). Perfect Absorber Metamaterial for Designing Low-RCS Patch Antenna. IEEE Antennas and Wireless Propagation Letters, 13.

Liu, Z. D., Hall, P., & Wake, D. (1997, October). Dual-frequency planar inverted-F antenna. *IEEE Transactions on Antennas and Propagation*, *45*(10), 1451–1458. doi:10.1109/8.633849

Louertani & Chio. (2014). Low-profile broadband spiral antenna with meta materials. *IEEE Antennas and Propagation Society International Symposium (APSURSI),* 1131-1332.

Lu, Z. Q., Zhao, M., Yang, Z. Y., Wu, L., & Zhang, P. (2013, August 15). Helical Metamaterial Absorbers: Broadband and Polarization-Independent in Optical Region. *Journal of Lightwave Technology*, *31*(16), 2762–2768. doi:10.1109/JLT.2013.2272632

Luna, D. R., Vasconcelos, C. F. L., & Cruz, R. M. S. (2013). *Using Natural Optimization Algorithms and Artificial Neural Networks in the Design of Effective Permittivity of Metamaterials*. IMOC. doi:10.1109/IMOC.2013.6646572

Luo, Y., Han, Z., & Toshiyoshi, H. (2014). A Novel J-Shape Active MEMS Metamaterial Antenna. *IEEE Antennas and Propagation Society International Symposium (APSURSI), 569-570.* 10.1109/APS.2014.6904615

Ma, X., Huang, C., Pan, W., Zhao, B., Cui, J., & Luo, X. (2014). A Dual Circularly Polarized Horn Antenna in Ku-Band Based on Chiral Metamaterial. *IEEE Transactions on Antennas and Propagation*, *62*(4), 2307–2311. doi:10.1109/TAP.2014.2301841

Mahmoud & Hu. (2012). A Compact Dual Band Meta-material Antenna for Wireless Applications. *Loughborough Antennas & Propagation Conference.*

Mahmoud & Ibrahim. (2013). Compact and Closely Spaced Metamaterial MIMO Antenna With High Isolation for Wireless Applications. IEEE Antennas and Wireless Propagation Letters, 12.

Majid, H.A., & Rahim, M.K.A. (2007). Investigation of LH-meta material in microstrip antenna. *Asia-Pacific Conference on Applied Electromagnetics*, 1-4.

Markley, L., & Eleftheriades, G. V. (2011, March). Meta-screens and near-field antenna-arrays: A new perspective on subwavelength focusing and imaging. *Metamaterials (Amsterdam)*, *5*(2-3), 97–106. doi:10.1016/j.metmat.2011.03.004

Mendhe, S. E., & Kosta, Y. P. (2013). Ultra Wide Band Three Layer Microstrip Patch Antenna Using Single Layer Helical Resonating Metamaterial Cover. *International Conference on Wireless and Optical Communications Networks (WOCN),* 1-5. 10.1109/WOCN.2013.6616250

Mercier, L. (2005). Steerable and tunable "EBG resonator antennas" using smart metamaterials. *IEEE Antennas and Propagation Society International Symposium, 4B*, 406-409.

Mishra, R. K., & Patnaik, A. (1998). Neural network-based CAD model for the design of square patch antennas. *IEEE Transactions on Antennas and Propagation*, *46*(12), 1890–1891. doi:10.1109/8.743842

Mishra, R. K., & Patnaik, A. (2000). Neurospectral Analysis of Coaxial Fed Rectangular Patch Antenna. *IEEE Antennas and Propagation Society International Symposium*, 2, 1062 – 1065. 10.1109/APS.2000.875405

Mittra, R. (2008). To Use or Not to Use the Effective Medium Approach for Designing Performance-Enhanced Small Antennas-That is the Question. *IEEE International Workshop on Antenna Technology: Small Antennas and Novel Metamaterials*, 55-58. 10.1109/IWAT.2008.4511289

Mookiah, P., & Dandekar, K. R. (2009). Metamaterial-Substrate Antenna Array for MIMO Communication System. *IEEE Transactions on Antennas and Propagation*, *27*(October), 3283–3292. doi:10.1109/TAP.2009.2028638

Mussetta, M. (2010). Characterization of microstrip reflectarray square ring elements by means of an Artificial Neural Network. *Proceedings of the Fourth European Conference on Antennas and Propagation (EuCAP),* 1-4.

Nakano, H. (2014). Natural and Metamaterial-based Spiral and Helical Antennas. *Loughborough Antennas and Propagation Conference (LAPC).* 10.1109/LAPC.2014.6996405

Odabasi, H., & Fernando, L. (2013). Complementary Electric-Field-Coupled (CELC) Based Resonator Antennas. *Antennas and Propagation Society International Symposium (APSURSI),* 772-773.

Oliaei, M., & Abrishamian, M. (2014). Corrugated SIW K band horn antenna. *AEÜ. International Journal of Electronics and Communications*, *68*(12), 1199–1204. doi:10.1016/j.aeue.2014.06.009

Ourir, A., Burokur, S. N., & de Lustrac, A. (2007). Phase-Varying Metamaterial for Compact Steerable Directive Antennas. *Electronics Letters*, *43*(9), 493–494. doi:10.1049/el:20070298

Pan, Z. Y. (2012). Terahertz Metamaterials Fabrication from Narrowband to Broadband. *IEEE Asia-Pacific Conference on Antennas & Propagation*, 68 - 69. 10.1109/APCAP.2012.6333151

Pandeeswari, R. (2012). Artificial Neural Network Model for MNG-Metamaterial Spiral Resonator. *Electromagnetics Research Symposium Proceedings*, 29-33.

Pandey, G. K., Singh, H. S., Bharti, P. K., & Meshram, M. K. (2014, August 28). Metamaterial-based UWB antenna. *Electronics Letters*, *50*(18), 1266–1268. doi:10.1049/el.2014.2366

Park & Lee. (n.d.). Omnidirectional circularly polarized antenna utilizing zeroth-order resonance of epsilon negative transmission line. *IEEE Trans. on Antennas and Propag., 59*, 2717-2720.

Patnaik, A. (1997). An Artificial Neural Network Model for Effective Dielectric Constant of Microstrip Line. IEEE Trans. Antennas Propag., 45.

Patnaik, A. (2005). A Frequency Reconfigurable Antenna Design Using Neural Netwoks. *IEEE Antennas and Propagation Society International Symposium*, 409-412. 10.1109/APS.2005.1551830

Paul, P.M. (2013). Miniaturization of Square Patch Antenna using Complementary Split Ring Resonators. *Advances in Computing and Communications (ICACC)*, 122 – 125.

Pelosi, G. (2004). Inter-Element Coupling In Large Arrays by an ANN Approach. *IEEE Antennas and Propagation Society International Symposium*, 1, 523 – 526. 10.1109/APS.2004.1329711

Peng, Y. (2010). Ultra thin low profile U band folded meta-material wideband dipole antenna for multi-Gb/s data transmission using 65 nm CMOS technology. *Antennas and Propag Society Inter Symposium (APSURSI)*, 1-4.

Pintos, J. F., Louzir, A., Minard, P., Perraudeau, J., Mattei, J. L., Souriou, D., & Queffelec, P. (2010). *Ultra Miniature UHF Antenna Using Magneto-Dielectric Material.* ANTEM. doi:10.1109/ANTEM.2010.5552480

Polemi, A., Maci, S., & Kildal, P.-S. (2011). Dispersion Characteristics of a Metamaterial-Based Parallel-Plate Ridge Gap Waveguide Realized by Bed of Nails. *IEEE Transactions on Antennas and Propagation, 59*(March), 904–1003. doi:10.1109/TAP.2010.2103006

Pyo, Baik, Cho, & Kim. (2009). Metamaterial-based antenna with triangular slotted ground for efficiency improvement. *IEEE Electronics Letters, 45*(3).

Quevedo-Teruel, O., Rajo-Iglesias, E., Inclan-Sanchez, L., & Vazquez-Roy, J. (2010). Reconfigurable loaded planar inverted F-antenna by making use of varactor diodes. *Proc. IEEE APSURSI*, 1–4. 10.1109/APS.2010.5560948

Qureshi, F., Antoniades, M. A., & George, V. (2005). A Compact and Low-Profile Metamaterial Ring Antenna With Vertical Polarization. IEEE Antennas and Wireless Propagation Letters, 4.

Ranjan, K., & Singh, G. (2013). Effect of low dielectric permittivity on microstrip antenna at terahertz frequency. *Optik (Stuttgart)*, *124*(22), 5777–5780. doi:10.1016/j.ijleo.2013.04.061

Robustillo, P. (2011). ANN Element Characterization for Reflectarray Antenna Optimization. *Proceedings of the 5th European Conference onAntennas and Propagation*, 957 – 960.

Sabban, A. (2013). New Printed Meta Material Antenna. *Antennas and Propagation Society International Symposium (APSURSI),* 1790 – 1791.

Sabban, A. (2014). New microstrip metamaterial antennas. In *Antennas and Propagation Society International Symposium (APSURSI).* IEEE.

Saeed Majedi. (2012). *Compact MNG zeroth order resonator antenna*. Elsevier B.V.

Sarmah, K., & Sarma, K. K. (n.d.). Design Optimization of Loop Antenna Using Competitive Leakning ANN. *National Conference on Emerging Trends and Applications in Computer.*

Shang. (2010). An Effective Method for RFID Tag Antenna Optimization Based on Artificial Neural Network. *IEEE Youth Conference on Information Computing and Telecommunications*, 383 – 386.

Shena, Y., & Gongc, L. (2014). Investigation of gain effect of multi-band patch antenna based oncomposite rectangular SRRs. *Optik (Stuttgart)*, *124*(22), 933–935.

Sievenpiper, D., Zhang, L., Broas, R. F. J., Alex'opoulos, N. G., & Yablonovitch, E. (1999). Artificial Magnetic conductor Surfaces with a forbidden frequency band. *IEEE Transactions on Microwave Theory and Techniques*, *47*(11), 2059–2074. doi:10.1109/22.798001

Simion, S., Marcelli, R., & Sajin, G. (2007, August 16). Small-size CPW silicon resonating antenna based on transmission-line meta-material approach. *Electronics Letters*, *43*(17), 908–909. doi:10.1049/el:20071639

Singh, N., & Singh, S. (2010). Effect of Photonic Band Gap Structure on Planar Antenna Configuration. *Mediterranean Microwave Symposium (MMS),* 81 – 85. 10.1109/MMW.2010.5605140

Singh, N., Singh, S., & Kumar, H. (2011). A study on Applications of Meta-Material based Antennas. *Conference on Electronics Computer Technology (ICECT)*, 192 - 196. 10.1109/ICECTECH.2011.5941588

Song, X.-H., Wu, W.-Y., Shen, T.-G., & Zhou, Y.-Q. (2011). Investigation of A Patch Antenna Based on I-shaped Left-handed Material. *Optik (Stuttgart)*, *122*(16), 1426–1429. doi:10.1016/j.ijleo.2010.09.021

Thakur, A., Thakur, P., & Hsu, J.-H. (2011). Novel magnetodielectric nano-materials with matching permeability and permittivity for the very high frequency applications. *Scripta Materialia*, *642*(2), 205–208. doi:10.1016/j.scriptamat.2010.09.045

Turpin, J. P. (2011). Anisotropic Metamaterial Realization of a Flat Gain-enhancing Lens for Antenna Applications. *Symposium on Antennas and Propagation (APSURSI), 2011 IEEE International*, 2886-2889. 10.1109/APS.2011.5997130

Turpin. (2012). Low Cost and Broadband Dual-Polarization Metamaterial Lens for Directivity Enhancement. *IEEE Transactions on Antennas and Propagation, 12*.

Venkatrami Reddy, S., Saurav, K., Sarkar, D., Singh, A., & Srivastava, K. V. (2014). Design of a Novel Dual Band Patch Antenna Loaded with CELC Resonator for WLAN/WiMAX Applications. *Twentieth National Conference on Communications (NCC),* 1-6.

Wang, J., Zhu, Z., Sun, Y., Shen, T., & Gong, L. (2013). Study on left-handed effect of composite helices and its application in square frame patch antennas. *Optik (Stuttgart)*, *124*(21), 5189–5192. doi:10.1016/j.ijleo.2013.03.052

Wang, S. (2009). Design of broadband array elements based on neuralnetwork and genetic algorithm. *IET Journals & Magazines, Volume*, *45*, 1011–1012.

Wang, S.-Y., & Chen, F.-C. (2012). The Multi-Band EBG Characteristic of The Mushroom Structure with The Suspending Microstrip line. *Proceedings of the 42nd European Microwave Conference*, 663-666. 10.23919/EuMC.2012.6459242

Werner, D. H., Haupt, R. L., & Werner, P. L. (1999, October). Fractal Antenna Engineering "The Theory and Design of Fractal Antenna Arrays". *IEEE Antennas & Propagation Magazine*, *41*(5), 37–59. doi:10.1109/74.801513

Wnuk, M. (2013). The Utilization of Neural Networks to Radiation Pattern of Microstrip Antenna Formation. *International Conference on Signal Processing and Communication Systems*, 1-5.

Wong, K.-L., Lee, G.-Y., & Chiou, T.-W. (2003, January). A Low-Profile Planar Monopole Antenna for Multiband Operation of Mobile Handset. *IEEE Transactions on Antennas and Propagation*, *51*(1), 121–125. doi:10.1109/TAP.2003.809044

Wu, C.-Y., Yeh, S. H., & Lu, T.-H. (2007). Planar High Gain Antenna for 5.8-GHz WiMAX Operation. *Antennas and Propagation Society International Symposium*, 3488 – 3491.

Xie, M. T., & Guo, Q. G. (2010). Design of a Novel Artificial Magnetic Conductor Plane and Its Application for Low-Profile Dipole. *International Conference on Microwave and Millimeter Wave Technology (ICMMT)*, 2085 – 2087. 10.1109/ICMMT.2010.5525228

Yang, S.-L., Kishk, A., & Lee, K.-F. (2008). Frequency reconfigurable U-slot microstrip patch antenna. *IEEE Antennas and Wireless Propagation Letters*, *7*, 127–129. doi:10.1109/LAWP.2008.921330

Yu, Z., & Qi Wang, A. (2007). Multi-Band Small Antenna Based on Deformed Split Ring Resonators of Left-Handed Meta-Materials. *International Symposium on Microwave, Antenna, Propagation and EMC*, 531-534. 10.1109/MAPE.2007.4393672

Zani, M. Z. M., Jusoh, M. H., Sulaiman, A. A., Baba, N. H., Awang, R. A., & Ain, M. F. (2010). Circular Patch Antenna On Metamaterial. *IEEE International Conference on Electronic Devices, Systems and Applications (ICEDSA)*, 313-316. 10.1109/ICEDSA.2010.5503053

Zhang, Y., Wei, K., Zhang, Z., Li, Y., & Feng, Z. (2015). A Compact Dual-Mode Metamaterial-Based Loop Antenna for Pattern Diversity. *IEEE Antennas and Wireless Propagation Letters*, *14*, PP394–PP397. doi:10.1109/LAWP.2014.2363847

Zhang, Z., Xie, H., & Zheng, L. (2008). V- Shape wired structure H.F bowtie antenna. *International Symposium on Antennas, Propagation and EM Theory*, 207–210.

Zhao, X., Lee, Y., Choi, J. H., & Jung, K. Y. (2012, August 21). Design of a metamaterial-inspired size-reduced wideband loop antenna with frequency scanning characteristic with frequency scanning characteristic. *IET Microwaves, Antennas & Propagation*, *6*(11), 1227–1235. doi:10.1049/iet-map.2012.0006

Zhi. (2014). A Low-Profile High-Gain Substrate-Integrated Waveguide Slot Antenna Enabled by an Ultrathin Anisotropic Zero-Index Metamaterial Coating. *IEEE Transactions On Antennas And Propagation*, *62*(March), 1173–1184.

Zhu, J., & Eleftheriades, G. V. (2009). A Compact Transmission-Line Metamaterial Antenna With Extended Bandwidth. *IEEE Antennas and Wireless Propagation Letters*, *8*, 295–298. doi:10.1109/LAWP.2008.2010722

Chapter 3
Prospect of Dynamic Metasurface Array Antenna System With Machine Learning

Rupesh Kumar
SRM University, India

ABSTRACT

This chapter presents the machine learning (ML) concept for standard RF component design in microwave frequency. It will explain the use of the deep machine learning concept for antenna and other RF components, such as RF filter, and all relevant analysis will be based on the CST simulations. The comparative study of the ML approach and the antenna design tool (such as CST) will be presented in the form of their performance. This chapter will explain the perspectives of ML in RF system design and analysis. This chapter will present the design of the antenna and filter as examples. The simulated results will be obtained by the CST MW Studio and the ML will be implemented in MATLAB.

INTRODUCTION

Traditionally the standard RF components (such as divider, filter, coupler, etc.) have been designed using different available 3D EM (Electro-Magnetic) Wave Simulator tools such as ADS, CST, HFSS, etc. For example, antenna simulation has been satisfactorily performed in CST or HFSS whereas the standard RF components, filter coupler, mixer, amplifier, etc., in ADS. These tools have gained reputation and trust from researchers and other Radio Frequency (RF) fraternities, and over the years, it has performed very well from an individual component design to a

DOI: 10.4018/978-1-6684-8287-2.ch003

more complex integrated system design and analysis through different simulation modalities. At the same time, the complexity scale has also grown exponentially with the advancement of wireless technologies for different applications such as wireless sensing and for establishing wireless communication. In this context, the array aperture antenna based on dynamic metasurface design has many applications related to remote sensing as well as wireless communication under 5G and 6G schemes (Araghi, 2022; Bianco, 2020; Chaudhuri & Pal, 2022; Hamdi & Aguili, 2021). Nevertheless, this new concept has brought new challenges for system design, especially concerning simulation tools because of their more complex structure compared to generally known antenna or RF component designs.

It becomes more challenging, in terms of computational cost as well as time, for a larger array size. This is evident in the case of Dynamic Metasurface Array antenna design because of the scarce increase in the number of radiating antenna elements which are known as meta-elements. Further, the optimization of such design becomes impossible with traditional computing resources because of the increased number of iteration steps involved in such approaches irrespective of known algorithms.

However, the recent development of Machine Learning (ML) (Alsharari, 2023; Huang et al., 2023; Koziel & Pietrenko-Dabrowska, 2023; Sendrea et al., 2023; Yang et al., 2023) has opened a new door across the domains such as statistics, decision-making, signal processing, optimization, prediction, identification, etc., while targeting different applications such as autonomous-driving, industrial autonomous, speech recognition, agriculture, and computer vision, where it is a challenge for researchers, scientists, and engineers to develop conventional algorithms to perform the needed tasks. Now, we need to understand machine learning before we go further. In simple terms, Machine Learning can be defined as a branch of Computer Science and Artificial Intelligence that focuses on the development of human-like learning as well as self-correction tendencies for attaining higher accuracy through data mining and algorithm developments. It is important to emphasize the two very special features of ML, that is, data mining can be related to establishing unknown mathematical relationships among data and doing meaningful data processing in the form of algorithms so that deviation from the expected values should remain minimal or within acceptable limits as an error. One keynote can be outlined here as these two special features always remain associated with repetitive or iterative tasks. Likewise, the available 3D Microwave Simulator tools are also solving Maxwell's equations at the mesh level in iterative patterns to compute the performance of an antenna or an RF component. Therefore, it is possible to use the ML concepts to achieve the expected results which are computed through different available EM Wave tools.

In this chapter, we will learn more about the dual-band antenna and RF filters as the standard RF component designs under a supervised learning approach, and the necessary steps requirements. We will learn more about the designing of RF standard components or antennas based on machine learning approaches. It is worth noting that prior fundamental knowledge of machine learning is required for a better understanding of the presented chapter!

MACHINE LEARNING CONCEPT FOR ANTENNA AND STANDARD RF COMPONENT DESIGN

The Machine Learning process is widely categorized into three ways of learning (Afacan et al., 2021; Guan et al., 2021; Mishra et al., 2022; Zhang et al., 2022):

- Supervised Learning
- Unsupervised Learning
- Reinforcement learning

In Supervised Learning, the input and its expected output information as labels are used to develop a learning algorithm for establishing a mapping between input and out. This approach can be utilized for designing an antenna or RF components. However, labels are missing in unsupervised learning, and it focuses more on hidden patterns or features. The last one, reinforcement learning, is used to solve the challenges related to dynamic conditions such as autonomous driving where it must safely & continuously drive according to the changing environment.

Usually, in supervised machine learning, we use two sets of data for the development of a trained network and its verification. The workflow of supervised learning, in general, is illustrated in Figure 1.

Figure 1. Workflow of supervised learning

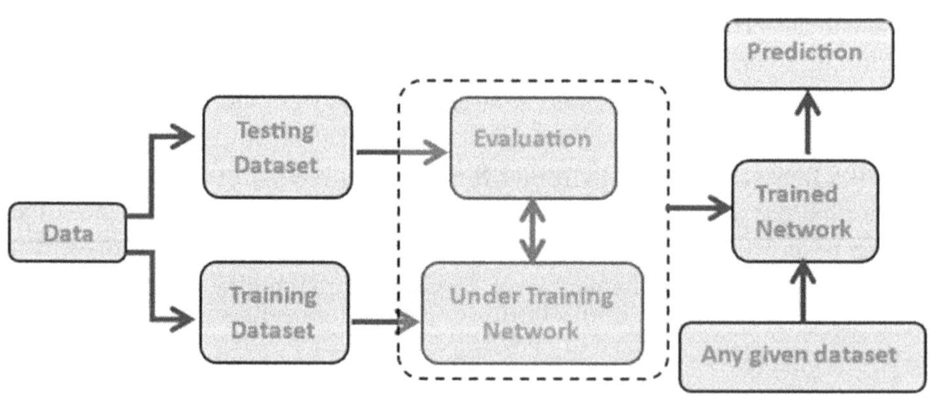

Supervised learning needs a well-labeled data set which is divided into two parts, training sets, and testing sets. These two sets are used to build a trained network with a continuous feedback mechanism between evaluation and the developing network. Later, the trained network becomes available for any given set of data as input, and its corresponding output is predicted by it. The trained network takes almost no processing time, and output data is expected to come in real-time which is impressive in comparison with the available 3D EM simulator tools.

However, it is important to know that the input data set must be defined in the proper way to facilitate the training step in Figure 1. There is not any known rule on how to decide the input dataset's format. This makes the situation slightly confusing for new researchers, but one basic rule that can be deployed here is to consider those datasets which have an impact on the output performance of the RF design. Towards the design modeling, if you take one example of antenna design then try to consider one or multiple parameters such as any dimensions of the antenna to which you want to see the variation in the output dataset. In case of multiple parameters, try to use all possible permutations and their corresponding output dataset. Moreover, all other parameters which are not expected to vary should be considered constant. Although there is no rule as of now regarding the design specifications, the input and output datasets can be prepared for the ML approach. In this sense, it is important, first, to understand the design requirements and then prepare the datasets, otherwise, there is a high chance that you will waste your time while preparing a not suitable dataset for training purposes.

After the design of the trained network, this can be used for the analysis of a new set of values for obtaining output results. Once the training period is over, a good, trained network will predict a close value to the expected without incurring time delay which is the biggest advantage one can have from machine learning approaches compared to 3D EM Wave simulator tools.

Figure 2 represents the workflow of the machine learning approach for RF design in general. The input sets may contain the design dimensions under investigation,

Figure 2. Workflow of supervised learning

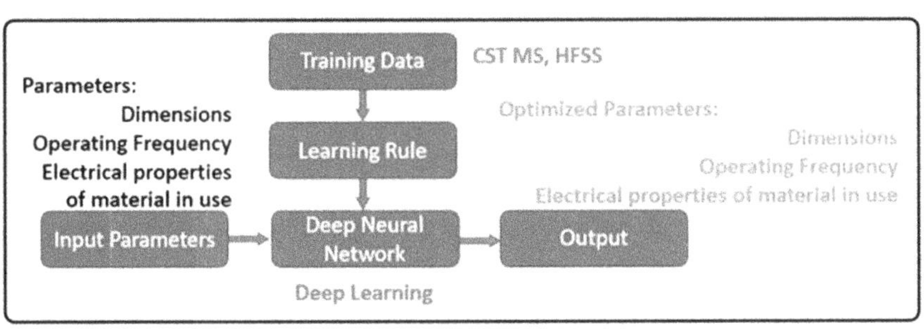

operating frequency, electrical properties, etc., and its corresponding output datasets will be radiation patterns, scattering parameters, etc. Each possible dataset is collected from the available simulator tools such as CST, and Ansys HFSS.

The prepared dataset can be used for developing the trained network based on the learning approach. Once the trained network is ready, it can be used for predicting the output from any given input data set.

DYNAMIC METASURFACE ANTENNA: MACHINE LEARNING APPROACH

Although a deep learning network needs to be trained which is a time-consuming process, however, once trained, the trained network can be used to achieve real-time performance. Hence, a Deep Learning Network may be able to provide a solution for real-time pattern synthesis. As presented in (Araghi, 2022), this approach can be used for any antenna analysis such as an array antenna.

The results of the modeled antenna in the 3D EM Wave Simulator will be used to train the neural network. The scheme for the network training is illustrated in Figure 2. Once the network is trained, it will be used to predict the radiation pattern in real-time for a given set of input information such as the number of antenna elements, size, etc. Some works based on machine learning approaches have been reported for the characterization of antennae (Kumar et al., 2022; Sharma et al., 2020). These reported works help in predicting the radiation patterns as a real-time solution. This can be further extended to validate the performance of antennae in more complex situations such as the presence of metallic appendages in a spacecraft environment where traditional approaches take more. Genuinely, there is a large scope for more active research work in this field. Moreover, the fast analysis of antenna performance will benefit from the scope of machine learning.

Many researchers and scientists are actively involved in the research on antenna array characterization based on the Machine Learning approach; this involves the RF Antenna system modeling and designing using the 3D electromagnetic simulation software such as Ansys HFSS, and CST Studio Suite. Further, the Deep Machine Learning (Deep ML) approach has been used to predict the radiation patterns, radiated powers, s-parameters, etc. of an antenna. For example, the dynamic metasurface antenna (DMA) aperture as an application for a Computational Millimeter-Wave Imaging system can be designed with a machine learning approach. The designed DMA is shown in Figure 3.

The antenna elements are designed for random radiation patterns to aid microwave imaging. Each radiating element which is called a meta-atom is used to radiate at their respective resonating frequency because of frequency selective nature, and

Figure 3. Dynamic metasurface antenna (DMA)
Source: Kumar et al. (2022)

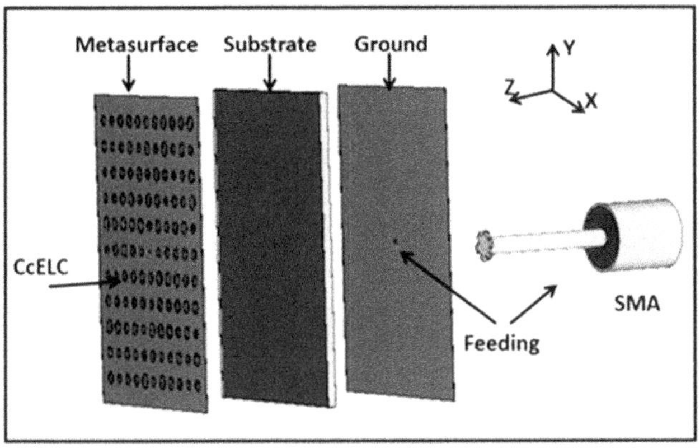

this way the collective radiation pattern of the design is random in nature and varies with sweeping frequency. The design reported in (Hamdi & Aguili, 2021) has 120 radiating meta-atoms, and each has a different radius in comparison to others. The reported design has used around 180 million mesh cells for full-wave analysis, and one simulation took around one hour and 30 minutes on a computing system with sixteen cores, 128 GB RAM, and a couple of CST accelerator tokens for simulation.

To optimize such a design, we need an extremely large number of simulation iterations, say, if each meta-element is considered three different radius values, then the required simulation time will increase exponentially for 3^{120} iterations, where 120 is the total number of meta-atoms. This is simply not possible to complete with the traditionally known computing resources. Perhaps! Quantum Computing can play an important role here!

However, the benefits of machine learning become quite evident for such challenges which are basically iterative.

ANTENNA AND RF COMPONENT DESIGN: MACHINE LEARNING APPROACH

As we have understood the constraints related to DMA system design and its optimization, it is impossiblc to solve this issue with limited computing resources. Let's see! What can be done with the Machine Learning approach for RF system design? First, we will see the scope of machine learning scope for designing a standard RF component as an RF band filter, then we will see its scope for dual band antenna design. The design of the RF filter and dual band RF filter are shown in Figure 4. Both designs are accomplished in CST Studio Suite. The RF filter is analyzed at the UHF band with a center frequency of 950 MHz. Likewise, a dual-band antenna is also completed in CST for ISM and WLAN bands, the lower band is around 2.45 GHz whereas the upper band is around 5.1 GHz, respectively. The performance s-parameters are presented in Figure 5.

Figure 4. a) RF filter design, b) dual band antenna
Source: Sharma et al. (2020)

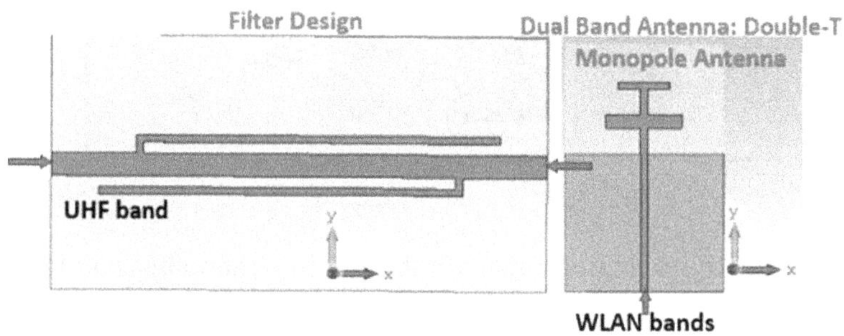

Figure 5. a) RF filter design, s-parameters, b) dual band antenna design, s-parameter
Source: Sharma et al. (2020)

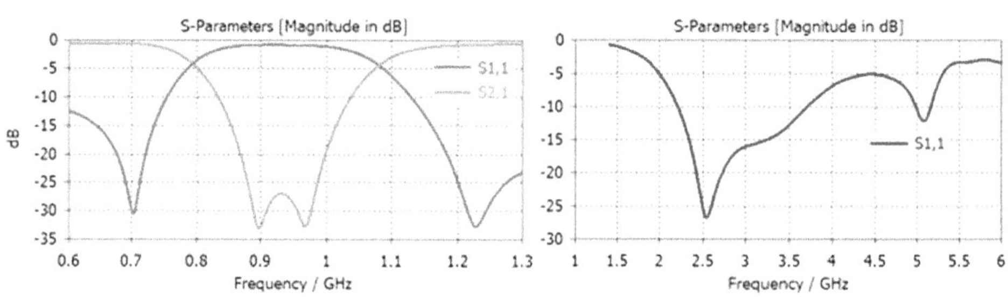

MACHINE LEARNING IN MATLAB

The training datasets are used to build the training network in MATLAB which involved the Artificial Neural Network (ANN) toolbox, and the steps are mentioned below:

Step 1. Open the Neural Network Start GUI by clicking Apps, and then opting for Neural Network Fitting or by entering the 'start' command and then clicking the Fitting app.

Step 2. From the Neural Network Fitting window, click on import data and then import the input dataset and output datasets.

Step3. Split the data into training, validation, and test sets.

Figure 6. Neural network toolbox, MATLAB

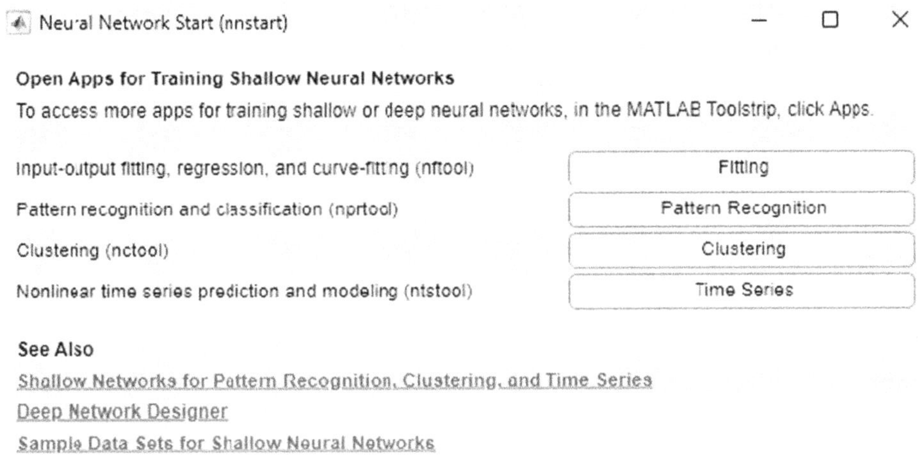

Figure 7. Neural network fitting, MATLAB

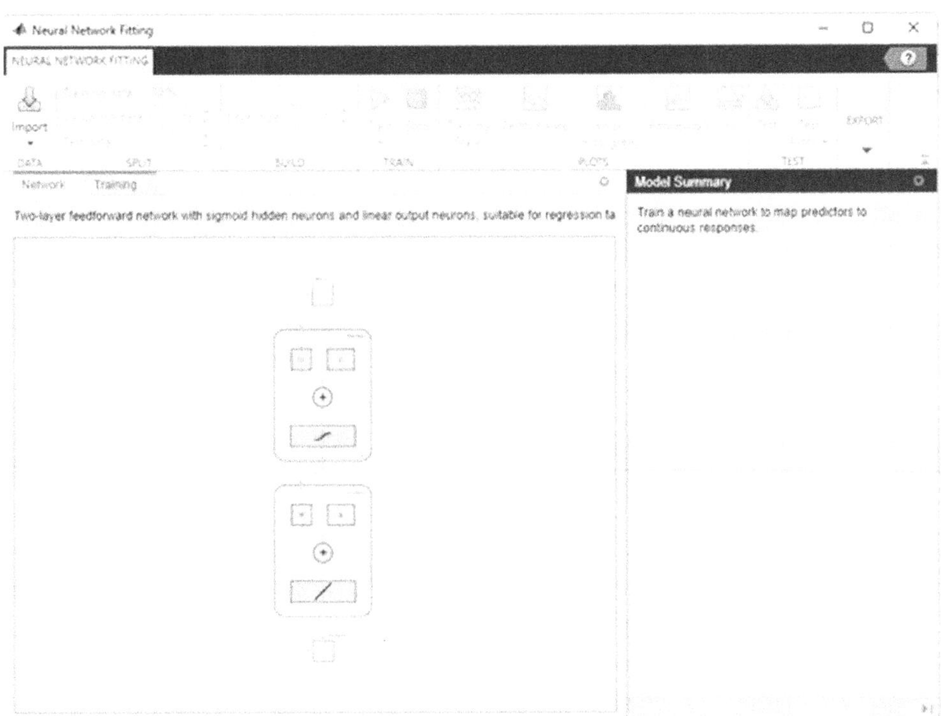

Step4. Define the neural network architecture and press the train option.

Step5. Check the performance using different analysis options such as mean squared, regression, etc.

Step6. Visualize the plots, such as regression fit, histogram of errors, etc.

Step7. Save the MATLAB function for use as a trained network.

DUAL BAND ANTENNA DESIGN: MACHINE LEARNING APPROACH

The antenna modeling is obtained by following the steps mentioned in the previous section. It has three main steps for deploying the Machine Learning approach:

First, the input and output datasets are obtained from the simulator tools, and they should be large enough to model an acceptable trained model. These datasets play an important role in the development of the model and therefore, they must be very specific, and a proper data collection method is required to make sure that the collected datasets are good for the training process. There is no specific rule which can define data collection. However, one can define certain boundaries for the collection of the dataset which should consider the ranges of different parameters over which the trained model is meant to predict the output dataset. It can only be learned from past experiences that are used for such dataset preparation.

Second, this extracts the hidden features in the training data by establishing the mapping relationship between the input and output datasets that should be able to predict the output data for any unknown input dataset within the defined boundaries. This is possible to achieve using a suitable ML algorithm such as linear regression, neural network, etc. This is an important step that establishes the mapping relationship between input and output datasets. After the successful modeling, this helps in predicting the output datasets for any unknown input dataset within the defined boundaries.

Third, after modeling of mapping relationship in the second step, the accuracy of the model should be tested using the testing dataset. This step should consider only those datasets which were not considered during the training stage, and the designed model is used to predict the expected output dataset value for each input dataset in the testing step. Its performance is compared with the true value of the output dataset in the training datasets. A good performance in this step means that the trained network is ready for general purpose and will have a similar performance with the predetermined boundary condition of the training dataset as mentioned in the Second step.

Traditionally, the antenna designs and analysis are mostly accomplished in 3D EM Wave Simulators, the ML approach is a potential candidate, especially in computing expensive situations as explained in the case of dynamic metasurface antenna design. In comparison to traditional simulators, a well-trained network can also provide the output performance parameters such as 10 dB frequency-bandwidth, gain, efficiency, etc. based on input parameters such as geometric and material dielectric properties in almost real-time.

However, the main constraint in the deployment of the ML approach is the time taken for the construction of the training input dataset and its corresponding output datasets information. Although, the training dataset can be constructed from either measurement data or simulation analysis. Some works have been reported in (Afacan et al., 2021; Guan et al., 2021; Mishra et al., 2022) which have used the ML approach for antenna design, and it seems to be better in terms of computational power and time. The biggest advantage of ML is the computational time reduced to a few seconds otherwise it would have taken several hours using 3D EM Wave simulations.

DUAL-BAND ANTENNA DESIGN

To demonstrate the capability of ML for an antenna design, a T-shaped monopole antenna design has been simulated in CST Studio Suits. The antenna structure is shown in Figure 4(b). The performance of this antenna mainly depends on its T-Shaped lengths and widths. In this analysis, we have considered the length of two upper horizontal stripes and their corresponding widths for the construction of the input datasets, and their corresponding output datasets are taken as the reflection coefficient. For this analysis, a total of 450 data samples are considered for the training. The predicted reflection coefficient is shown in the blue-colored curve in Figure 8. The predicted reflection coefficients have good agreement with the CST simulation. The two resonating frequencies, the WLAN band, perfectly match with the CST simulation.

The defined Neural Network for this antenna design has ten hidden layers, four input parameters, and one corresponding output dataset. The architecture and its performance parameters are shown in Figure 9. Details about the defined network are shown here, including time taken to train, gradient descent, number of validations, etc. The input dataset has been randomly divided for training purposes. Also, it uses the 'Levenberg-Marquardt' algorithm for training purposes.

The performance graph is shown in Figure 10. The horizontal axis and vertical axis are represented by the epochs and mean squared error, respectively. The best performance is considered from the epoch with the lowest validation error, that is, the mean squared error. As it is mentioned the best validation performance is 0.5605 at epoch 1000.

Figure 8. Dual band antenna: a) Broken curve corresponds to CST studio suits, b) solid continuous solid line represents the predicted curves by the trained neural network in MATLAB

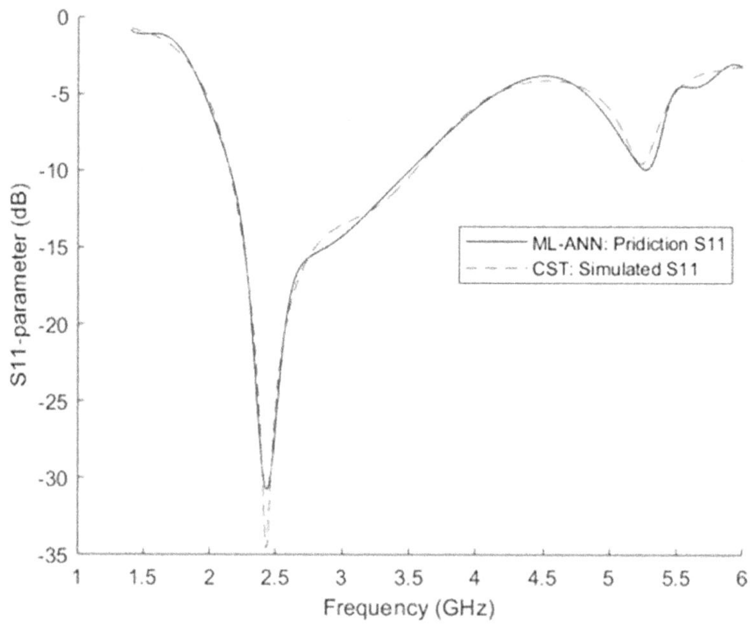

The training, testing, and validation graphs are shown in Figure 14. These are the regression plots that represent the output data values present during the training process for the expected output values as 'Target'. The fitting is good if the data align along a 45-degree slope which is equal to 1. This means that the under-training network outputs are equal to the expected output values. All curves are above 0.99 or above which is a sign of a well-trained network. Re-training can be initiated by clicking 'Train' in the neural network toolbox, see Figure 7 for reference. The true value and the output value are represented by the horizontal axis and the vertical axis, respectively. If we see Figure 11 clockwise, the first one represents 'Training', followed by 'Validation', 'Test', and 'Target-All', respectively. Further, the Regression (R) value is mentioned for each plot.

RF BAND FILTER: MACHINE LEARNING APPROACH

The RF filter is designed in CST Studio Suits which is shown in Figure 4(a). The designed passive RF band filter has two ports and two layers. The top layer has two arms which are meant to tune the resonating frequency. For machine learning

Figure 9. Neural network architecture for WLAN antenna analysis

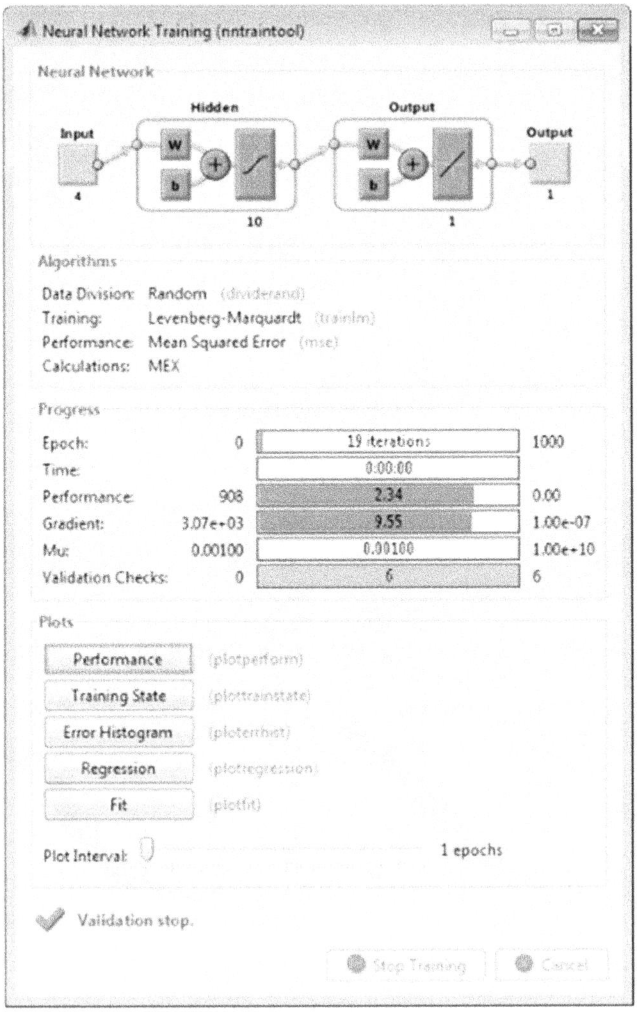

modeling, the same approach has been used as explained in the previous sections. Input and its corresponding output datasets are constructed from 750 samples. The s-parameters from CST Studio Suits are presented in broken curves in Figure 12. The reflection coefficient (|S11|) and transmission coefficient (|S21|) are presented in red and blue colors, respectively. The predicted values from the trained network are plotted as a solid line in Figure 12. The performance curves are not a good match with the CST simulations. The predicted curves from the trained model are deemed to be deviating from the true values. The reason behind such deviation could be

Figure 10. Neural network architecture for RF filter analysis

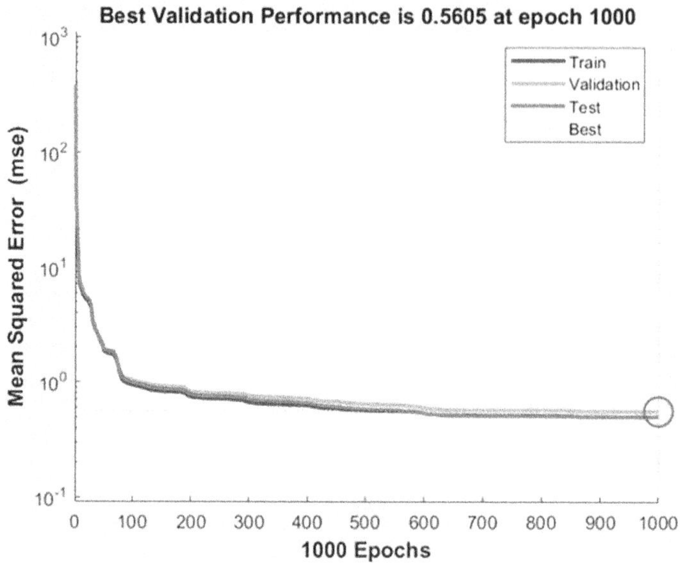

Figure 11. RF filter-training, testing, and validation plots of the trained network

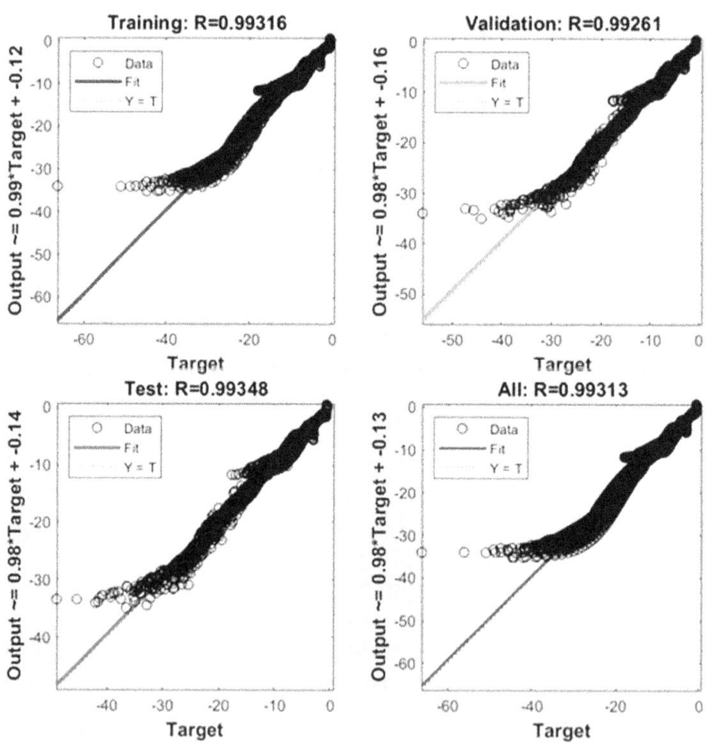

Figure 12. RF filter: a) Broken curves correspond to CST studio suits, b) solid continuous solid represents the predicted curves by the trained neural network in MATLAB

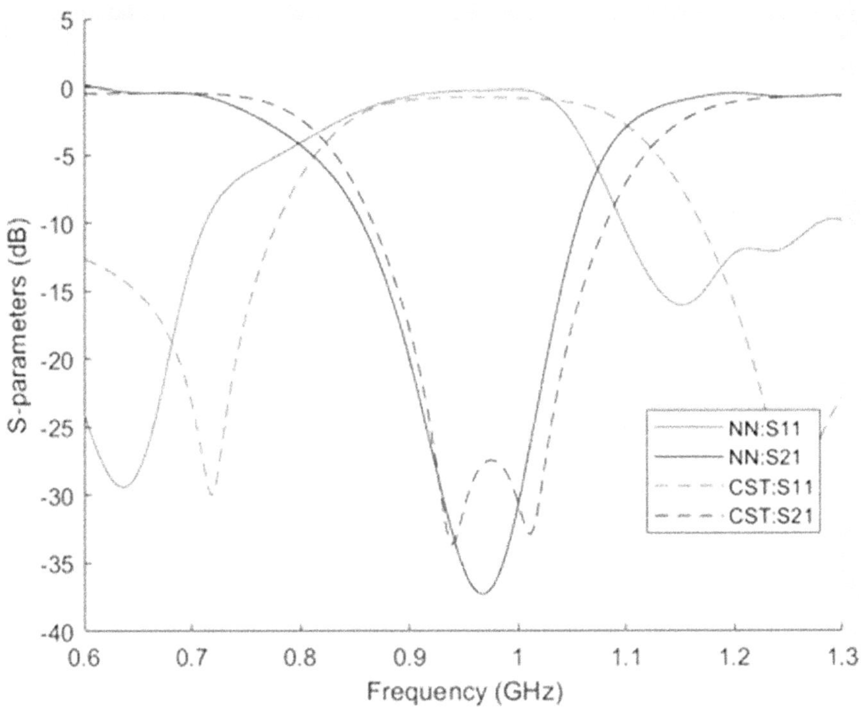

the underperformance of the trained model, improper training datasets, or neural network architecture.

The trained neural network architecture is shown in Figure 13. It has 10 hidden networks with 4 input parameters, mainly representing the length of the arms, its widths along with the frequency points. The output parameters are a reflection coefficient (|S11|) and a transmission coefficient (|S21|), respectively.

Figure 13. Neural network architecture for RF filter analysis

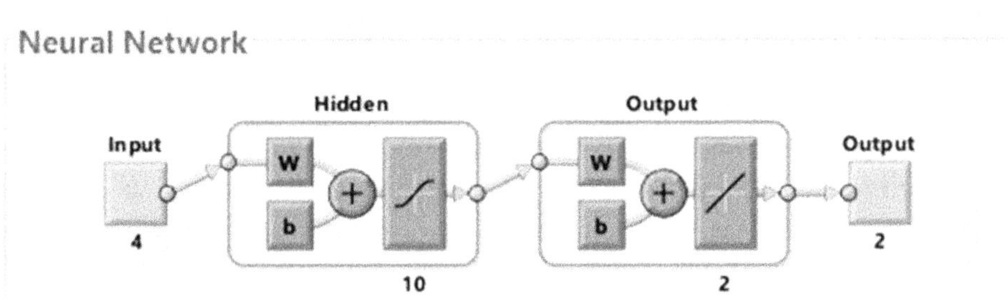

The performance graph of the RF filter is shown in Figure 14. The horizontal axis and the vertical axis are represented by the epochs and the mean squared error values, respectively. The best performance is considered from the epoch with the lowest validation error, that is, the mean squared error. As it is mentioned the best validation performance is 2.6384 at epoch 668. Compared to the dual-band antenna, this performance is not good, and it is evident from the predicted values in Figure 12.

Figure 14. Neural network architecture for RF filter analysis

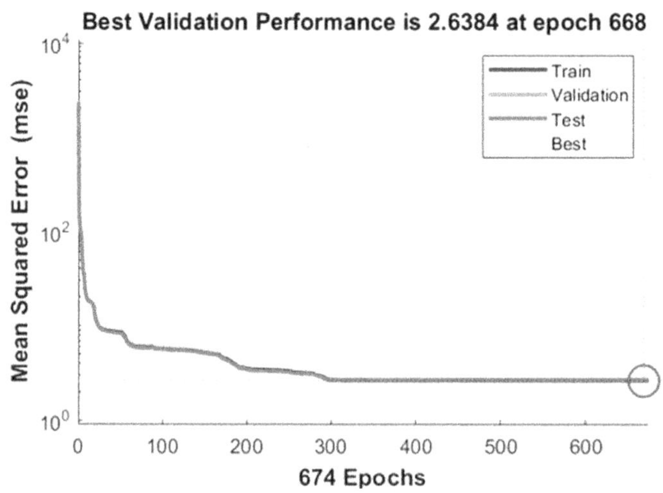

Further, training, testing, and validation graphs of the trained network are shown in Figure 15. As we know, the fitting is good if the data align along a 45-degree slope which is equal to 1. However, all curves are around 0.98 which is less than the previously seen case of antenna, see Figure 11. Hence, we can say that the trained network is not good, and while doing the training we should consider enough data for attaining the proper accuracy.

CONCLUSION

Researchers and Scientists are known for using the 3D EM Wave simulator to do the antenna and RF component design and its analysis. So far, such simulators have performed well, however, in recent years complex design such as dynamic metasurface antenna (DMA) is facing some issues because of the huge complexity

Figure 15. RF filter: Training, testing, and validation plots of the trained network

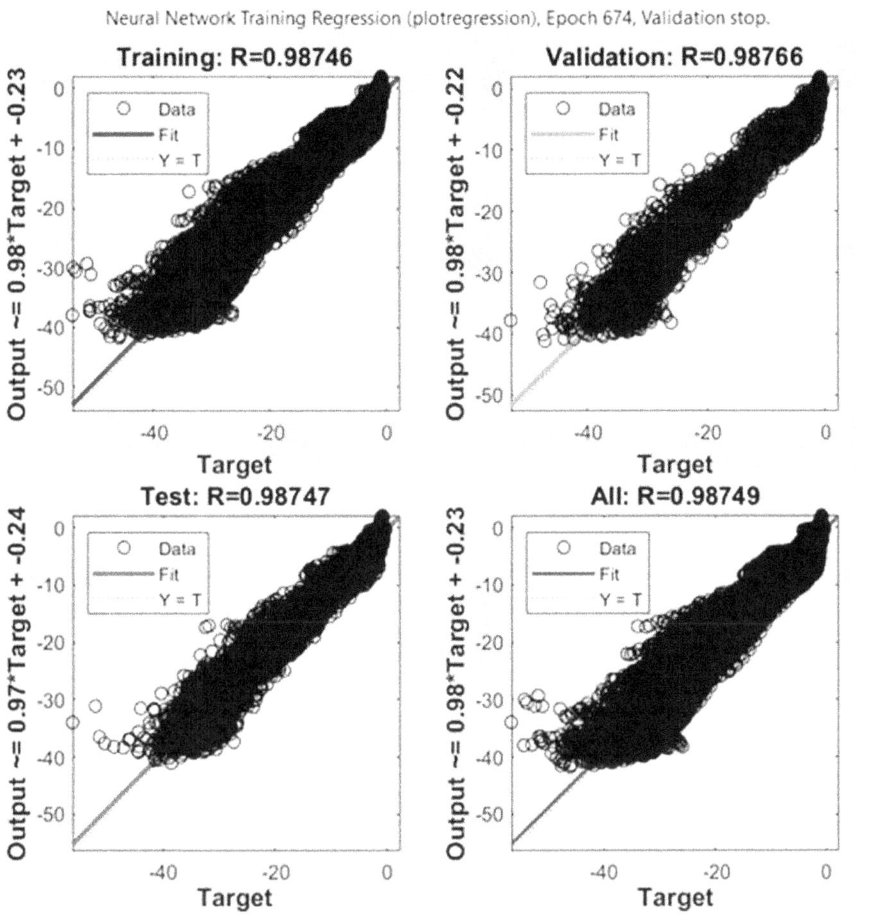

of their design and it incurs an unavoidable computing cost. However, the use of machine learning for RF component design has been reported by researchers (Afacan et al., 2021; Guan et al., 2021; Mishra et al., 2022; Sharma et al., 2020) which shows its potential to solve some pertinent issues associated with the traditional tools like simulation time.

In this chapter, we have understood the prospect of dynamic metasurface antenna and its complexity to envisage the potential use under 5G and 6G technologies for different modalities such as wireless sensing & communication, etc. However, the analysis of the design in traditional tools is a challenging job in terms of computational resources and simulation time. Although machine learning has shown good prospects, it still has a long way to go before it will get a formidable acceptance by industry as well as academia.

REFERENCES

Afacan, E., Lourenço, N., Martins, R., & Dündar, G. (2021). Review: Machine learning techniques in analog/RF integrated circuit design, synthesis, layout, and test. *Integration, 77*, 113-130. https://doi.org/ doi:10.1016/j.vlsi.2020.11.006

Alsharari, M. (2023). *Monopole Antenna Modeling and Optimization for 5G Communication System using Machine Learning. In 2023 International Wireless Communications and Mobile Computing*. IWCMC.

Araghi, M. (2022, October). Systematic Design of a Holographic-Based Metasurface Reflector in the Sub-6 GHz Band. *IEEE Antennas and Wireless Propagation Letters, 21*(10), 1960–1964. doi:10.1109/LAWP.2022.3186922

Bianco, S. (2020). AESA Adaptive Beamforming Using Deep Learning. *Proceedings of 2020 IEEE Radar Conference*.

Chaudhuri, & Pal. (2022). Design and Analysis of Programmable Cross-Polarization Converter In Terahertz Band Using Photo-excited Split-ring Resonator Based Metasurface. *2022 IEEE International Conference on Emerging Electronics (ICEE)*, 1-4. doi: 10.1109/ICEE56203.2022.10117692

Guan, Z., Zhao, P., Wang, X., & Wang, G. (2021). Modeling Radio-Frequency Devices Based on Deep Learning Technique. *Electronics (Basel), 10*, 1710. doi:10.3390/electronics10141710

Hamdi & Aguili. (2021). Spectral Floquet analysis devoted to meta-surface applied for 5G and planned 6G antenna designs. *2021 IEEE 19th International Symposium on Antenna Technology and Applied Electromagnetics (ANTEM),* 1-2. doi: 10.1109/ANTEM51107.2021.9518709

Huang, H., Yang, X.-S., & Wang, B.-Z. (2023). Machine-Learning-Based Generative Optimization Method and Its Application to an Antenna Decoupling Design. *IEEE Transactions on Antennas and Propagation, 71*(7), 6243–6248.

Koziel, S., & Pietrenko-Dabrowska, A. (2023). Improved-Efficacy EM-Driven Optimization of Antenna Structures Using Adaptive Design Specifications and Variable-Resolution Models. *IEEE Transactions on Antennas and Propagation, 71*(2), 1863–1874.

Kumar, Fusco, & Yurduseven. (2022). Preliminary Study of Breast Cancer Detection Using A Computational Microwave Imaging System. *2022 52nd European Microwave Conference (EuMC),* 195-198. doi: 10.23919/EuMC54642.2022.9924492

Mishra, Elbir, & Zaghloul. (2022). *Machine Learning for Metasurfaces Design and Their Applications.* Academic Press.

Sendrea, R. E., Zekios, C. L., & Georgakopoulos, S. V. (2023). Multifidelity Surrogate Modeling Based on Analytical Eigenfunction Expansions. *IEEE Transactions on Antennas and Propagation*, *71*(2), 1673–1683.

Sharma, Zhang, & Xin. (2020). Machine Learning Technique for Optimizing Design of Double T-Shaped Monopole Antenna. *IEEE Trans. on Ant. and Prop.*, *68*(7).

Yang, X., Chen, Y., Zhao, Y., Pan, J., Guo, J., & Yang, D. (2023). Application of KNN for Linear Array Pattern Prediction Based on the Active Element Pattern Method. *IEEE Antennas and Wireless Propagation Letters*, *22*(5), 1134–1138.

Zhang, J., Feng, F., & Na, W. (2022). Machine Learning Technologies for RF/Microwave CAD: Past, Present and Future Perspectives. *2022 52nd European Microwave Conference (EuMC),* 16-16. doi: 10.23919/EuMC54642.2022.9991872

Chapter 4
Design and Fabrication Techniques for Nonlinear Metamaterials and Metasurfaces for Wireless Communication

J. Mangaiyarkkarasi
N.M.S.S. Vellaichamy Nadar College, India

J. Shanthalakshmi Revathy
https://orcid.org/0000-0003-1724-7117
Velammal College of Engineering and Technology, India

ABSTRACT

The chapter explores the development and manufacturing of tailored nonlinear metamaterials and metasurfaces for wireless communication. It covers design and fabrication techniques, material selection and characterization, and design considerations. The chapter also examines simulation and modelling techniques, advanced fabrication methods, integration into wireless communication systems, and characterization and testing processes. It addresses reliability and scalability challenges, the role of nonlinear effects, and the significance of reliability testing and performance characterization. Economic, scalability, and sustainability considerations are discussed, along with case studies and future trends in design and fabrication techniques. The chapter strongly emphasizes the potential of nonlinear metamaterials for enhancing wireless communication and aids as a valuable source for experts in the field.

DOI: 10.4018/978-1-6684-8287-2.ch004

INTRODUCTION

In the realm of wireless communication, the demand for faster, more efficient, and reliable systems continues to grow exponentially. To meet these ever-increasing requirements, researchers and engineers are continuously exploring novel materials and designs that can revolutionize the way we transmit and receive information wirelessly. Among these ground breaking innovations, nonlinear metamaterials and metasurfaces have emerged as promising candidates, offering unprecedented capabilities in manipulating electromagnetic waves.

This chapter delves into the design and fabrication techniques for nonlinear metamaterials and metasurfaces, focusing on their application in the field of wireless communication. By harnessing the unique properties of metamaterials and metasurfaces, we can enhance the performance of antennas, improve signal quality, and enable new functionalities that were once unimaginable.

Metamaterials possess remarkable electromagnetic properties as a result of their subwavelength unit cells, which are deliberately designed and created. These materials possess exotic characteristics not found in nature, such as negative refractive index, effective medium behavior, and strong light-matter interactions. Nonlinear metamaterials take these capabilities further by incorporating nonlinear components, enabling active control and manipulation of electromagnetic waves.

On the other hand, metasurfaces are two-dimensional structures composed of subwavelength scatterers arranged in a specific pattern. These surfaces can tailor the amplitude, phase, and polarization of incident electromagnetic waves with exceptional precision, effectively moulding the wavefront according to desired specifications. By judiciously designing the unit cells and arranging them in a precise configuration, metasurfaces can exhibit a wide range of functionalities, including beam steering, polarization conversion, and focusing, among others.

This chapter provides an in-depth exploration of the design principles and fabrication techniques for nonlinear metamaterials and metasurfaces, with a strong emphasis on their applications in wireless communication systems. We discuss various approaches to engineering nonlinear responses in metamaterials, including the utilization of active elements such as diodes, varactors, and transistors. Moreover, we delve into the intricacies of metasurface design, including the choice of unit cell geometry, material selection, and optimization methods.

Furthermore, this chapter explores the fabrication techniques used to realize these sophisticated structures, considering both top-down and bottom-up approaches. We discuss conventional manufacturing methods, such as photolithography and electron-beam lithography, as well as emerging techniques like nanoimprint lithography and self-assembly processes. We also touch upon the challenges and limitations associated with each method, providing insights into the practical considerations of fabricating nonlinear metamaterials and metasurfaces.

As wireless communication technologies continue to advance, the integration of nonlinear metamaterials and metasurfaces into wireless devices holds great promise for achieving unprecedented performance and functionality. By mastering the design and fabrication techniques presented in this chapter, researchers and engineers can pave the way for a new generation of wireless communication systems that are faster, more efficient, and more adaptable to the ever-evolving demands of the digital age.

In the subsequent sections of this chapter, we delve into the design principles, fabrication methods, and potential applications of nonlinear metamaterials and metasurfaces in wireless communication, offering a comprehensive guide for those interested in exploring the frontiers of this exciting field.

OVERVIEW OF DESIGN AND FABRICATION TECHNIQUES

To understand the design and fabrication techniques of nonlinear metamaterials and metasurfaces for wireless communication, it is essential to have the knowledge of their classifications (Quevedo-Teruel & Chen, 2021). Some common classification categories for nonlinear metamaterials and metasurfaces based on the properties, in the context of wireless communication are,

Nonlinear Response Mechanism

Kerr-type nonlinearity: These metamaterials exhibit nonlinear responses based on the Kerr effect, where the refractive index changes with the intensity of the incident electromagnetic field.

Nonlocal response: Nonlocal nonlinear metamaterials exhibit a nonlinear response that is dependent on the local electromagnetic field as well as the fields in neighboring regions.

Other nonlinear mechanisms: Some metamaterials utilize other nonlinear effects, such as electro-optic, magneto-optic, or saturable absorption, to achieve their nonlinear responses.

Nonlinear Modulation Function

Phase modulation: Metamaterials and metasurfaces that modulate the phase of the incident electromagnetic waves nonlinearly fall under this category.

Amplitude modulation: Nonlinear metamaterials that modulate the amplitude or intensity of the incident waves nonlinearly are classified as amplitude modulators.

Frequency conversion: These metamaterials convert the frequency of the incident waves nonlinearly, allowing for up conversion or down conversion of wireless signals.

Operating Frequency Range

Microwave and RF range: Nonlinear metamaterials and metasurfaces designed to operate in the microwave and radio frequency (RF) range, typically used in wireless communication systems.

Terahertz range: Metamaterials and metasurfaces operating in the terahertz frequency range, which has applications in high-speed wireless communication and imaging.

Integration and Device Configuration

Metamaterial sheets: These are planar metamaterial structures that can be fabricated on thin sheets or substrates, making them suitable for integration into various wireless communication devices (Tadesse et al., 2020).

Metasurface antennas: Metasurface-based antennas are designed to replace conventional antennas and provide enhanced beam steering, beam shaping, and polarization control functionalities.

Waveguide-based devices: Metamaterials and metasurfaces integrated into waveguides, such as rectangular or circular waveguides, for guiding and manipulating wireless signals.

Specific Application

Beamforming and beam steering: Nonlinear metamaterials and metasurfaces designed for dynamically controlling the direction and focusing of wireless signals.

Frequency conversion and mixing: Metamaterials and metasurfaces used for nonlinear frequency conversion and mixing, enabling signal processing, frequency translation, and band extension.

Nonlinear signal modulation: Metamaterials and metasurfaces employed for nonlinear modulation schemes, offering advantages in spectral efficiency, signal-to-noise ratio, and interference resistance (Dai et al., 2020).

Sensing and detection: Nonlinear metamaterials and metasurfaces integrated into wireless sensors for enhanced sensitivity, selectivity, and dynamic range.

Energy harvesting: Nonlinear metamaterials and metasurfaces used for converting ambient wireless signals into electrical energy for low-power wireless devices and sensors.

These classification categories help in categorizing and understanding the different types and applications of nonlinear metamaterials and metasurfaces in wireless communication systems. It is important to mention that these categories are not mutually exclusive, and there can be instances where they overlap or combine with each other.

DESIGN PRINCIPLES

The design and fabrication of nonlinear metamaterials and metasurfaces have opened up exciting possibilities for enhancing wireless communication systems. These advanced materials and structures offer unique properties that can manipulate electromagnetic waves in unprecedented ways, enabling improved signal quality, increased transmission efficiency, and innovative functionalities (Vellucci et al., 2022). In this overview, we provide a glimpse into the design and fabrication techniques employed in the development of nonlinear metamaterials and metasurfaces for wireless communication. Figure 1 explains the Wireless information and power transmission using metamaterials.

Figure 1. Wireless information and power transmission using metamaterials

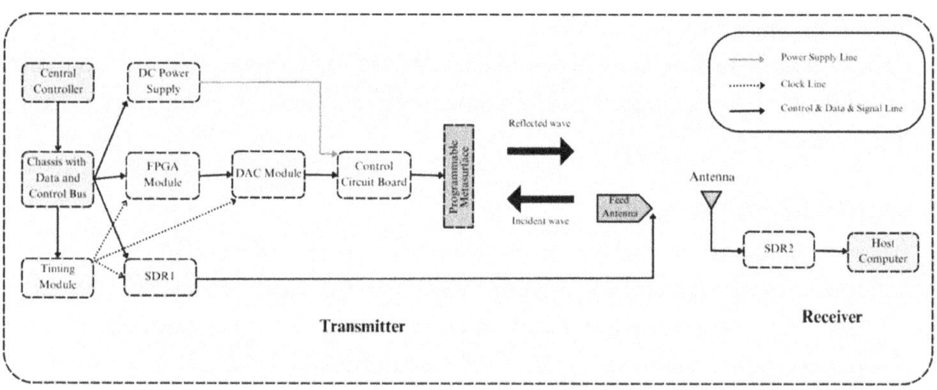

The design of nonlinear metamaterials and metasurfaces for wireless communication involves careful consideration of various factors to achieve the desired electromagnetic response. The design principles revolve around the manipulation of the unit cell geometry, material selection, and the incorporation of nonlinear elements (H.-T. Chen et al., 2016). By tailoring the unit cell arrangement and its dimensions, researchers can control the effective electromagnetic properties of the metamaterial or metasurface. The choice of materials, including dielectrics, metals, and active components, significantly influences the desired response. Nonlinear elements, such as diodes or transistors, are introduced to enable active control of the metamaterial or metasurface, allowing dynamic tuning and reconfigurability.

FABRICATION TECHNIQUES

Fabricating nonlinear metamaterials and metasurfaces requires specialized techniques capable of producing intricate structures with subwavelength features. Various fabrication methods have been employed, each with its advantages and challenges. Traditional top-down techniques, such as photolithography and electron-beam lithography, offer high precision but may be limited by time and cost (Yang et al., 2008). Nanoimprint lithography, another top-down approach, can provide large-area patterning at reduced costs. Bottom-up techniques, such as self-assembly processes and colloidal lithography, offer simplicity and scalability but may have limitations in terms of feature size and control. Hybrid approaches that combine different fabrication methods are also explored to overcome the limitations of individual techniques.

The design and fabrication of nonlinear metamaterials and metasurfaces have revolutionized the field of wireless communication. By manipulating electromagnetic waves at the subwavelength scale, these advanced materials and structures offer unprecedented control and functionality (Franco-Martínez et al., 2022). The design principles encompass careful consideration of unit cell geometry, materials, and nonlinear components, while fabrication techniques range from traditional top-down methods to emerging bottom-up and hybrid approaches. With their applications in antenna design, polarization control, frequency-selective surfaces, and more, nonlinear metamaterials and metasurfaces are poised to shape the future of wireless communication systems, enabling faster, more efficient, and adaptable wireless technologies (Yoon et al., 2016).

Applications in Wireless Communication

Nonlinear metamaterials and metasurfaces find numerous applications in wireless communication systems. One prominent application is in antenna design, where these advanced materials and structures can enhance radiation patterns, improve directivity, and enable beam steering capabilities (Esmail et al., 2022). Metasurfaces, in particular, enable the control of polarization, enabling efficient polarization conversion or manipulation. Additionally, nonlinear metamaterials and metasurfaces can be utilized for frequency-selective surfaces, filters, and absorbers, enabling the design of compact and multifunctional wireless devices (Hu et al., 2020). The integration of active elements in nonlinear metamaterials opens up opportunities for signal modulation, switching, and reconfigurability in wireless communication systems. Nonlinear metamaterials and metasurfaces have the potential to revolutionize wireless communication systems by enabling advanced functionalities and performance improvements. Some specific applications of nonlinear metamaterials and metasurfaces in wireless communication are discussed below (Tadesse et al., 2020).

(i) ***Beam steering and focusing***: Nonlinear metasurfaces can manipulate the phase of electromagnetic waves in a nonlinear fashion, allowing for dynamic control over the direction and focusing of wireless signals. This capability enables beam steering and beamforming techniques, which are crucial for next-generation wireless communication systems, such as 5G and beyond.
(ii) ***Frequency conversion***: Nonlinear metamaterials can be engineered to exhibit frequency mixing and conversion properties. By exploiting the nonlinear responses of these materials, it becomes possible to perform frequency up conversion or down conversion of wireless signals. This capability can be used for frequency translation, signal processing, and frequency band extension in wireless communication systems.
(iii) ***Nonlinear signal modulation***: Nonlinear metamaterials can be employed to modulate wireless signals in a nonlinear fashion, enabling advanced modulation schemes beyond traditional linear modulation techniques. Nonlinear signal modulation techniques offer advantages such as enhanced spectral efficiency, improved signal-to-noise ratio, and resistance to certain types of interference.
(iv) ***Enhanced wireless sensing***: Nonlinear metamaterials can be utilized to improve wireless sensing capabilities. By integrating nonlinear materials into sensor devices or antennas, it becomes possible to enhance the sensitivity, selectivity, and dynamic range of wireless sensors. This can be beneficial in various applications, including environmental monitoring, medical sensing, and security systems.
(v) ***Nonlinear frequency selective surfaces***: Metasurfaces composed of nonlinear metamaterials can be designed to exhibit nonlinear frequency selective properties. These surfaces can selectively control the transmission or reflection of wireless signals based on their frequency and intensity. Nonlinear frequency selective surfaces find applications in filter design, signal routing, and frequency multiplexing in wireless communication systems.
(vi) ***Nonlinear energy harvesting***: Nonlinear metamaterials can be utilized in energy harvesting applications. By exploiting the nonlinear responses of these materials to incoming electromagnetic waves, it becomes possible to convert ambient wireless signals into useful electrical energy. This concept, known as nonlinear energy harvesting, has the potential to power low-power wireless devices and sensors without the need for external power sources (K. Li et al., 2019).

These are just a few examples of how nonlinear metamaterials and metasurfaces can be applied in wireless communication systems. As research and development in this field progress, we can expect to see even more innovative applications and improvements in wireless communication technologies.

MATERIALS SELECTION AND CHARACTERIZATION FOR NONLINEAR METAMATERIALS

The selection of appropriate materials is a crucial aspect in the design and fabrication of nonlinear metamaterials for wireless communication. The properties of the chosen materials directly impact the electromagnetic response and nonlinear behavior of the metamaterial, influencing its performance and functionality (Huang et al., 2011). In this section, we discuss the considerations involved in materials selection and the characterization techniques used to evaluate their suitability for nonlinear metamaterials in wireless communication applications. Some important properties to be considered for the materials selection are mentioned below. Figure 2 represents chart for material selection.

Figure 2. Chart for material selection

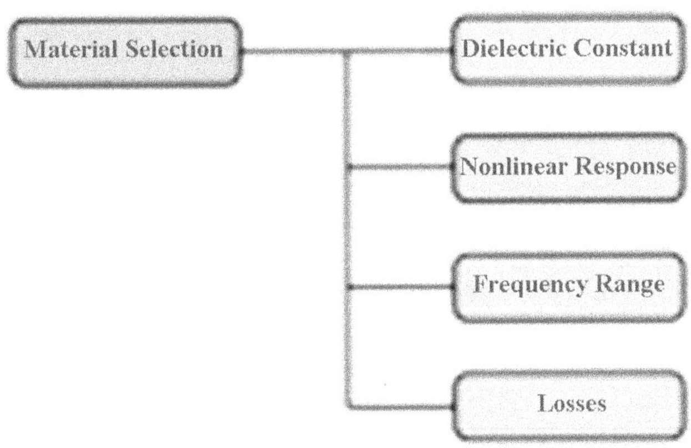

(i) ***Dielectric Constant***: The dielectric constant of a material determines its ability to manipulate electromagnetic waves. For nonlinear metamaterials, materials with high dielectric constants are often preferred as they offer stronger interaction with electromagnetic fields. Materials such as ferroelectrics, semiconductors, and certain polymers exhibit high dielectric constants and can be utilized for achieving nonlinear responses.

(ii) ***Nonlinear Response***: Nonlinear metamaterials rely on the presence of nonlinear components to enable active control and manipulation of electromagnetic waves. Materials with nonlinear optical or electrical properties, such as certain semiconductors or nonlinear crystals, are sought after for nonlinear metamaterial applications. The nonlinear response of these materials can be characterized by parameters like the nonlinear refractive index or nonlinear conductivity.

(iii) ***Frequency Range***: The selection of materials is heavily influenced by the desired operating frequency range of the wireless communication system. Different materials exhibit different electromagnetic properties at various frequency ranges. Therefore, it is essential to choose materials that possess the desired characteristics within the target frequency range(Huang et al., 2011). For example, metals like gold or silver are frequently utilized in the visible or infrared range because of their advantageous plasmonic properties.

(iv) ***Losses***: Minimizing losses is crucial for efficient wireless communication systems. Losses can occur due to absorption, scattering, or other factors within the metamaterial structure. Materials with low inherent losses are preferred to maintain signal integrity and maximize transmission efficiency. Understanding the material's absorption and scattering characteristics is important in evaluating its suitability for a particular application.

CHARACTERIZATION TECHNIQUES

Some important characterization techniques to be carried out to study the nonlinear meta materials are given below (Elnaggar & Milford, 2017) . Figure 3 explains the block diagram for characterization techniques.

Figure 3. Block diagram for characterization techniques

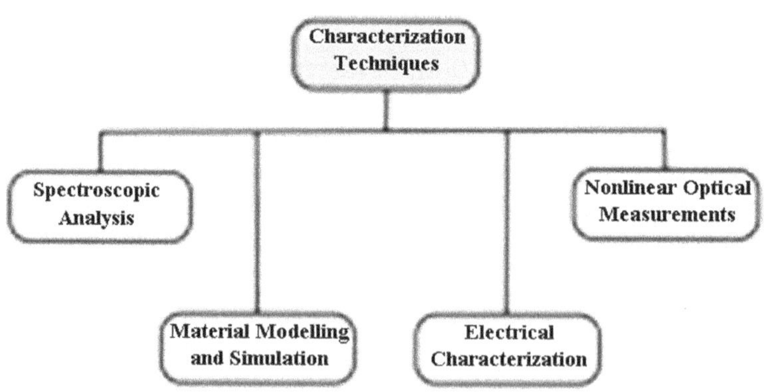

(i) ***Spectroscopic Analysis***: Commonly used spectroscopic techniques like Fourier-transform infrared (FTIR) spectroscopy or terahertz time-domain spectroscopy are employed to analyze the optical properties of materials. These techniques provide insights into parameters like the complex refractive index, absorption coefficient, and dispersion characteristics, which are vital for evaluating the suitability of materials for specific frequency ranges

(ii) ***Nonlinear Optical Measurements***: Nonlinear optical characterization methods, such as Z-scan or pump-probe techniques, are utilized to assess the nonlinear properties of materials. These techniques allow the measurement of parameters like the nonlinear refractive index or nonlinear absorption coefficient, providing valuable information on the material's nonlinear response and its potential for applications in nonlinear metamaterials(Huang et al., 2011).

(iii) ***Electrical Characterization***: Electrical characterization techniques, such as current-voltage (IV) measurements or capacitance-voltage (CV) measurements, are employed to analyze materials with electrical nonlinearities. These techniques enable the determination of parameters like the nonlinear conductivity or the voltage-dependent capacitance, aiding in the evaluation of the material's electrical nonlinear behavior.

(iv) ***Material Modelling and Simulation***: Computational techniques, such as finite element analysis (FEA) or rigorous coupled-wave analysis (RCWA), can be utilized to model and simulate the behaviour of materials within metamaterial structures. These simulations provide valuable insights into the material's electromagnetic properties, allowing for optimization and design improvements before fabrication (Salhi et al., 2016) .

By carefully considering the dielectric constant, nonlinear response, frequency range, and losses, coupled with appropriate characterization techniques, researchers can select and evaluate materials that are well-suited for nonlinear metamaterials in wireless communication applications. The integration of these materials into metamaterial structures opens up possibilities for achieving enhanced functionalities, active control of electromagnetic waves, and improved wireless communication performance.

KEY DESIGN CONSIDERATIONS FOR NONLINEAR METAMATERIALS AND METASURFACES

Designing nonlinear metamaterials and metasurfaces for wireless communication requires careful consideration of various factors to achieve desired functionalities and performance enhancements. These considerations encompass both the structural aspects of the metamaterials/metasurfaces and their integration into wireless communication systems. In this section, we discuss some key design considerations to keep in mind when developing nonlinear metamaterials and metasurfaces for wireless communication applications (Shadrivov et al., 2015). Figure 4 represents block diagram for design considerations.

Figure 4. Block diagram for design considerations

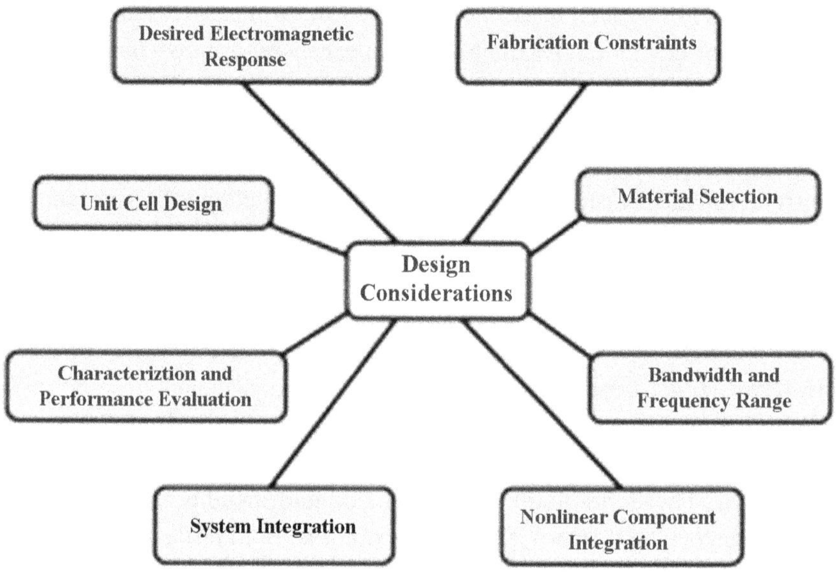

(i) ***Desired Electromagnetic Response***: Clearly defining the desired electromagnetic response is the first step in the design process. Determine the specific functionalities required for the wireless communication system, such as beam steering, polarization control, or frequency selectivity. This will guide the selection of appropriate design parameters and help optimize the performance of the metamaterial/metasurface.

(ii) ***Unit Cell Design***: The unit cell is the fundamental building block of the metamaterial/metasurface structure. Its geometry, size, and arrangement significantly influence the electromagnetic response. Tailor the unit cell parameters to achieve the desired electromagnetic properties, such as resonance frequencies, dispersion characteristics, or effective medium behavior. Numerical simulations and optimization techniques can assist in fine-tuning the unit cell design.

(iii) ***Nonlinear Component Integration***: To enable active control and manipulation of electromagnetic waves, nonlinear components need to be integrated into the metamaterial/metasurface structure. Consider the type of nonlinear element, such as diodes, varactors, or transistors, those best suits the desired functionality. Optimize the placement and configuration of these components to achieve effective nonlinear responses and facilitate dynamic tuning and reconfigurability (Minovich et al., 2015).

(iv) ***Material Selection***: The design of nonlinear metamaterials/metasurfaces heavily relies on the careful selection of materials. Consider materials with suitable properties for the desired functionality, such as high dielectric constant for strong field manipulation or materials with inherent nonlinearities. Pay attention to material losses, as they can impact signal integrity and efficiency. Additionally, consider compatibility with fabrication techniques and environmental factors, such as temperature stability or durability.

(v) ***Bandwidth and Frequency Range***: Wireless communication systems operate within specific frequency bands, and the metamaterial/metasurface design should be tailored to these frequency ranges. Consider the bandwidth requirements of the system and ensure that the metamaterial/metasurface can provide the desired functionality across the entire operating frequency range. Broadband designs may require trade-offs between performance and complexity (Jung et al., 2020) .

(vi) ***Fabrication Constraints***: Take into account the available fabrication techniques and their limitations when designing nonlinear metamaterials/ metasurfaces. Consider factors such as feature size, scalability, cost, and compatibility with chosen materials. Hybrid approaches that combine different fabrication methods may be necessary to overcome specific challenges and achieve desired structures.

(vii) ***Characterization and Performance Evaluation***: Develop appropriate characterization techniques to assess the performance of the designed nonlinear metamaterial/metasurface. Measure and analyze key parameters such as transmission/reflection coefficients, resonance frequencies, phase response, polarization conversion efficiency, and nonlinear responses. Compare experimental results with simulations to validate the design and identify areas for improvement.

(viii) ***System Integration*** Consider the integration of nonlinear metamaterials/ metasurfaces into the wireless communication system. Account for factors such as size, weight, power requirements, and compatibility with existing technologies. Ensure seamless integration without compromising the overall system performance or introducing additional complexity (Ramaccia et al., 2021). By considering these design considerations, engineers and researchers can develop nonlinear metamaterials and metasurfaces that meet the specific requirements of wireless communication systems. Iterative design processes, incorporating simulations, prototyping, and performance evaluations, can aid in refining and optimizing the design to achieve desired functionalities and performance enhancements.

SIMULATION AND MODELLING TECHNIQUES FOR DESIGN OPTIMIZATION

Simulation and modelling techniques play a crucial role in the design optimization of metamaterials and metasurfaces for wireless communication. Metamaterials are man-made materials specifically engineered to showcase electromagnetic properties that are not typically observed in natural materials (Franco-Martínez et al., 2022) . Metasurfaces, on the other hand, are two-dimensional arrangements of subwavelength structures that manipulate electromagnetic waves with extraordinary precision. To harness the full potential of these advanced materials, researchers employ simulation and modelling techniques to analyze and optimize their performance. These techniques involve utilizing electromagnetic simulation software to simulate the behavior of metamaterials and metasurfaces under different conditions and configurations (Abadal et al., 2017). By adjusting parameters such as material composition, unit cell geometry, and incident wave properties, designers can explore a vast design space to achieve desired functionalities, including enhanced signal manipulation, beamforming, and antenna miniaturization. Furthermore, advanced optimization algorithms are employed to automate the design process and identify optimal configurations that meet specific performance criteria. These techniques enable researchers to expedite the development of metamaterial-based devices and systems, facilitating the advancement of wireless communication. Simulation and modelling techniques are essential tools in the design optimization of metamaterials and metasurfaces for wireless communication. Metamaterials are artificially designed structures that possess exceptional electromagnetic properties, surpassing those found in natural materials. They are composed of subwavelength unit cells that interact with incident waves in unique ways, allowing for precise control and manipulation of electromagnetic fields. Metasurfaces, on the other hand, are two-dimensional arrangements of subwavelength structures that can manipulate the phase, amplitude, and polarization of electromagnetic waves with remarkable accuracy (Abadal et al., 2017).

To fully harness the potential of these advanced materials, researchers rely on simulation and modelling techniques to analyze and optimize their performance. To analyze the behavior of metamaterials and metasurfaces under different conditions and configurations, electromagnetic simulation software like finite-difference time-domain (FDTD) or finite-element method (FEM) is commonly employed. These simulations consider factors like material composition, unit cell geometry, incident wave properties, and the desired functionality of the device or system.

By adjusting these parameters and performing iterative simulations, designers can explore a vast design space and identify optimal configurations that meet specific performance criteria. For example, they can optimize metamaterials and metasurfaces to achieve enhanced signal manipulation, such as creating artificial magnetic or electric responses, focusing or steering electromagnetic waves, or achieving broadband absorption or reflection characteristics. Moreover, these techniques allow for the miniaturization of antennas and the development of compact and efficient wireless communication devices.

The design process can be automated by employing advanced optimization algorithms such as genetic algorithms, particle swarm optimization, or Bayesian optimization.. These algorithms search through the design space, iteratively refining the parameters to converge on the best possible solutions. This approach not only accelerates the design process but also ensures that complex design goals are met effectively (Holloway & Kuester, 2020). Furthermore, simulation and modelling techniques enable researchers to gain insights into the underlying physics and mechanisms governing the behaviour of metamaterials and metasurfaces. They can analyze how different design parameters affect the electromagnetic response, investigate the impact of fabrication imperfections, and explore the robustness of the designed devices to variations in operating conditions.

Simulation and modelling techniques are indispensable in the design optimization of metamaterials and metasurfaces for wireless communication. They facilitate the exploration of the design space, automate the design process, and provide valuable insights into the behavior and performance of these advanced materials. By leveraging these techniques, researchers can advance the development of innovative devices and systems that revolutionize wireless communication technologies.

ADVANCED FABRICATION METHODS FOR NONLINEAR METAMATERIALS AND METASURFACES

Advanced fabrication methods are crucial for realizing nonlinear metamaterials and metasurfaces with tailored functionalities for wireless communication applications. Nonlinear metamaterials and metasurfaces possess nonlinear response characteristics, enabling them to manipulate electromagnetic waves in a nonlinear manner, such as frequency conversion, harmonic generation, and nonlinear modulation (McClintock et al., 2021). To fabricate these complex structures, several cutting-edge techniques have been developed.

Nanofabrication techniques like focused ion beam (FIB) milling or electron beam lithography (EBL) are utilized for the purpose of creating precise structures at the nanoscale (P. Li et al., 2021). Figure 5 explains the focussed ion beam technique and figure 6 shows the electron beam lithography process.

Figure 5. Focused ion beam technique

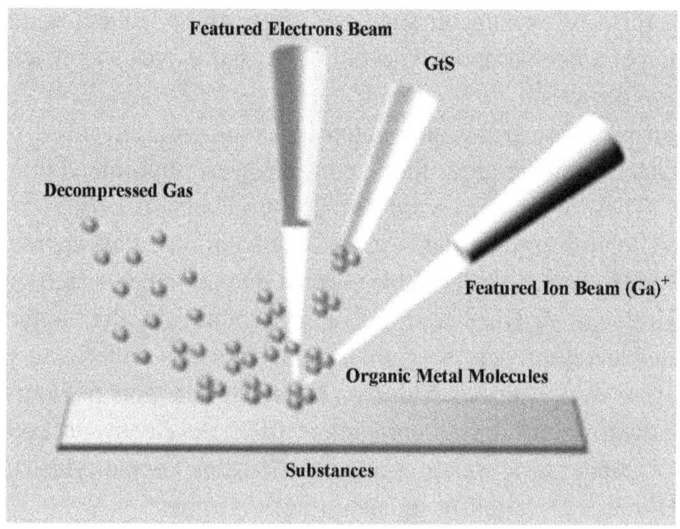

Figure 6. Electron beam lithography process

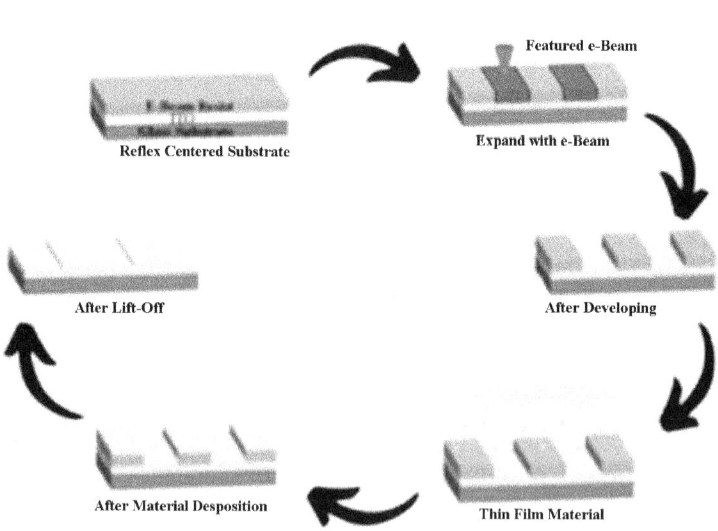

These methods allow for precise patterning of subwavelength structures on the surface of metamaterials or metasurfaces. By carefully designing the geometry and arrangement of these structures, nonlinear effects can be induced, enhancing the overall functionality of the devices.

Another advanced fabrication method is the use of nanoscale self-assembly techniques, such as colloidal self-assembly or block copolymer lithography. These methods exploit the inherent properties of materials at the nanoscale to spontaneously arrange themselves into desired patterns and structures. This approach enables the fabrication of metamaterials and metasurfaces with intricate geometries and precise control over their nonlinear properties.

Furthermore, additive manufacturing techniques, such as 3D printing, have gained attention in the fabrication of metamaterials and metasurfaces. Additive manufacturing allows for the direct printing of complex three-dimensional structures with subwavelength features, offering great design flexibility. By incorporating nonlinear materials into the printing process, it becomes possible to create functional metamaterials and metasurfaces with enhanced nonlinear response.

Techniques like nanoimprint lithography (NIL) and nanocontact printing (NCP) have shown promise in fabricating metamaterials and metasurfaces. These methods utilize molds or stamps to transfer nanoscale patterns onto the surface of materials, enabling large-scale fabrication with high resolution and repeatability (Y. Chen, 2015). Through precise control of the stamp design, nonlinear elements can be incorporated into the patterns, resulting in enhanced nonlinear behaviour.

Moreover, advancements in materials synthesis and deposition techniques have played a significant role in fabricating nonlinear metamaterials and metasurfaces. Techniques such as chemical vapour deposition (CVD), atomic layer deposition (ALD), and molecular beam epitaxy (MBE) enable the controlled growth of thin films with precise composition and atomic-level precision. These techniques facilitate the integration of nonlinear materials, such as ferroelectrics or semiconductors, into the metamaterial or metasurface structures, enabling efficient nonlinear response.

In conclusion, advanced fabrication methods are essential for the realization of nonlinear metamaterials and metasurfaces for wireless communication. These methods, including nanofabrication, self-assembly techniques, additive manufacturing, imprint lithography, and advanced deposition techniques, enable the creation of complex structures with tailored nonlinear functionalities. By combining these fabrication techniques with simulation and modelling approaches, researchers can advance the development of nonlinear metamaterial-based devices and systems, leading to transformative advancements in wireless communication technologies.

INTEGRATION OF NONLINEAR METAMATERIALS INTO WIRELESS COMMUNICATION SYSTEMS

The integration of nonlinear metamaterials into wireless communication systems holds immense potential for enhancing their performance and enabling new functionalities. Nonlinear metamaterials exhibit nonlinear response characteristics, allowing for the manipulation and control of electromagnetic waves in unique ways. By incorporating these materials into wireless communication systems, several benefits can be achieved (Keiser et al., 2013).

One significant advantage is the ability to achieve frequency conversion and harmonic generation. Nonlinear metamaterials can convert signals from one frequency to another, enabling frequency up conversion or down conversion. This capability is particularly useful in applications such as signal mixing, frequency translation, and heterodyne detection. By integrating nonlinear metamaterials into the system, these frequency conversion processes can be achieved more efficiently and with improved accuracy.

Moreover, nonlinear metamaterials enable nonlinear modulation of signals. This means that the amplitude, phase, or frequency of a signal can be modulated nonlinearly, providing enhanced control over the transmitted or received signals. Nonlinear modulation techniques, such as four-wave mixing or parametric amplification, can be employed to achieve functionalities like signal amplification, modulation format conversion, or signal regeneration. Integrating nonlinear metamaterials into wireless communication systems allows for more efficient and versatile modulation schemes.

Furthermore, nonlinear metamaterials can enhance the sensitivity and selectivity of wireless communication systems. By exploiting the nonlinear response of these materials, it becomes possible to design highly sensitive and selective sensors for wireless communications. Nonlinear sensors can detect and analyse weak signals, enabling applications such as low-power communication, spectrum sensing, and signal detection in noisy environments. These sensors can provide improved signal-to-noise ratios, enhanced detection limits, and better overall system performance (Harutyunyan et al., 2012).

Additionally, nonlinear metamaterials can be utilized to achieve efficient signal control and manipulation. By designing metamaterial structures with tailored nonlinear responses, it becomes possible to control the propagation, scattering, and absorption of electromagnetic waves. This capability opens up opportunities for beam steering, wave front shaping, and antenna miniaturization. Nonlinear metamaterial-based devices can exhibit tunable properties, allowing for dynamic control of signal directionality, polarization, and focusing.

Integration of nonlinear metamaterials into wireless communication systems also facilitates the development of reconfigurable and adaptive devices. Nonlinear materials can exhibit tunable properties through external stimuli such as electric fields, magnetic fields, or optical pumping. By leveraging these tuneable characteristics, it becomes possible to design wireless communication systems that can adapt to changing operating conditions, optimize their performance, or switch between different functionalities on-demand. Reconfigurable devices offer flexibility, improved system efficiency, and adaptability to diverse communication scenarios.

In conclusion, the integration of nonlinear metamaterials into wireless communication systems offers significant advantages in terms of frequency conversion, modulation, sensitivity, selectivity, signal control, and reconfigurability. These materials enable advanced functionalities that enhance system performance, provide greater control over transmitted or received signals, and enable the development of novel wireless communication applications. By leveraging the unique properties of nonlinear metamaterials, researchers and engineers can drive innovation in wireless communication technology, leading to more efficient, versatile, and adaptive communication systems.

CHARACTERIZATION AND TESTING OF NONLINEAR METAMATERIALS AND METASURFACES

Characterization and testing of nonlinear metamaterials and metasurfaces are essential steps in understanding their behaviour and optimizing their performance for wireless communication applications. These materials exhibit nonlinear response characteristics, which necessitate specific techniques for accurate characterization and testing. Several approaches are employed to evaluate the nonlinear properties and validate the functionality of these advanced materials.

One crucial aspect of characterization is the measurement of the nonlinear response of metamaterials and metasurfaces. Nonlinear optical techniques like second-harmonic generation (SHG), third-harmonic generation (THG), or four-wave mixing (FWM) are widely employed to investigate and analyse the nonlinear behaviour of materials (Gentile et al., 2011).

These techniques involve illuminating the material with laser beams of different frequencies and analysing the generated nonlinear signals. By measuring the intensity, polarization, or phase of the generated signals, the nonlinear response of the material can be quantified and studied.

To evaluate the performance of nonlinear metamaterials and metasurfaces in wireless communication systems, characterization techniques also include measuring their electromagnetic properties. This involves measuring parameters such as reflection, transmission, absorption, or scattering of electromagnetic waves at different frequencies and incident angles. Techniques like terahertz time-domain spectroscopy, vector network analyzers, or free-space measurements are used to analyse the electromagnetic response of these materials and verify their designed functionalities.

In addition to characterization, testing nonlinear metamaterials and metasurfaces involves validating their performance in practical wireless communication scenarios. This includes testing their capabilities in real-world environments and assessing their behaviour under various operating conditions. Performance metrics such as signal-to-noise ratio, power efficiency, bandwidth, and range are evaluated to ensure the functionality and effectiveness of the materials in wireless communication systems.

Testing may also involve assessing the durability and reliability of the nonlinear metamaterials and metasurfaces. This includes subjecting the materials to environmental conditions, such as temperature variations, humidity, or mechanical stress, to evaluate their stability and performance under different circumstances. Accelerated aging tests and reliability studies are conducted to ensure the longevity and robustness of the materials in practical applications.

Furthermore, computational modelling and simulation techniques are employed as complementary tools to characterize and test nonlinear metamaterials and metasurfaces. These techniques involve mathematical modelling, finite element analysis, or numerical simulations to predict and understand the behaviour of these materials. Simulation results can provide insights into the nonlinear response, electromagnetic properties, and overall performance of the materials, aiding in the design optimization process and guiding experimental characterization efforts.

In summary, characterization and testing of nonlinear metamaterials and metasurfaces for wireless communication involve measuring their nonlinear response, electromagnetic properties, and performance in practical scenarios. Techniques such as nonlinear optical measurements, electromagnetic measurements, and computational modelling are employed to evaluate the behaviour and validate the functionalities of these materials. Through rigorous characterization and testing, researchers can gain a comprehensive understanding of the materials' capabilities, optimize their design, and ensure their effectiveness in wireless communication applications.

RELIABILITY AND SCALABILITY CHALLENGES IN FABRICATION

The fabrication of metamaterials and metasurfaces for wireless communication poses significant challenges in terms of reliability and scalability. These challenges arise due to the intricate nature of these materials and the need for precise control over their properties. Several factors contribute to the reliability and scalability challenges in their fabrication (Yoon et al., 2016).

One key challenge is the fabrication of subwavelength structures with high precision and uniformity. Metamaterials and metasurfaces often consist of subwavelength features that require nanoscale fabrication techniques. Achieving high-resolution patterning over large areas with uniformity and repeatability is a complex task. Fabrication imperfections, such as variations in feature size, shape, or position, can lead to deviations in the desired electromagnetic response, affecting the overall performance of the material. Ensuring reliable and scalable fabrication methods that can consistently reproduce the desired structures is a significant challenge.

Another challenge lies in the selection and integration of suitable materials for metamaterial and metasurface fabrication. These materials should possess specific electromagnetic properties and mechanical stability to ensure reliable performance. However, finding materials that meet the desired criteria, are compatible with the fabrication processes, and exhibit long-term stability can be challenging. Many metamaterials and metasurfaces rely on nanoscale or thin-film structures, which can be susceptible to environmental factors such as temperature, humidity, and mechanical stress. Developing materials with improved reliability and scalability is crucial to ensure long-term performance and practical implementation.

Scalability is a significant challenge in the fabrication of metamaterials and metasurfaces. Although fabrication techniques like electron beam lithography or focused ion beam milling provide high resolution and precision, they often face limitations in terms of scalability. These techniques are time-consuming, expensive, and not suitable for large-scale production. As a result, alternative fabrication methods that can achieve high throughput and scalability, such as nanoimprint lithography, roll-to-roll printing, or self-assembly techniques, are being explored. However, ensuring the reliability and uniformity of the fabricated structures at a larger scale remains a challenge (Martín, 2012).

The integration of metamaterials and metasurfaces into practical wireless communication devices and systems also poses reliability challenges. The performance of these materials can be sensitive to factors such as fabrication imperfections, material degradation, or variations in operating conditions. Reliability testing under realistic conditions is crucial to ensure the robustness and durability of the fabricated devices. Moreover, ensuring compatibility with existing fabrication processes and integration with standard wireless communication components adds complexity to the fabrication process and requires careful consideration.

To address these reliability and scalability challenges, a multidisciplinary approach is required. Collaboration between material scientists, device engineers, and fabrication experts is crucial to develop novel fabrication techniques, materials, and characterization methods that enhance reliability and scalability. The advancement of additive manufacturing processes, the development of more stable and reliable materials, and the establishment of rigorous quality control measures are areas of active research. Additionally, improvements in simulation and modeling techniques can aid in predicting and optimizing the performance of fabricated metamaterials and metasurfaces, reducing the reliance on extensive trial-and-error fabrication processes.

In summary, ensuring reliability and scalability in the fabrication of metamaterials and metasurfaces for wireless communication remains a significant challenge. Overcoming these challenges requires advancements in fabrication techniques, material selection, integration strategies, and reliability testing. Addressing these challenges will pave the way for the widespread adoption of metamaterial-based devices and systems in wireless communication applications.

CASE STUDIES: DESIGN AND FABRICATION OF NONLINEAR METAMATERIALS FOR SPECIFIC WIRELESS COMMUNICATION APPLICATIONS

Case Study 1: Nonlinear Metamaterial for Frequency Upconversion in 5G Communications

In this case study, a nonlinear metamaterial was designed and fabricated to enable frequency upconversion in 5G wireless communication systems. The goal was to convert lower frequency signals to higher frequency bands for efficient spectrum utilization. The design involved incorporating nonlinear elements into the metamaterial structure to achieve frequency mixing and upconversion.

The design process started with simulation and modeling techniques to optimize the metamaterial structure. The unit cell geometry, material composition, and nonlinear properties were iteratively adjusted to achieve the desired frequency conversion characteristics. The simulation results guided the selection of suitable materials and fabrication techniques (Fischer et al., 2020).

The fabrication utilized innovative nanofabrication methods like electron beam lithography and reactive ion etching. Subwavelength features were patterned with high precision on a substrate, allowing for the desired nonlinear response. The choice of materials with appropriate nonlinear properties, such as nonlinear optical materials or ferroelectrics, ensured efficient frequency upconversion.

After fabrication, the nonlinear metamaterial was characterized using nonlinear optical measurements. Second-harmonic generation (SHG) and mixing experiments were performed to verify the frequency upconversion capability. Electromagnetic measurements were conducted to evaluate the metamaterial's act in the process of transmission, reflection, and absorption at different frequencies.

The fabricated nonlinear metamaterial demonstrated efficient frequency upconversion, enabling the conversion of lower frequency signals to higher frequency bands suitable for 5G communications. The system achieved improved spectrum utilization and enhanced signal transmission, contributing to the advancement of wireless communication technologies.

Case Study 2: Nonlinear Metasurface for Beam Steering in Millimeter-Wave Communications

In this case study, a nonlinear metasurface was designed and fabricated for beam steering in millimeter-wave wireless communication systems. The objective was to develop a metasurface-based device that could dynamically control the direction of millimeter-wave beams for improved communication link quality and coverage (Milias et al., 2021).

The design process involved numerical simulations and optimization algorithms to determine the metasurface structure that would achieve the desired beam steering functionality. The metasurface was designed to exhibit nonlinear properties, allowing for active mechanism of the phase and amplitude of the transmitted or reflected waves. The simulations considered factors such as incident angle, polarization, and frequency to optimize the metasurface performance.

The fabrication of the nonlinear metasurface utilized techniques such as nanoimprint lithography or self-assembly methods to pattern subwavelength structures with high precision. The choice of suitable materials with tunable nonlinear properties facilitated dynamic control of the metasurface response. Integration with external control mechanisms, such as electrical or optical stimuli, was also considered to enable real-time beam steering.

After fabrication, the metasurface was tested for beam steering performance. The device was characterized using experimental setups that included millimeter-wave sources, antennas, and beam scanning systems. The transmitted or reflected beam direction was measured at different control settings and compared with the simulation results.

The fabricated nonlinear metasurface demonstrated effective beam steering capabilities, enabling dynamic control of millimeter-wave beams. The device showcased enhanced communication link quality, improved coverage, and adaptability to changing communication scenarios. This case study exemplified the potential of nonlinear metasurfaces for advanced beamforming and beam steering in millimeter-wave wireless communication system (Walia et al., 2015).

In conclusion, these case studies highlight the design and fabrication of nonlinear metamaterials for specific wireless communication applications. By incorporating nonlinear elements and utilizing advanced fabrication techniques, these materials offer enhanced functionalities and performance in wireless communication systems. The success of these case studies demonstrates the potential of nonlinear metamaterials and metasurfaces in enabling novel wireless communication technologies and addressing specific communication challenges.

FUTURE TRENDS IN DESIGN AND FABRICATION TECHNIQUES FOR NONLINEAR METAMATERIALS AND METASURFACES

Future trends in the design and fabrication techniques for nonlinear metamaterials and metasurfaces for wireless communication are poised to drive advancements in the field and unlock new possibilities. Several key trends are anticipated to shape the future of these technologies (Jung et al., 2020):

(i) ***Multifunctional and reconfigurable designs***: Future nonlinear metamaterials and metasurfaces will likely exhibit multifunctional and reconfigurable properties. By incorporating tunable or switchable elements, such as phase-change materials, liquid crystals, or electroactive polymers, these materials can dynamically adapt their electromagnetic response to varying communication scenarios. This trend will enable greater versatility, adaptability, and optimization in wireless communication systems.

(ii) ***Integrated and miniaturized devices***: The integration of nonlinear metamaterials and metasurfaces with existing wireless communication components is a growing trend. Future fabrication techniques will focus on developing methods for seamlessly integrating these materials with antennas, waveguides, or other communication devices. This integration will lead to compact, miniaturized, and multifunctional devices, enabling advancements in wearable technology, Internet of Things (IoT), and beyond.

(iii) ***Advances in fabrication scalability***: Addressing the scalability challenge is an ongoing trend in the field. Future fabrication techniques will continue to explore high-throughput and scalable methods, such as roll-to-roll printing, nanoimprint lithography, or self-assembly techniques. These techniques will enable the large-scale production of nonlinear metamaterials and metasurfaces, paving the way for their widespread adoption in wireless communication systems.

(iv) ***Nanoscale and 2D materials***: The use of nanoscale and two-dimensional (2D) materials in nonlinear metamaterials and metasurfaces is an emerging trend. Materials such as graphene, transition metal dichalcogenides (TMDs), or topological insulators offer unique electromagnetic properties and can be integrated into the metamaterial structures. Future research will focus on exploring the potential of these materials for achieving enhanced nonlinear effects, improved device performance, and novel functionalities.

(v) ***Advances in characterization and testing***: Future trends will involve advancements in characterization and testing techniques for nonlinear metamaterials and metasurfaces. This includes the development of more accurate and efficient measurement techniques to quantify their nonlinear response, electromagnetic properties, and performance in practical wireless communication scenarios. Moreover, computational modeling and simulation techniques will continue to improve, aiding in the design optimization process and reducing the reliance on extensive experimental characterization.

(vi) ***Hybrid and metasurface-enhanced systems***: The combination of nonlinear metamaterials with other emerging technologies, such as plasmonics, photonic crystals, or quantum systems, is an exciting trend for future wireless communication systems. Hybrid systems can leverage the unique properties of different materials and platforms, leading to enhanced functionality, improved efficiency, and novel applications. Metasurfaces, with their ability to control wavefronts and manipulate electromagnetic fields, will play a crucial role in realizing these hybrid systems(Jung et al., 2020).

(vii) ***Application-driven design and optimization***: Future trends will see an increasing emphasis on application-driven design and optimization of nonlinear metamaterials and metasurfaces for specific wireless communication scenarios. The customization of these materials for specific frequency bands, communication protocols, or system requirements will allow for targeted solutions and improved performance. This trend will involve a close collaboration between material scientists, communication engineers, and system integrators to design and fabricate materials tailored to the specific needs of wireless communication applications.

In summary, future trends in the design and fabrication techniques for nonlincar metamaterials and metasurfaces in wireless communication are expected to focus on multifunctionality, integration, scalability, nanoscale materials, characterization advancements, hybrid systems, and application-driven design. These trends will shape the development of advanced wireless communication technologies, offering improved performance, increased functionality, and novel applications in areas.

CONCLUSION

This book chapter has provided a comprehensive overview of the design and fabrication techniques for nonlinear metamaterials and metasurfaces in the context of wireless communication. We have explored the fundamental principles of these advanced materials and their potential for revolutionizing wireless communication technologies.

Throughout the chapter, we discussed various design considerations, including the incorporation of nonlinear elements, optimization algorithms, and simulation techniques. The fabrication techniques employed in the production of nonlinear metamaterials and metasurfaces were also examined, highlighting the challenges of reliability, scalability, and integration with existing communication systems.

We delved into case studies that demonstrated the practical application of these materials in specific wireless communication scenarios. From frequency upconversion in 5G communications to beam steering in millimeter-wave systems, these case studies showcased the capabilities and advantages of nonlinear metamaterials and metasurfaces in enhancing communication performance.

Furthermore, we explored future trends in the field, identifying areas of growth and potential breakthroughs. Multifunctional and reconfigurable designs, integration with miniaturized devices, scalability advancements, utilization of nanoscale and 2D materials, and application-driven design and optimization were identified as key trends that will shape the future of nonlinear metamaterials and metasurfaces in wireless communication.

As the demand for higher data rates, improved signal quality, and increased spectrum efficiency continues to grow, the design and fabrication techniques discussed in this book chapter offer promising avenues for meeting these challenges. By harnessing the unique properties of nonlinear metamaterials and metasurfaces, we can unlock new possibilities for wireless communication systems, enabling advancements in areas such as 5G, IoT, and beyond.

It is our hope that this chapter aids as a valuable source for researchers, engineers, and practitioners in the wireless communication field. With a comprehensive understanding of the design principles, fabrication techniques, and future trends, we aim to inspire further innovation and collaboration in the development of nonlinear metamaterials and metasurfaces for the next generation of wireless communication technologies.

REFERENCES

Abadal, S., Liaskos, C., Tsioliaridou, A., Ioannidis, S., Pitsillides, A., Sole-Pareta, J., Alarcon, E., & Cabellos-Aparicio, A. (2017). Computing and Communications for the Software-Defined Metamaterial Paradigm: A Context Analysis. *IEEE Access : Practical Innovations, Open Solutions*, *5*, 6225–6235. doi:10.1109/ACCESS.2017.2693267

Chen, H.-T., Taylor, A. J., & Yu, N. (2016). A review of metasurfaces: Physics and applications. *Reports on Progress in Physics, 79*(7), 076401. doi:10.1088/0034-4885/79/7/076401 PMID:27308726

Chen, Y. (2015). Nanofabrication by electron beam lithography and its applications: A review. *Microelectronic Engineering, 135*, 57–72. doi:10.1016/j.mee.2015.02.042

Dai, J. Y., Tang, W., Yang, L. X., Li, X., Chen, M. Z., Ke, J. C., Cheng, Q., Jin, S., & Cui, T. J. (2020). Realization of Multi-Modulation Schemes for Wireless Communication by Time-Domain Digital Coding Metasurface. *IEEE Transactions on Antennas and Propagation, 68*(3), 1618–1627. doi:10.1109/TAP.2019.2952460

Elnaggar, S. Y., & Milford, G. N. (2017). Description and stability analysis of nonlinear transmission line type metamaterials using nonlinear dynamics theory. *Journal of Applied Physics, 121*(12), 124902. doi:10.1063/1.4979022

Esmail, B. A. F., Koziel, S., Golunski, L., Majid, H. B. A., & Barik, R. K. (2022). Overview of Metamaterials-Integrated Antennas for Beam Manipulation Applications: The Two Decades of Progress. *IEEE Access : Practical Innovations, Open Solutions, 10*, 67096–67116. doi:10.1109/ACCESS.2022.3185260

Fischer, S. C. L., Hillen, L., & Eberl, C. (2020). Mechanical Metamaterials on the Way from Laboratory Scale to Industrial Applications: Challenges for Characterization and Scalability. *Materials (Basel), 13*(16), 3605. doi:10.3390/ma13163605 PMID:32824029

Franco-Martínez, F., Grasl, C., Kornfellner, E., Vostatek, M., Cendrero, A. M., Moscato, F., & Lantada, A. D. (2022). Hybrid design and prototyping of metamaterials and metasurfaces. *Virtual and Physical Prototyping, 17*(4), 1031–1046. doi:10.1080/17452759.2022.2101009

Gentile, M., Hentschel, M., Taubert, R., Guo, H., Giessen, H., & Fiebig, M. (2011). Investigation of the nonlinear optical properties of metamaterials by second harmonic generation. *Applied Physics. B, Lasers and Optics, 105*(1), 149–162. doi:10.100700340-011-4766-y

Harutyunyan, H., Volpe, G., Quidant, R., & Novotny, L. (2012). Enhancing the Nonlinear Optical Response Using Multifrequency Gold-Nanowire Antennas. *Physical Review Letters, 108*(21), 217403. doi:10.1103/PhysRevLett.108.217403 PMID:23003302

Holloway, C. L., & Kuester, E. F. (2020). Electromagnetic metamaterials and metasurfaces: historical overview, characterization, and the effect of length scales. In *Dielectric Metamaterials* (pp. 1–38). Elsevier. doi:10.1016/B978-0-08-102403-4.00006-2

Hu, Y., Wang, X., Luo, X., Ou, X., Li, L., Chen, Y., Yang, P., Wang, S., & Duan, H. (2020). All-dielectric metasurfaces for polarization manipulation: Principles and emerging applications. *Nanophotonics*, *9*(12), 3755–3780. doi:10.1515/nanoph-2020-0220

Huang, D., Poutrina, E., Zheng, H., & Smith, D. R. (2011). Design and experimental characterization of nonlinear metamaterials. *Journal of the Optical Society of America. B, Optical Physics*, *28*(12), 2925. doi:10.1364/JOSAB.28.002925

Jung, J., Park, H., Park, J., Chang, T., & Shin, J. (2020). Broadband metamaterials and metasurfaces: A review from the perspectives of materials and devices. *Nanophotonics*, *9*(10), 3165–3196. doi:10.1515/nanoph-2020-0111

Keiser, G. R., Fan, K., Zhang, X., & Averitt, R. D. (2013). Towards Dynamic, Tunable, and Nonlinear Metamaterials via Near Field Interactions: A Review. *Journal of Infrared, Millimeter and Terahertz Waves*, *34*(11), 709–723. doi:10.100710762-013-9993-3

Li, K., He, X., Wang, X., & Jiang, S. (2019). A Nonlinear Electromagnetic Energy Harvesting System for Self-Powered Wireless Sensor Nodes. *Journal of Sensor and Actuator Networks*, *8*(1), 18. doi:10.3390/jsan8010018

Li, P., Chen, S., Dai, H., Yang, Z., Chen, Z., Wang, Y., Chen, Y., Peng, W., Shan, W., & Duan, H. (2021). Recent advances in focused ion beam nanofabrication for nanostructures and devices: Fundamentals and applications. *Nanoscale*, *13*(3), 1529–1565. doi:10.1039/D0NR07539F PMID:33432962

Martín, F. (2012). Metamaterials for Wireless Communications, Radiofrequency Identification, and Sensors. *ISRN Electronics*, *2012*, 1–29. doi:10.5402/2012/780232

McClintock, H. D., Doshi, N., Iniguez-Rabago, A., Weaver, J. C., Jafferis, N. T., Jayaram, K., Wood, R. J., & Overvelde, J. T. B. (2021). A Fabrication Strategy for Reconfigurable Millimeter-Scale Metamaterials. *Advanced Functional Materials*, *31*(46), 2103428. Advance online publication. doi:10.1002/adfm.202103428

Milias, C., Andersen, R. B., Lazaridis, P. I., Zaharis, Z. D., Muhammad, B., Kristensen, J. T. B., Mihovska, A., & Hermansen, D. D. S. (2021). Metamaterial-Inspired Antennas: A Review of the State of the Art and Future Design Challenges. *IEEE Access : Practical Innovations, Open Solutions*, *9*, 89846–89865. doi:10.1109/ACCESS.2021.3091479

Minovich, A. E., Miroshnichenko, A. E., Bykov, A. Y., Murzina, T. V., Neshev, D. N., & Kivshar, Y. S. (2015). Functional and nonlinear optical metasurfaces. *Laser & Photonics Reviews*, *9*(2), 195–213. doi:10.1002/lpor.201400402

Quevedo-Teruel, O., & Chen, Q. (2021). Classification and opportunities of metasurfaces for antenna designs. *2021 International Symposium on Antennas and Propagation (ISAP)*, 1–2. 10.23919/ISAP47258.2021.9614437

Ramaccia, D., Toscano, A., & Bilotti, F. (2021). Time-varying metamaterials and metasurfaces for antennas and propagation applications. *Journal of Physics: Conference Series*, *2015*(1), 012121. doi:10.1088/1742-6596/2015/1/012121

Salhi, R., Laabidi, M., & Choubani, F. (2016). Modeling and simulation of practical metamaterial structures: SRR and HIS structures. *2016 IEEE International Conference on Computational Electromagnetics (ICCEM)*, 259–261. 10.1109/COMPEM.2016.7588663

Shadrivov, I. V., Lapine, M., & Kivshar, Y. S. (Eds.). (2015). *Nonlinear, Tunable and Active Metamaterials* (Vol. 200). Springer International Publishing. doi:10.1007/978-3-319-08386-5

Tadesse, A. D., Acharya, O. P., & Sahu, S. (2020). Application of metamaterials for performance enhancement of planar antennas: A review. *International Journal of RF and Microwave Computer-Aided Engineering*, *30*(5). Advance online publication. doi:10.1002/mmce.22154

Vellucci, S., Monti, A., Barbuto, M., Longhi, M., Toscano, A., & Bilotti, F. (2022). Metasurface Coatings Enabling Scattering, Frequency, and Radiation Tunability for Next-Generation Antenna Systems. *2022 Microwave Mediterranean Symposium (MMS)*, 1–5. 10.1109/MMS55062.2022.9825540

Walia, S., Shah, C. M., Gutruf, P., Nili, H., Chowdhury, D. R., Withayachumnankul, W., Bhaskaran, M., & Sriram, S. (2015). Flexible metasurfaces and metamaterials: A review of materials and fabrication processes at micro- and nano-scales. *Applied Physics Reviews*, *2*(1), 011303. doi:10.1063/1.4913751

Yang, Y., Li, Q., & Wang, G. P. (2008). Design and fabrication of diverse metamaterial structures by holographic lithography. *Optics Express*, *16*(15), 11275. doi:10.1364/OE.16.011275 PMID:18648445

Yoon, G., Kim, I., & Rho, J. (2016). Challenges in fabrication towards realization of practical metamaterials. *Microelectronic Engineering*, *163*, 7–20. doi:10.1016/j.mee.2016.05.005

Chapter 5
Metamaterial-Based Absorber

Mehaboob Mujawar
https://orcid.org/0000-0002-8260-7062
Glocal University, India

Subuh Pramono
Universitas Sebelas Maret, Indonesia

ABSTRACT

In this chapter, the authors discuss metamaterial absorbers. The need for active absorbers and also different types of active absorbers have been listed in detail such as switchable absorber/reflector, switchable absorber, and tunable absorbers. The recent trends and the work investigated by researchers around the globe have also been mentioned in brief, which can set a benchmark for further research in the field of metamaterial absorbers. The proposed MTM structure is made up of four-square patches. The structure was designed on an FR4 substrate measuring 11 mm by 11 mm with a thickness of 1.60 mm. Other properties of the substrate include relative permittivity of 4.4 and loss tangent of 0.019. The structure has a full ground plane to minimize transmission. The structure's primary resonance for absorber mode configuration is at 8 GHz. The simulations of the proposed structure were carried out using CST.

INTRODUCTION

The high frequency region has been basically quantified into two regions, microwave and optical range. Optical region covers infra-red, visible and ultra-violet region. The microwave range lies between 300 MHz to 300 GHz and in terms of wavelength it lies between 1m to 1mm (Mujawar et al., 2022). For all the conventional users,

DOI: 10.4018/978-1-6684-8287-2.ch005

microwave frequency ranges from 1 to 40 GHz, over which we can also consider the mmwave band from 30 to 300 GHz and the operating wavelength is between 1mm to 10mm or 1 cm. The optical range lies between 30 to 3000 THz. There is a THz gap of 0.3 to 30 THz between microwave and optical range. According to carson's rule, as carrier frequency increases the bandwidth requirement increases by 10% of the carrier frequency. The channel capacity as per shannon's law depends on bandwidth and signal-to-noise ratio. $C = B\log_2(1+SNR)$, As SNR remains constant and bandwidth increases, the channel capacity will increase. Therefore, the number of channels that can be accommodated over a particular medium increases. Hence the researchers have opted for higher frequencies for communication purpose.

Introduction to Metamaterial: The word "Meta" means "beyond". Metamaterials are basically structures which are homogeneous in nature. They are the periodic structures and when their periodicity is excited by external waves or signals. It exhibits unusual electromagnetic properties. There are different type of materials and each material has its own properties. The property of the material never changes at any cost. When the properties of the material are changed artificially, it is called as metamaterial. All natural materials like glass, or diamond have positive electrical permittivity, magnetic permeability and an index of refraction. When the properties of these materials are changed making them negative, it is called metamaterial (Smith et al., 2005). Metamaterial is not a natural material; it is artificial material by changing the character so that they can change the property of the electromagnetic waves. A basic meta-atom consists of unit cell, the size of unit cell is less than $\lambda/4$ (effective homogeneity limit). Meta-surface is the two dimensional analog of the metamaterial. It basically comes from frequency selective surface, where we are having 2 dimensional array, which can be made of metallic patches, slots etc. The thickness of the structure is very thin in nature that is less than $\lambda/10$. When the thickness is less than $\lambda/10$, the structure is ultra-thin in nature. Whenever the periodicity is operating in the sub-wavelength region ($\leq\lambda/4$), the structure is a meta-surface. Ripin et al. (2012) came up with the idea for an MSA that makes use of EBG structures at the ground and incorporates the dielectric from Rogers RO3003. As compared to an antenna that does not have an EBG structure, this model was resonated at 7.3 GHz, which resulted in an increased bandwidth of 39.63% while simultaneously lowering its size by 22.38%. It is put to use in analyzing a wide variety of MSA features. The Nicolson-Ross-Weir (NRW) method was used in order to do research on the metamaterial characteristics of the proposed antenna.

Wearable sensors equipped with antennas that can monitor a person's health, as well as their performance in sports, during exercise, and other activities, are gaining a significant amount of popularity. These types of applications could make use of the ON-body and OFF-body sensors that are a part of Wireless Body Area Networks (WBANs). The ISM band is regarded as the band that is most ideally suited for WBANs since it does not need a license and it provides a sufficient amount of bandwidth. In order to communicate local data from the sensor nodes to the central module, these sensor nodes make use of antennas that are worn on the body. In addition to being bent and crumpled, the functionality of an antenna may be negatively affected when it is in close proximity to a human body. When a person is exposed to radiation from an antenna back-lobe, the amount of power that is absorbed by their body in relation to their total mass also has an effect on the health of their biological tissues. The term "Specific Absorption Rate," which is abbreviated as "SAR," is used to evaluate the level of exposure in terms of the amount of power that is taken in by the human body in relation to its unit mass. The limit has been set by the American National Standards Institute (ANSI) at 1.6 W/Kg absorbed per 1 g of tissue, while the International Commission on Non-Ionizing Radiation Protection (ICNIRP) has set it at 2 W/Kg absorbed per 10 g of tissue. Both of these organizations have established the standard. Omnidirectional antennas often produce a high specific absorption rate (SAR) inside the human body. This is as a result of the fact that they need strict control over the high back-lobe radiations that they emit toward the human body. According to Sievenpiper et al. (1999), the use of a metamaterial-based Electromagnetic Band Gap (EBG) as a ground plane that functions as a High Impedance Surface (HIS) for body worn antennas has the potential to lower the intensity of these radiations and, as a result, the specific absorption rate (SAR). As a ground plane, a dual band concentric square patch 3 x 3 EBG structure has been suggested in order to lessen the effects of bending on antenna driving point impedance and SAR. Both a dual band Jerusalem cross 3 x 3 HIS structure and a dual band hexagonal type concentric patch EBG layer have been included. Both of these structures use dual bands.

Meta-surface Absorber: When a wave is incident on an object, a part of it will reflect back and a part will transmit. If a lossy media is considered, the rest of the wave will be absorbed (Mujawar, 2021).

$$\Lambda(\omega) = 1 \quad |S_{11}|^2 \quad |S_{21}|^2 \tag{1}$$

S_{11} is the reflected power (R) and S_{21} is the transmitted power (T). with complete metallic backing $|S_{21}|^2=0$

$$A(\omega) = 1 - |S_{11}|^2 \tag{2}$$

Absorption (A) = 1 – R

As the reflected power decreases, the absorption will increase. The input impedance (eq. 3) from the conventional equation is given by:

$$Z(\omega)=\eta_0\sqrt{\frac{(1+S_{11})^2-S_{21}{}^2}{(1-S_{11})^2-S_{21}{}^2}} \tag{3}$$

With complete metallic backing, $|S_{21}|^2=0$

$$Z(\omega)=\eta_0\frac{1+S_{11}}{1-S_{11}} \tag{4}$$

When $S_{11}=0$

$$Z(\omega)=\eta_0 \tag{5}$$

$$\eta_0\sqrt{\frac{\mu_m}{\varepsilon_m}}=\eta_0 \tag{6}$$

$$\mu_m=\varepsilon_m \tag{7}$$

Metamaterial Based Microwave Absorber

A microwave absorber exhibits absorption at microwave frequencies. They can be either single or multiband absorber. The metamaterial absorbers are ultra-thin having thickness of $\lambda/20$. The periodicity is of the order of $\lambda/4$ to $\lambda/6$, but the thickness is very small compared to wavelength. The metamaterial absorbers are mostly based on meta-surface. The effective electromagnetic parameters (ε_{eff} and μ_{eff}) can be regulated and absorbers can be made scalable from microwave, terahertz, infra-red to optical. Microwave absorbers can be easily fabricated using PCB technology (Chen et al., 2004).

Active Absorbers

Need for active absorbers: Conventional passive absorbers offer no flexibility once fabricated. E.g: once antenna is designed for 4 to 8 GHz frequency, it cannot be re-designed for 8 to 9 GHz. which means that the structure cannot be re-used in passive absorbers. Active absorbers can be re-used by controlling through external stimuli. They can be realized by using magnetically/mechanically/electronically. Electronically controlled absorbers utilize active components such as p-i-n diode, varactor diode. These components offer high tuning speed, wideband tuning, compact size and low cost (Mujawar & Gunasekaran, 2022).

Types of Active Absorbers

The Switchable absorber/reflectors use stealth technology. Aircraft can be detected by its own radar during reflective state, whereas it will act as an absorber under the complementary state. Switchable absorbers are used in electromagnetic compatibility test, radar camouflage and deception roles, tagging and asset tracking system. Tunable absorbers are used for reduction of RCS in radomes. Tunable absorbers can be used to replace broadband absorber. They can compensate fabrication and installation error.

1. **Switchable absorber/reflectors:** A switchable absorber exhibits absorption and reflection responses, in two different conditions, regulated by an external control. Consider a simple electric resonator structure for which electric field is connected. Due to the electric field in the horizontal direction, there is surface current distribution. This structure offers reflection at 10 GHz and narrowband absorption at 8GHz. Due to the cut at the middle, it is not giving any surface current and a complete reflection at the entire frequency. Both the structures are offering reflection and absorption within the frequency range of 8 to 12 GHz. p-i-n diode is the ideal solution to realize this switchable operation from single geometry. P-i-n diode offers small resistance in ON state and large capacitance in OFF state. A DC voltage source needs to be connected across the p-i-n diode, such that the diode can be turned ON or OFF by regulating the external voltage level. A diode mounted across a single unit cell can be easily biased. But in a metamaterial absorber structure, unit cells are arranged in periodic pattern in order to simultaneously bias all diodes, they are connected in series and a large voltage is applied across the two ends. The inductor allows direct current (DC) and blocks alternating current (AC). p-i-n diode model: The p-i-n diode is inserted in the middle portion of the structure. During ON state, p-i-n diode offers small resistance and it is equivalent to short circuit. During OFF state, it is giving capacitive effect and basically it is similar to gap. Therefore, by making the p-i-n diode ON/OFF, two different structures can be represented by single geometry. Simulation of p-i-n diode in EM software includes the following steps: 1) Consider an air box/vacuum with definite dimensions. 2) Assign boundary then select lumped RLC. 3) During ON state, assign small resistance (R_{ON}). 4) During OFF state, assign

large capacitance (C_{OFF}). 5) Indicate the current flow line in proper direction. We need to note that all these R, L, C values are connected in parallel and be careful while selecting.

2. **Switchable Absorber:** The switchable absorber exhibits absorption response that can be switched between two frequency bands by external stimuli. P-i-n diode is mostly used in the geometries. The absorption is happening at both ON/OFF state, but the frequency is changing due to the diode action. When the geometries are asymmetrical, polarization is sensitive. In circular split ring geometry, 4 diodes can be connected. During the ON state, diode is functional and it will operate at 3.41 GHz. during the OFF state, due to the diode being working like capacitor, it will operate at 5.35 GHz band. The diodes are in series connection, hence large supply voltage is required to turn ON the diodes. This disadvantage is overcome by using parallel biasing network in cross dipole geometry. During the ON state, when the diode is offering small resistance, the entire dipole arm gets activated. During the OFF state, when the diode is offering high resistance, the entire dipole arm is not activated. Absorption frequency changes based on the effective length of the cross dipole geometry. The positive voltage supply is given to the left side metal line and the negative voltage supply is given to the ground plane. Metallic vias are used to supply the bias voltages across the diodes. P-i-n diodes are in parallel connection, hence the overall supply voltage is small.
3. **Tunable Absorbers:** A tunable absorber exhibits variable absorption frequency that can be continuously regulated by an external bias voltage. The structure has infinite number of states. Most of the tunable absorbers are realized by varactor diodes. A varactor when operated under reverse voltage exhibits tunable junction capacitance represented by:

$$C_J = \frac{C_{J0}}{\left(1+\frac{V_R}{V_J}\right)^M} + C_P \tag{8}$$

C_J is the junction capacitance(eq. 8) which depends on the reverse voltage, V_R is the reverse voltage, V_J is the junction voltage, C_{J0} is the zero bias junction capacitance, C_p is the package capacitance and M is grading coefficient. Junction capacitance can be controlled by varying the reverse voltage. The absorption frequency will change based on changes in junction capacitance. In polarization independent single band tunable absorber, when the reverse bias voltage is 0V and 8V. The response frequency is 1.14 GHz and 1.94 GHz respectively with a tuning range of 70.18%. in this structure, the varactor diodes operate under reverse bias voltage unlike the p-i-n diodes. All the varactors need to be biased under parallel combination, otherwise the overall voltage required will be very high.

LITERATURE REVIEW

In Landy et al. (2008), they have designed single band meta-surface absorber. The patches have been printed over 1mm thin FR4 substrate. It has been found that the absorption is approximately 99.6% at 7.46 GHz. k is the scaling factor multiplied to each geometrical dimension of the structure. The scaling up and scaling down of the structure basically means that periodicity remains constant but the top metallic patches will be optimized. As the dimensions are increased, the frequency decreases and wavelength increases. If the scaling factor was one then the thickness was $\lambda/4$. In Li et al. (2011), they have designed dual band absorber. Two sub-cells have been used, sub-cell 1 has been scaled up by factor of 1.2 and sub-cell 2 by a factor of 1.3. The fabricated structure has dimension of $240\text{m}m\times240\text{m}m$. The fabricated structure is placed in the anechoic chamber, which is illuminated using a pair of horn antenna. The antenna measurements are carried out using vector network analyzer. Initially they have taken copper surface of identical dimension and reflection coefficient is measured. Then the sample structure is kept and reflection coefficient is measured again. The difference between the two provides the actual reflection coefficient from the structure and accordingly absorption coefficient can be determined. The meta-surface antenna has been described in Bilotti et al. (2006). Array of cells has been used on the other side of the patch. VNA is used for the measurement of antenna parameters. By employing this type of structure dual band response is obtained and one of the bandwidth subsequently increased. The simulated and measured results of radiation pattern for co-pol and cross-pol are in good matching. In Schurig et al. (2006), they have designed filter in THz domain. Gold is deposited on the top of silicon layer. As the number of layers are increased, the bandwidth is enhanced in the 1.7 to 2.2 THz region. The cut-off can be increased by increasing the number of layers. But compactness is lost in the third dimension. The algorithm is developed using binary search technique (Chen, 2012). The whole search space is divided into two halves and the cut-off frequency is determined. Based on the frequency obtained, one of the two halves is discarded and the whole search space is divided into half. This continues until both the left extreme and right extreme have their corresponding frequencies difference less than the acceptable precision error.

Design of the Structure

The proposed MTM structure is shown in the figure 1 and figure 2 shows the basic unit cell. It is made up of four-square patches. The structure was designed on an FR4 substrate measuring 11 mm by 11 mm with a thickness of 1.60 mm. Other properties of the substrate include relative permittivity of 4.4 and loss tangent of 0.019. The structure has a full ground plane to minimize transmission. The structure's primary resonance for absorber mode configuration is at 8 GHz. The simulations of the proposed structure were carried out using CST. Figure 3 shows the Absorptivity response for the proposed structure.

Figure 1. The proposed metamaterial absorber

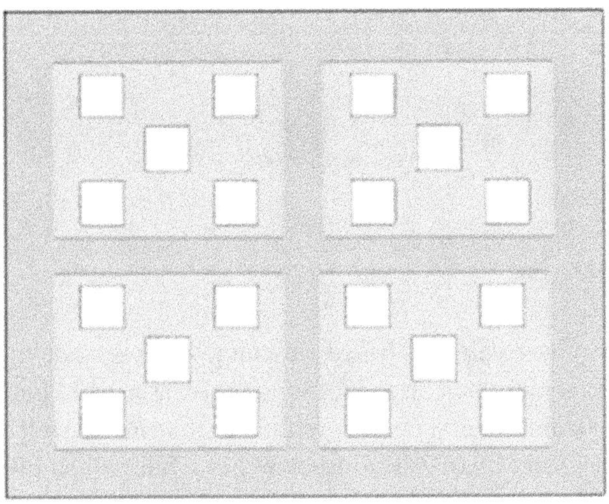

Figure 2. The basic unit cell

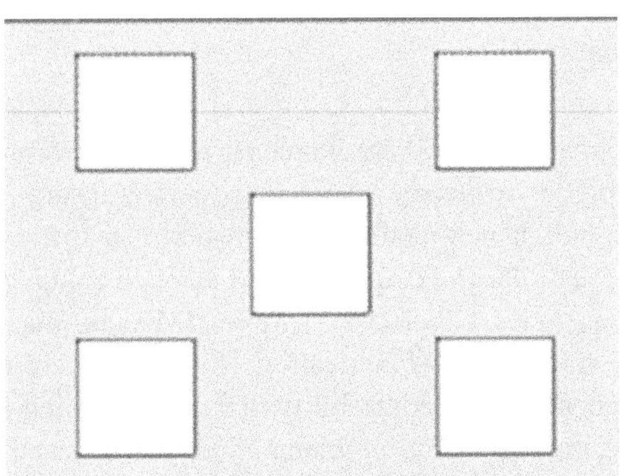

Figure 3. Absorptivity response

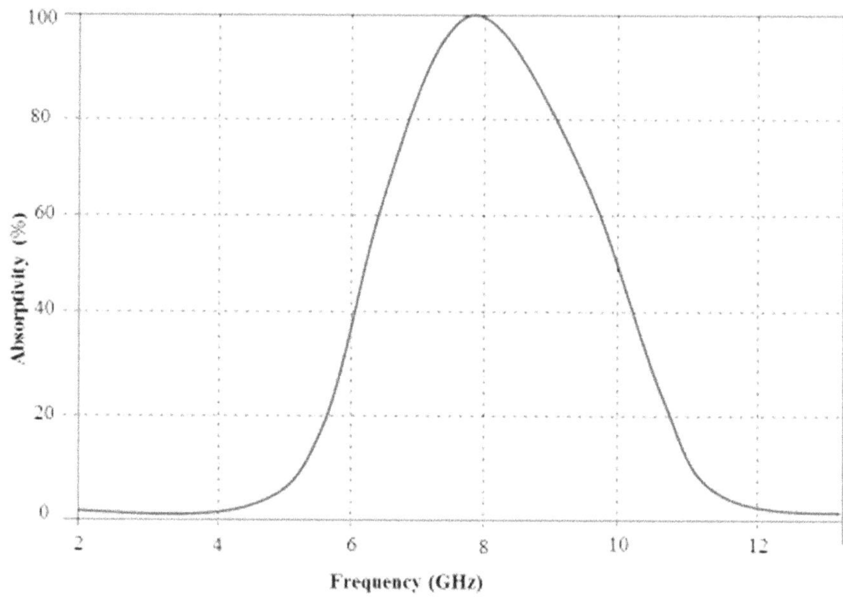

CONCLUSION

In this chapter, we have designed MTM structure which is made up of four-square patches. The structure was built on an FR4 substrate measuring 11 mm by 11 mm with a thickness of 1.60 mm. Other properties of the substrate include relative permittivity of 4.4 and loss tangent of 0.019. The structure has a full ground plane to minimize transmission. The structure's primary resonance for absorber mode configuration is at 8 GHz. The simulations of the proposed structure were carried out using CST.

FUTURE SCOPE

The suggested metamaterial absorbers, which have straightforward designs and can operate at lower or higher frequencies, can achieve perfect absorption. These designs can be changed to highlight new elements that are pertinent to the intended practical applications. The size of the unit cell is the most difficult component of MMAs for GHz and lower frequencies. Otherwise, flexible MMAs for microwave frequency can be made using straightforward fabrication. However, for absorbers functioning at other higher frequencies, this adaptability to various working environments and sophisticated fabrication still pose problems. Spectroscopy and imaging are two areas where the THz-regime-optimized MMAs could be used. For the future use of MMAs at optical frequency, a better manufacturing tolerance is also required. Similar to visible-range absorbers, MMAs in the GHz and THz frequencies have simpler designs, less complicated manufacturing processes, and lower production costs.

REFERENCES

Bilotti, F., Nucci, L., & Vegni, L. (2006). An SRR based microwave absorber. *Microwave and Optical Technology Letters*, *48*(11), 2171–2175. doi:10.1002/mop.21891

Chen, H. (2012). Interference theory of metamaterial perfect absorbers. *Optics Express*, *20*(7), 7165. doi:10.1364/OE.20.007165 PMID:22453398

Chen, X., Grzegorczyk, T. M., Wu, B., Pacheco, J., & Kong, J. (2004). Robust method to retrieve the constitutive effective parameters of metamaterials. *Physical Review. E*, *70*(1), 016608. Advance online publication. doi:10.1103/PhysRevE.70.016608 PMID:15324190

Landy, N. I., Sajuyigbe, S., Mock, J. J., Smith, D. J., & Padilla, W. J. (2008). Perfect Metamaterial Absorber. *Physical Review Letters*, *100*(20), 207402. Advance online publication. doi:10.1103/PhysRevLett.100.207402 PMID:18518577

Li, H., Yuan, L., Zhou, B., Shen, X., Cheng, Q., & Cui, T. J. (2011). Ultrathin multiband gigahertz metamaterial absorbers. *Journal of Applied Physics*, *110*(1), 014909. doi:10.1063/1.3608246

Mujawar, M. (2021). Compact Microstrip Patch Antenna Design with Three I-, Two L-, One E- and One F-Shaped Patch for Wireless Applications. CRC Press eBooks. doi:10.1201/9781003093558-7

Mujawar, M., & Gunasekaran, T. (2022). Multiband Slot Microstrip Antenna for Wireless Applications. In EAI/Springer Innovations in Communication and Computing (pp. 23–34). Springer International Publishing. doi:10.1007/978-3-030-76636-8_3

Mujawar, M., Saradhi, D. V., Naz, M. Y., & Muduli, A. (2022). Design and performance comparison of arrays of circular, square and hexagonal meta-material structures for wearable applications. *Journal of Magnetism and Magnetic Materials*, *553*, 169235. doi:10.1016/j.jmmm.2022.169235

Ripin, N., Awang, R., Sulaiman, A., Baba, N., & Subahir, S. (2012). Rectangular microstrip patch antenna with EBG structure. *SCOReD 2012 - 2012 IEEE Student Conference on Research and Development,* 266-271. . doi:10.1109/SCOReD.2012.6518651

Schurig, D., Mock, J. J., & Smith, D. J. (2006). Electric-field-coupled resonators for negative permittivity metamaterials. *Applied Physics Letters*, *88*(4), 041109. doi:10.1063/1.2166681

Smith, D. J., Vier, D. C., Koschny, T., & Soukoulis, C. M. (2005). Electromagnetic parameter retrieval from inhomogeneous metamaterials. *Physical Review. E*, *71*(3), 036617. Advance online publication. doi:10.1103/PhysRevE.71.036617 PMID:15903615

Chapter 6
Study of Novel Design of Multi-Band and Broad-Band Metamaterial Microwave Absorber

Alkesh Agrawal
Shri Ramswaroop Memorial University, India

Bhagwant Singh
Shri Ramswaroop Memorial University, India

ABSTRACT

The artificially engineered materials termed as metamaterials (MMs) possess unnatural electromagnetic (EM) properties with μ < 0 and ε < 0 due to their geometrical and structural uniqueness. This resulted in diversified study of MMs from the GHz to THz range with one of the potential application as microwave absorber. In this chapter, a case study has been taken up with a unique design of a MM microwave absorber that exhibits multi-band as well as broad-band absorptance within 5GHz - 40GHz. The proposed structure consists of a square ring and a "+" shaped patch at its centre and the scaled "+" shaped patches at four corners. The proposed MM absorber exhibited absorptance peaks of 99.74%, 98.58%, 90.00%, 90.04%, 84.74%, and 94.94% at 6.33 GHz, 14.08 GHz, 28.92 GHz, 32.48 GHz, 33.84 GHz, and 37.69 GHz, respectively, with a broad-band lying within frequency range of 32 GHz to 37 GHz. The MM behavior is studied in terms of normalized matched impedance, effective permeability, and surface current distribution with potential application in stealth technology.

DOI: 10.4018/978-1-6684-8287-2.ch006

INTRODUCTION

Metamaterials

A material's electromagnetic (EM) properties are described in terms of permittivity and permeability values, which can be either negative or positive. Table 1 depicts the classification of the materials based on positive or negative values of and .

Table 1. Material classification based on the values of $\tilde{\varepsilon}$ and $\tilde{\mu}$

$\tilde{\varepsilon}$	$\tilde{\mu}$	Classification of Material
+	+	"double-positive" (DPS)
-	+	"single-negative" (SNG) or "epsilon-negative" (ENG)
+	-	"single-negative" (SNG) or "mu-negative" (MNG)
-	-	"double-negative" (DNG) or "metamaterial" (MM)

Source: Kshetrimayum (2004)

The artificially engineered ENG material was combined with MNG material to generate the metamaterial (Smith et al., 1999). The term "metamaterial" was given by (Walser, 2001). A new class of materials known as metamaterials (MMs) have been artificially synthesized and are not present in nature. These artificially engineered materials possess unusual electromagnetic (EM) characteristics that are absent from materials that are present in nature, such as effective permittivity and permeability values that are less than zero (Smith et al., 1999). Unusual electromagnetic properties result from the structure's geometry, periodic arrangement, and orientation. The composition of the material has nothing to do with these peculiar electromagnetic properties. Metamaterials are sub-wavelength metallic structures that are arranged in a periodic or irregular pattern and their EM material properties are derived from the geometry of the device rather than directly from the material's composition or band structure.

The periodically structured microwires exhibit negative value of effective permittivity $\tilde{\varepsilon}$ below plasma frequency. A plasma is a density wave of electron gas. Plasmons exist in metals. The electrons in metals are loosely coupled to the atoms and are so free to move. A plasmon is a wave of synchronised oscillations made up of billions of electrons in an electron gas. The electrons oscillate violently at a resonance frequency known as the plasmon frequency ω_p. Metal electrons respond to an electric field by producing an opposite screening field, therefore an electric field cannot exist inside a metal. Since an electromagnetic wave cannot pass through metal, it is reflected back instead. However, there comes a point where the external field oscillates too quickly for the electrons to follow at high frequency (= high photon energy). A metal stops reflecting above this frequency. The plasmon energy $E_p = \hbar\omega_p$ is the equivalent energy. The periodic structure of microwires lowers the average concentration of electrons and increases the effective electron mass by self inductance (Pendry et al., 1996).

Split ring resonators (SRRs), arranged in a periodic array, exhibit negative effective permeability $\tilde{\mu}_{eff}$. A split ring resonator is made out of concentric copper rings with splits at the ends (Pendry et al., 1999). Depending on the resonance characteristics of the SRR, surface currents are created when an SRR is exposed to EM wave along it's axis. The magnetic field created by the induced surface currents either strengthens or weakens the incident field (Smith et al., 1999). An essential factor in the SRR's ability to have resonance at a wavelength much longer than the ring diameter is the presence of splits in the SRR. High capacitance between the concentric rings is produced by the split of the inner concentric ring, which also lowers the resonating frequency and confines the electric field. The SRR's dimensions are much lower than the wavelength of free space, which lowers radiation losses and raises the Q factor significantly (Smith et al., 1999).

The metamaterials find applications in perfect-lenses (Pendry, 2000), super-lenses (Fang et al., 2005), super-resolution imaging with hyper-lenses (Liu et al., 2007), EM cloaks (Schurig et al., 2006), EM concentrators (Rahm et al., 2008), and MM based perfect absorbers (Landy et al., 2008).

EM Absorber

A device, under impedance matched conditions, that absorbs all electromagnetic radiations incident on it is known as an electromagnetic (EM) wave absorber. EM absorbers were developed by two scientists, "J. Jaunmann" and "W. W. Salisbury," in the United States to enhance radar performance and give stealth technology (Salisbury, 1954). Resonant absorbers and non-resonant (or broad-band) absorbers are two broad categories of EM absorbers (Lee et al., 2016). The non-resonant absorbers are further classified as geometric transition and low density absorbers (Lee et al., 2016). The resonant absorbers are classified as Salisbury screen (Salisbury, 1954), circuit analog absorbers (Munk, 2000), and metamaterial perfect absorbers (Landy et al., 2008), as shown in Figure 2.

The broad-band absorbers are capable of absorbing the incident EM radiation over a wide range of frequencies. The surface impedance of a broad-band absorber is feebly dependent on the frequency of EM waves in a wide range of EM frequencies, and it matches to the impedance of free space Z_o. The incident EM waves will therefore not be reflected, and are completely absorbed for a wide range of frequencies. The absorbed EM waves must be dissipated inside the material. Therefore, the material used in non-resonant absorbers must have high electric and magnetic losses.

First and foremost, the salisbury screen's performance is constrained by its large size. As dielectric lossy layers cascade, the thickness rises for broad-band absorption. (Jaumann screen), as shown in Figure 1 (b). The dielectric layer's thickness needs to be $\left(\frac{\lambda/4}{\eta}\right)$, in order to fulfil completely destructive interference. The dielectric layer's thickness must be quite high if the EM wave's wavelength (λ) is large. Second, because of impedance matching, the Salisbury screen's performance is restricted to microwave frequencies.

Figure 1. (a) Salisbury screen, (b) Jaumann screen

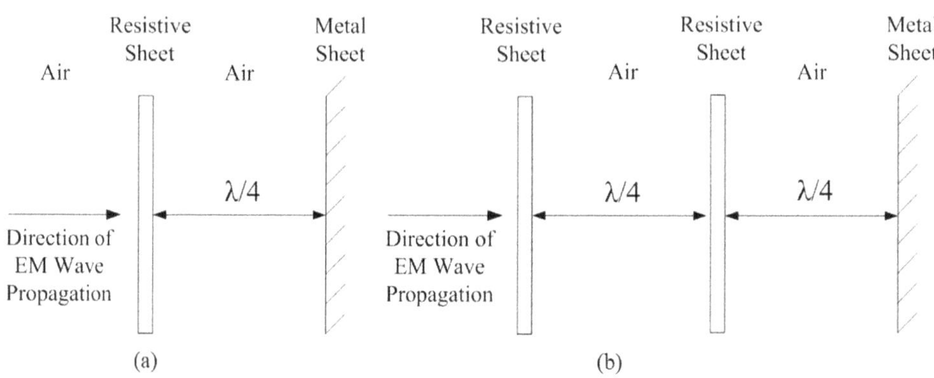

Figure 2. Classification of absorbers

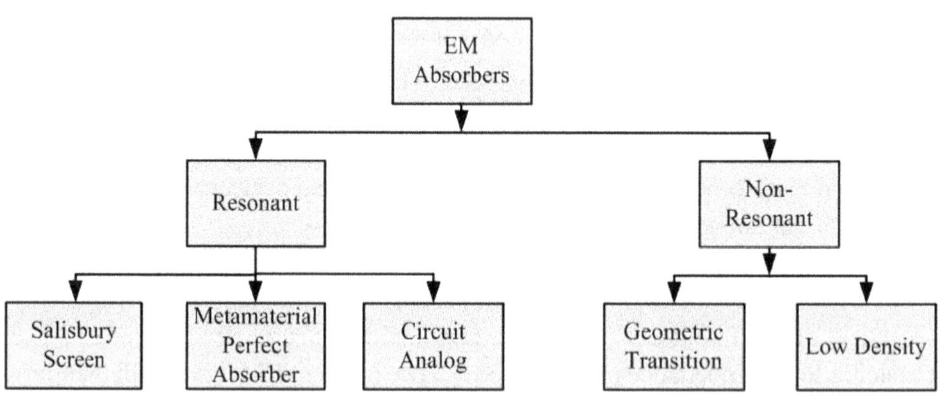

Metamaterial Perfect Absorber

Typically, an MPA has three layers (Agrawal & Misra, 2021). It is made up of a bottom layer with a metal laminated ground plane, layer in the middle which is dielectric layer (lossy or lossless), and a top layer with a periodic array of unit cells making up a metamaterial high impedance surface. MPA's layered structure is depicted in Figure 3. In order to match the impedance of frequency selective top layer of an MPA with that of free space, the structural properties in terms of shape and size of the patches, periodicity and the orientation of the unit cell are to be optimised. The EM waves when encounter the frequency selective surface (FSS) having very high impedance do not reflect back but travel through the FSS into the dielectric layer. The EM waves are entirely blocked and reflected back by the continuous metal laminated bottom layer. The dielectric layer within the MPA enables incident electromagnetic waves to persist and get attenuated within the layer. High permittivity in the dielectric layer is ideal since it allows for a reduction in thickness while retaining the optical path. Additionally, the loss tangent (tan δ) of the dielectric layer should be large.

Figure 3. Layered structure of MPA

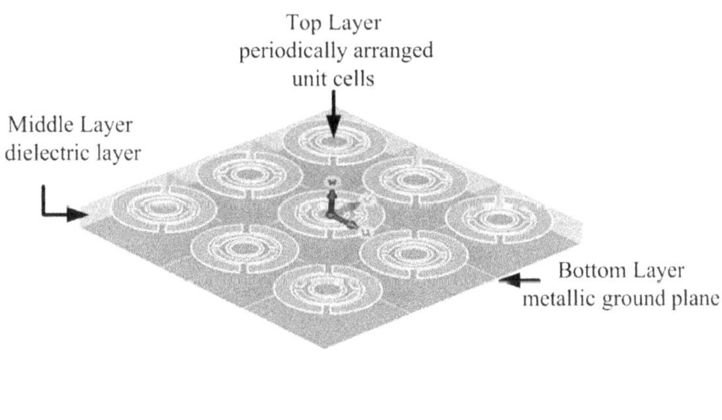

In conventional MPA, the reflecting boundary seems to be invisible to the incident EM wave due to suitable boundary conditions; hence, there is no or minimum reflection at the top layer of MPA, given by Equation (1), where A(f) is absorptance, $|S_{11}(f)|^2$ is reflectance, and $|S_{21}(f)|^2$ is transmittance.

$$A(f) = 1 - |S_{11}(f)|^2 - |S_{21}(f)|^2 \quad (1)$$

The effective value of the impedance of the top layer of the MPA is given by Equation (2), where $\tilde{Z}_{eff}$ (f), $\tilde{\mu}_{eff}$ (f), and $\tilde{\varepsilon}_{eff}$ (f) are the effective impedance, permeability, and permittivity of the frequency selective surface.

$$\tilde{Z}_{eff}(f) = \left[\tilde{\mu}_{eff}(f) / \tilde{\varepsilon}_{eff}(f)\right]^{1/2} \tag{2}$$

The impedance of free space is given by Equation (3), where Z_o, μ_o, and ϵ_o are the impedance, permeability, and permittivity of free space.

$$Zo = [\mu_o / \epsilon_o]^{1/2} \tag{3}$$

For A(f)=1, the $|S_{11}(f)|^2$ and $|S_{21}(f)|^2$ should be reduced to zero, which can be obtained if, $\widetilde{Z} = Z_o$. This situation is known as "perfectly matched condition" (Landy et al., 2008; Tao et al., 2008). The MPAs should fulfil the conditions of very high absorptance (Landy et al., 2008), perfectly matched condition (Landy et al., 2008; Tao et al., 2008), multi-band (Yan et el., 2020) and broad-band absorptance (Agrawal & Misra, 2021), polarisation insensitive absorptance (Singh et al., 2019), absorptance at wide angle of incidence (Agrawal et al., 2016).

For the first time, the MM narrow-band microwave absorber was realised in 2008 (Landy et al., 2008).

According to Figure 4, the unit cell was made up of an Electric Ring Resonator (ERR) atop a micro-structured cut wiring section that was divided from it by a dielectric layer. FR-4 substrate with a thickness of 0.2mm is used as the dielectric layer. The copper layers in Figure 4 that are 17 μm thick were etched to create the top layer that is ERR and bottom layer that is micro-structured cut wiring portion. The MMA structure was made up of a 2D array of unit cells that were 0.72 mm apart. The photosensitized approach was used to create the MPA.

In contrast to the absorptance of 88% achieved experimentally at 11.5 GHz, the results of the simulation in the FDTD solver with Computer Simulation Technology (CST) microwave studio, indicated 96% absorptance at a frequency of 11.48 GHz. Fabrication errors were the key factor in explaining the discrepancy between the simulated and experimental values. At the resonance frequency of 11.48 GHz, the ERR top layer and micro-structured cut wire segment bottom layer, tightly coupled the incident EM waves, producing a powerful electric response. Magnetic coupling was caused by the induction of anti-parallel currents in the ERR and the micro-structured cut wire section. Electric and magnetic responses were therefore caused by the interaction of ERR and the micro-structured cut wire section. According to the authors, changing the geometry of the ERR caused a subtle deviation in the

Figure 4. Experimentally demonstrated top view of narrow-band MPA reported in 2008 by Landy et al.

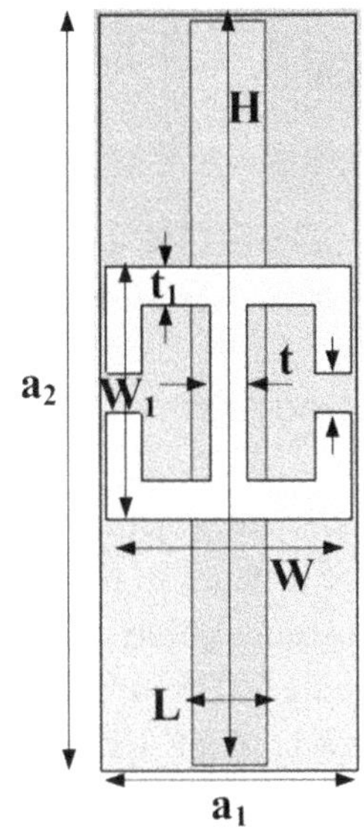

absorptance strength of absorptance resonance frequency, whereas altering the distance of the two metallic structures-the top layer of the ERR and the micro-structured cut wire section bottom layer, modified the magnetic response. It was claimed that the suggested MPA may be used in devices like bolometers.

By using MPA dimensions in μm and a polyamide dielectric substrate layer in place of the FR-4 substrate layer, the same work was extended in the THz domain by (Tao et al., 2008). With a potential use as a low thermal mass absorber with thermal sensing, the simulation results indicated 98% absorptance at 1.12 THz in contrast to experimental measured absorptance of 70% at 1.3 THz.

The first MM absorber in the mid-IR regime was reported by (Liu et al., 2010) and the first demonstrations of MM absorbers in the near-IR regime were simultaneously reported by (Hao et al., 2010; Liu et el., 2010). Since then, MM absorbers have been investigated in the fields of multi-band (Agrawal et al., 2017) and broad-band (Agrawal & Misra, 2021) MM absorbers.

The ability to demonstrate high absorptance at a wide range of resonant frequencies makes multi-band microwave absorbers practically more useful. Applications for multi-band microwave absorbers that are insensitive to condition of polarisation and wide angle of incident EM waves are expanding, whereas the broad-band MMAs are preferred over multi-band MMAs with the advantage of absorbing any band of frequencies in the entire high absorptance broad-band. The absorption capability of broad-band MMA is enhanced by a highly symmetric design with a stable and insensitive response of MMA for different polarisations and wide angular incident microwaves, with potential applications in reducing radar cross-section, electromagnetic interference, providing stealth technology, and improving metamaterial based micro-strip antennas.

CASE STUDY OF A NOVEL DESIGN OF MULTI-BAND AND BROAD-BAND MM ABSORBER

Design of Proposed MM Absorber

A three-layered metamaterial unit cell makes up the proposed MM absorber structure. The top layer has a square ring and a patch in the shape of "+" at its centre. The scaled "+" shaped patches are placed in all four quadrants, as depicted in Figure 5. The top as well as the bottom layer are 0.035 mm thick copper layers (conductivity of 5.8×10^7S/m). The middle layer is of FR4 substrate (ε_r=4.4, tanδ=0.02) with thickness 0.8 mm. Table 2, shows the optimum unit cell dimensions of the proposed MM absorber.

Figure 5. (a) The unit cell's perspective view in three dimensions, (b) the suggested MM absorber's dimension and geometry

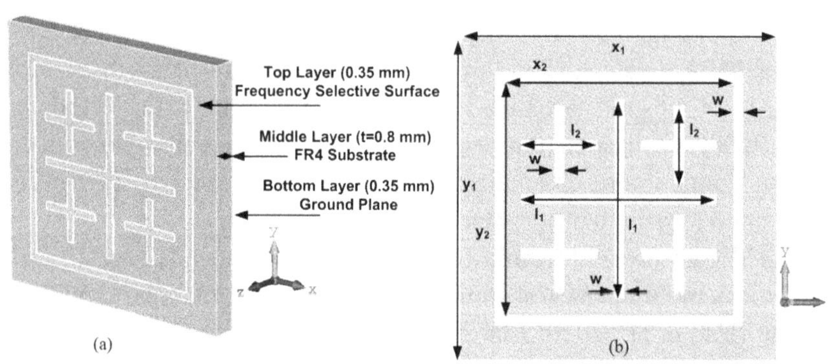

The Proposed MM Absorber Unit Cell Simulation Results

The dimensions of the design are optimised as depicted in Table 2. Utilising periodic boundary conditions, the optimised unit cell is simulated using CST Microwave Studio software. Figure 6 shows the reflection coefficient and the absorptance. Under perfectly impedance matched conditions, the proposed MM

Table 2. Unit cell Measurements for the proposed MM absorber

Element	Dimension (mm)
x_1	9
y_1	9
x_2	6.58
y_2	6.58
w	0.26
l_1	5.58
l_2	2.16
t	0.8

Figure 6. The graph for simulated $S_{11}(f)$, reflection coefficient and A(f), absorptance of MM absorber unit cell structure at 6.33 GHz, 14.08 GHz, 28.92 GHz, 32.48 GHz, 33.84 GHz, and 37.69 GHz

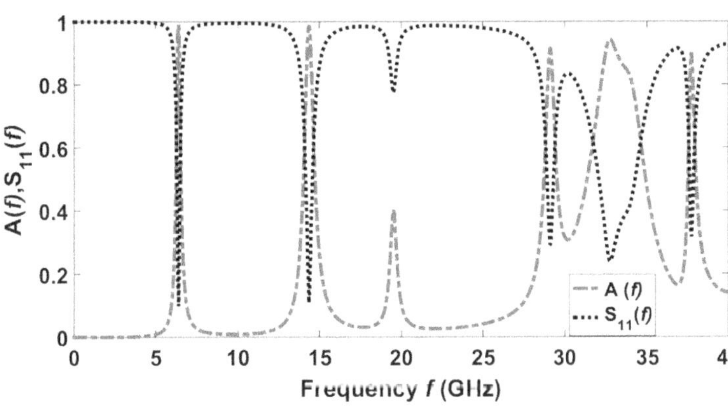

absorber exhibits six absorptance bands at 6.33 GHz, 14.08 GHz, 28.92 GHz, 32.48 GHz, 33.84 GHz, and 37.69 GHz, with the fourth and fifth absorptance bands merged to generate a broad-band.

The normalised matched impedance Z is simulated using Equation (4), where S_{11} depicts reflection coefficient and S_{21} depicts transmission coefficient.

$$Z = \sqrt{\frac{(1+S_{11})^2 - S_{21}^2}{(1-S_{11})^2 - S_{21}^2}} \tag{4}$$

Table 3. The values of $S_{11}(f)$, reflection coefficient and A(f), absorptance of MM absorber unit cell structure at resonance frequencies

Resonance frequency (GHz)	6.33	14.08	28.92	32.48	33.84	37.69
Reflection coefficients (dB)	-20.10	-19.29	-10.67	-12.45	-12.45	-9.86
Absorptance (%)	99.74	98.58	90.00	90.04	84.74	94.94

It can be verified from Figure 7, that the real value of (Z_{eff}(f)) close to 1 and the imaginary value of (Z_{eff}(f)) is close to 0, which verifies the impedance matched conditions and further adds to the maximum absorptance at resonance frequencies.

Figure 7. Normalized matched impedance of proposed MM absorber

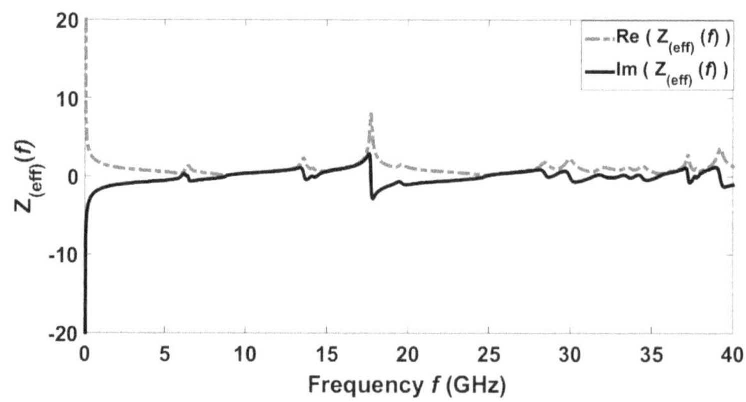

The simulated values of μ_{eff} are depicted in Figure 8. It can be clearly verified that the imaginary values at resonance frequencies are positive and contribute to the attenuation of EM waves inside the MM absorber whereas the imaginary values transit at resonance frequencies termed as Lorentzian transitions (Agrawal & Misra, 2021), thus showing metamaterial behavior.

Surface Current for Proposed MA at Frequency Where High Absorptance Is Obtained

The upper layer of the MM absorber is the frequency selective surface. The closed ring and the "+" shaped patches engineered over the top layer contribute to the

Figure 8. Complex natured permeability of proposed MM absorber

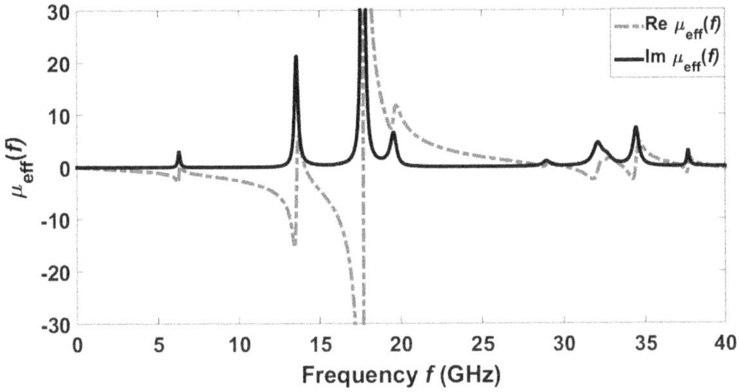

absorptance bands. The surface current distribution at various resonant frequencies is investigated in order to analyse the absorptance behaviour of the proposed MM absorber. Surface current concentration along the structure reflects how susceptible that portion is to absorption at a specific resonance frequency (Smith et al., 1999). According to (Dincer et al., 2013), the rotating surface currents in parallel direction are caused by electric response, whereas the rotating surface currents in anti-parallel direction are induced by magnetic response. The currents induced on the surface caused the incident electromagnetic radiation field to couple with the electric field and magnetic field responses, which established resonance frequencies for the increase of the localised EM field. Under normalised impedance matching conditions, the simultaneous magnetic and electric resonance at resonance frequencies leads to full absorption of incident EM radiation. As a result, the incident electromagnetic radiation (EMR) within 5 to 40 GHz frequency range was contained within the MPA structure with minimal reflection and maximal absorption. The frequency-selective surface's metallic patches serve as an inductive component on which the magnetic field is concentrated. Magnetic resonance is caused by the inductive component. According to (Smith et al., 1999), the electric field is centered within the metallic patches, whereas the spaces within the patches function as capacitive elements and cause electric resonance. As a result, the absorption peak at a given resonance frequency is influenced by the magnetic and electric resonance at each metallic patch. From Figure 9, the current distribution on the surface is concentrated at the outer square ring, which results absorption at the first resonance frequency at 6.33 GHz. The centered '+' shaped patch is responsible for the second absorptance band at 14.08 GHz, for which the surface current is concentrated. For the third absorptance band at 28.92 GHz, the distribution of current on the surface is along the outer square ring and

Figure 9. Distribution of surface current at resonance frequency (a) 6.33GHz, (b) 14.08GHz, (c) 28.92GHz, (d) 32.48GHz, (e) 33.84GHz, (f) 37.69GHz

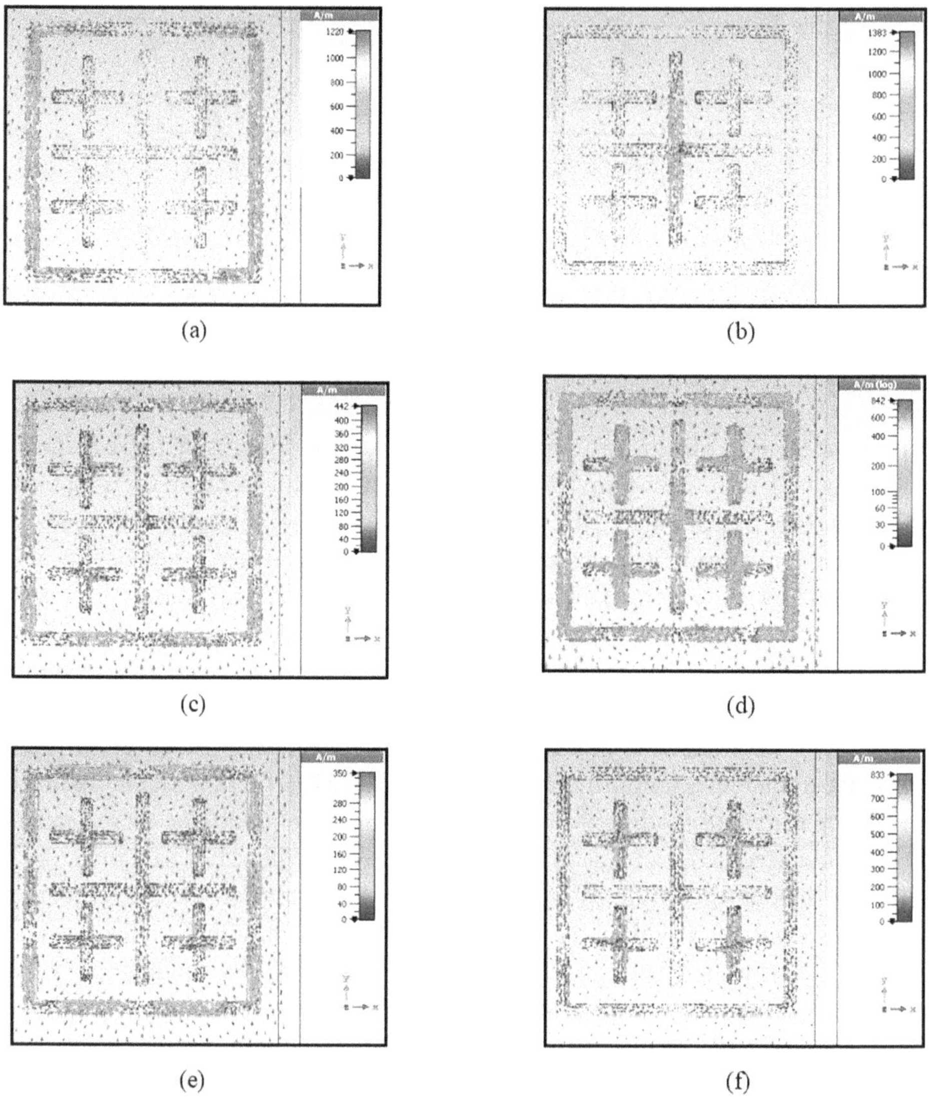

edges of the four inner "+" shaped patches. For the fourth and fifth absorptance bands, the surface current is distributed along the inner "+" shaped patches and the outer ring. A broad-band is generated by merger of both bands from 32.48 GHz to 33.84 GHz. At 37.69 GHz, the surface current is distributed along all the "+" shaped patches, that contribute to the sixth absorptance band.

CONCLUSION

The materials found naturally have positive values of εe_{ff} and $\mu ef_{f,}$ but artificially engineered metamaterials possess unnatural EM properties with negative values of $\varepsilon eff_{an}d$ $\mu eff_{tha}t$ make these materials unique. One of the potential applications of metamaterials is MM based microwave absorbers. In this chapter, a case study of a novel MM absorber structure is taken up that exhibits six resonant frequency bands with absorptance greater than 90%. The fourth and fifth absorptance bands merged to exhibit broad-band absorptance. Such a proposed design is unique and finds potential application in stealth technology.

REFERENCES

Agarwal, M., Behera, A. K., & Meshram, M. K. (2016). Wide-angle quad-band polarization insensitive metamaterial absorber. *Electronics Letters*, *52*(5), 340–342. doi:10.1049/el.2015.4134

Agrawal, A., & Misra, M. (2021). Angle of Incidence and Polarization Insensitive Ultra-Thin Ultra Wide-Band Metamaterial Absorber based on Novel Non-Linearity in Unit Cell Design. *Journal of Microwaves, Optoelectronics and Electromagnetic Applications*, *20*(2), 556–569. doi:10.1590/2179-10742021v20i31207

Agrawal, A., Misra, M., & Singh, A. (2017). Oblique Incidence and Polarization Insensitive Multiband Metamaterial Absorber with Quad Paired Concentric Continuous Ring Resonators. *Progress in Electromagnetics Research M. Pier M*, *60*, 33–46. doi:10.2528/PIERM17061302

Dincer, F., Karaaslan, M., Unal, E., & Sabah, C. (2013). Dual-band polarization independent metamaterial absorber based on omega resoanator and octa-starstrip configuration. *Electromagnetic Waves*, *141*, 219–231. doi:10.2528/PIER13061105

Fang, N., Lee, H., Sun, C., & Zhang, X. (2005). Sub-diffraction-limited optical imaging with a silver super lens. *Science*, *308*(5721), 534–537. doi:10.1126cience.1108759 PMID:15845849

Hao, J., Wang, J., Liu, X., Padilla, W. J., Zhou, L., & Qui, M. (2010). High performance optical absorber based on a plasmonic metamaterial. *Applied Physics Letters*, *96*(25), 251104. doi:10.1063/1.3442904

Kshetrimayum, R. S. (2004). A brief intro to metamaterials. *IEEE Potentials*, *23*(50), 44–46. doi:10.1109/MP.2005.1368916

Landy, N. I., Sajuyigbe, S., Mock, J. J., Smith, D. R., & Padilla, W. J. (2008). Perfect metamaterial absorber. *Physical Review Letters, 100*(20), 207402. doi:10.1103/PhysRevLett.100.207402 PMID:18518577

Lee, Y. P., Rhee, J. Y., Yoo, Y. J., Kim, K. W. (2016). Metamaterials for Perfect Absorption. *Springer Series in Materials Science,* 236.

Liu, N., Mesch, M., Weiss, T., Hentschel, M., & Giessen, H. (2010). Infrared perfect absorber and it's application as plasmonic sensor. *Nano Letters, 10*(7), 2342–2348. doi:10.1021/nl9041033 PMID:20560590

Liu, X., Starr, T., Starr, A. F., & Padilla, W. J. (2010). Infrared spatial and frequency selective metamaterial with near-unity absorbance. *Physical Review Letters, 104*(20), 207403. doi:10.1103/PhysRevLett.104.207403 PMID:20867064

Liu, Z., Lee, H., Xiong, Y., Sun, C., & Zhang, X. (2007). Far-field optical hyperlens magnifying sub-diffraction limited objects. *Science, 315*(5819), 1686. doi:10.1126cience.1137368 PMID:17379801

Munk, B. A. (2000). *Frequency Selective Surfaces: Theory and Design*. Wiley. doi:10.1002/0471723770

Pendry, J. B. (2000). Negative refraction makes a perfect lens. *Physical Review Letters, 85*(18), 3966–3969. doi:10.1103/PhysRevLett.85.3966 PMID:11041972

Pendry, J. B., Holden, A. J., Robbins, D. J., & Stewart, W. J. (1999). Magnetism from conductors and enhanced nonlinear phenomena. *IEEE Transactions on Microwave Theory and Techniques, 47*(11), 2075–2084. doi:10.1109/22.798002

Pendry, J. B., Holden, A. J., Stewart, W. J., & Youngs, I. (1996). Extremely low frequency plasmons in metallic microstructures. *Physical Review Letters, 76*(25), 4773–4776. doi:10.1103/PhysRevLett.76.4773 PMID:10061377

Rahm, M., Schurig, D., Roberts, D. A., Cummer, S. A., Smith, D. R., & Pendry, J. B. (2008). Design of electromagnetic cloaks and concentrators using form-invariant coordinate transformations of Maxwells equations. *Photonics and Nanostructures, 6*(1), 87–95. doi:10.1016/j.photonics.2007.07.013

Salisbury, W. W. (1954). *Absorbent body for electromagnetic waves.* United States Patent 2599944.

Schurig, D., Mock, J. J., Justice, B. J., Cummer, S. A., Pendry, J. B., Starr, A. F., & Smith, D. R. (2006). Metamaterial electromagnetic cloak at microwave frequencies. *Science, 314*(5801), 977–980. doi:10.1126cience.1133628 PMID:17053110

Singh, A. K., Pandurang, M., Shiban, A., & Koul, K. (2019). A Triple Band Polarization Insensitive Ultrathin Metamaterial Absorber for s- C- and X-Bands. *Progress in Electromagnetics Research M. Pier M*, *77*, 187–194. doi:10.2528/PIERM18110601

Smith, D. R., Padilla, W. J., Vier, D. C., Nemat-Nasser, S. C., & Schultz, S. (1999). Composite Medium with Simultaneously Negative Permeability and Permittivity. *Physical Review Letters*, *84*(10), 4184–4187. doi:10.1103/PhysRevLett.84.4184 PMID:10990641

Tao, H., Landy, N. I., Bingham, C. M., Zhang, X., Averitt, R. D., & Padilla, W. J. (2008). A metamaterial absorber for the terahertz regime: Design, fabrication and characterization. *Optics Express*, *16*(10), 7181–7188. doi:10.1364/OE.16.007181 PMID:18545422

Veselago, V. G. (1968). The electrodynamics of substances with simultaneously negative values of ε and μ. *Soviet Physics - Uspekhi*, *10*(4), 509–514. doi:10.1070/PU1968v010n04ABEH003699

Walser, R. M. (2001). Electromagnetic Metamaterials. *Proceeding SPIE Complex Mediums II beyond Linear Isotropic Dielectrics, 4467*, 1-15. https://doi.org/10.1117/12.432921

Yan, T. W., Ma, M. J., Li, C., & Wang, L. (2020). A Low Profile Quadruple-Band Polarization Insensitive Metamaterial Absorber. *Progress In Electromagnetics Research M, 90*, 69-79. http://www.jpier.org/PIERM/pier.php?paper=20010505

Chapter 7
Advanced Engineering Design of the Metamaterial Absorbers

Jingda Wen
https://orcid.org/0000-0001-5827-4360
Beihang University, China

Yinpeng Wang
Beihang University, China

Yuhang Wang
Sichuan International Studies University, China

ABSTRACT

In this chapter, the authors introduce the concept of MMA through reviewing the development of conventional absorbers. To illustrate the important operating mechanism, design process, and functionalization of the metamaterial absorber, they show the absorption relation of the absorber, impedance matching condition, equivalent circuit model, the evaluation of thickness, and oblique incidence condition in the first part. Then, they show the basic design method of the absorber, including parameter sweep method, equivalent circuit method, and deep learning method. Next, the planar integration strategy, vertical stacking strategy, and dispersive material strategy for broadband absorption bandwidth are discussed. In addition, they illustrate the tunable design of absorber, involving lumped element design method, thermal tunable design method, electrical tunable design method, mechanical tunable method, and reconfigurable tunable method. Finally, they talk about the basic engineering application of the metamaterial absorber in practical engineering application.

DOI: 10.4018/978-1-6684-8287-2.ch007

1. INTRODUCTION

The advance of modern electronic science and technology comes with a dramatic increase in the number of electronic devices, such as communication equipment, mobile phones and base stations. However, in order to achieve higher transmission efficiency, communication equipment needs to cover a wider range of electromagnetic wave spectra, which consequently results in a higher amount of electromagnetic radiation energy (Jeong, Nguyen, & Lim, 2018). Excessive electromagnetic radiation has negative effects in everyday life, industrial production and military defence. When the energy exceeds a certain range, the biological level interferes with the body's DNA and brain waves, which may cause fetal malformation or miscarriage in pregnant women, and may have adverse effects such as headaches in normal individuals. In industrial production, strong electromagnetic radiation energy may interfere with electronic equipment and prevent it from functioning properly, thus affecting normal industrial production activities. Besides, some electronic functional devices that require precise measurements cannot be measured accurately in environments with high levels of electromagnetic radiation interference due to the presence of large interference (Shen, Zhai, & Zheng, 2014). For example, the complex electromagnetic radiation environment that prevails requires artificial "electromagnetic clear zones" by means of electromagnetic absorbing materials. The locations with "electromagnetic clear zones" are often found in remote and mountainous areas, which is inconvenient for experimental functional tests. On the defence and military front, military radars emit spatial electromagnetic waves at certain frequencies and then analyze the echoes by measuring them to construct a radar reflection target. Radar detection stealth by absorbing space electromagnetic waves emitted by radar is an effective method to keep military targets out of detection and surveillance (Z. Li, Li, Zhao, & Zhou, 2020). In summary, realizing electromagnetic energy absorption in the microwave regime is of great importance in everyday life, industrial production and defence and military areas. Therefore, we mainly discuss the MMA in the microwave regime in this chapter to the scope of practical engineering application for wireless systems.

In order to address absorption issues in engineering, traditional absorption materials are limited by the preparation conditions and parameter characteristics of the materials themselves, making it difficult to achieve perfect absorption. Therefore, achieving perfect impedance matching conditions usually requires exploring improvements in materials composition, making it difficult to achieve effective perfect absorption design (Jones & Wooding, 1964). Compared to traditional absorption materials, MMA can accurately control the equivalent dielectric constant and magnetic permeability of the material surface, achieve perfect impedance matching with free space, reduce surface reflectivity, and combine with loss media in the structure to achieve perfect absorption (Wen, Ren, Peng, & Zhao, 2021). In 2008, Landy (Landy, Sajuyigbe, Mock, Smith, & Padilla, 2008) first introduced the metamaterial method into the field of absorption to achieve perfect absorption in a microwave band using a super surface. Figure 1(a) shows the structure of this absorbing super surface. The top is the (b) classic aperture resonant ring structure, the middle is the (c) dielectric layer, and the bottom is the (d) metal wire. When the incident wave enters the super surface, the aperture resonant ring above dominates the magnetic resonance mode that affects the equivalent magnetic permeability, the middle dielectric layer serves as the loss structure, and the metal wire at the bottom dominates the electric resonance mode that affects the equivalent dielectric constant. By optimizing the structural parameters separately, impedance matching conditions that satisfy both the equivalent dielectric constant and the equivalent magnetic permeability can be obtained.

Figure 1. The basic element of the perfect MMA: (a) Perspective view, (b) metallic split ring, (c) dielectric layer, and (d) metallic bar

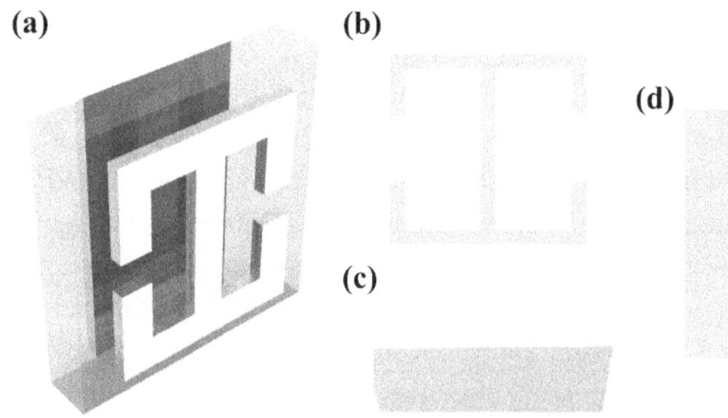

2. THE ABSORPTION THEORETICAL RELATION IN ABSORBER

2.1 The Energy Relation in the Absorber

Due to the increasing significance of problems such as electromagnetic pollution and interference, electromagnetic wave pollution has become the fourth major type of pollution after noise, water, and air pollution. To tackle electromagnetic pollution and achieve efficient electromagnetic absorption, previous researchers have conducted preliminary explorations into the absorption properties of materials. The figure 2 shows a schematic diagram of the interaction between electromagnetic waves and materials, which can be divided into transmission, reflection, and absorption.

When electromagnetic waves are incident on an absorbing material, a portion of the incident wave reflects back from the surface, another portion transmits through the material, and the rest is absorbed by the material. The frequency-dependent reflection coefficient is denoted as $R(\omega)$, the transmission coefficient as $T(\omega)$, and the absorption coefficient as $A(\omega)$. According to the conservation of energy and the normalization of the incident wave's energy, the relationship between them can be derived as follows:

$$R(\omega) + A(\omega) + T(\omega) = 1 \quad (1)$$

Introducing the two-port scattering parameter S, the absorption coefficient $A(\omega)$ can be expressed as:

$$A(\omega) = 1 - |S_{11}(\omega)|^2 - |S_{12}(\omega)|^2 \quad (2)$$

Here, $S_{11}(\omega)$ represents the reflection parameter that returns from the incident port, while $S_{12}(\omega)$ represents the transmission parameter that passes through the absorbing material. To achieve maximum absorption, $S_{11}(\omega)$ and $S_{12}(\omega)$ should be minimized. Low transmission can be achieved by adding a metal substrate, such as a perfect electric conductor (PEC) boundary. To achieve low reflection and impedance matching conditions, more incident waves will penetrate into the absorbing material, reducing the reflection on the surface (Zhang, Yan, Du, Wang, & Zhang, 2020).

Figure 2. The basic absorption relation in the absorber

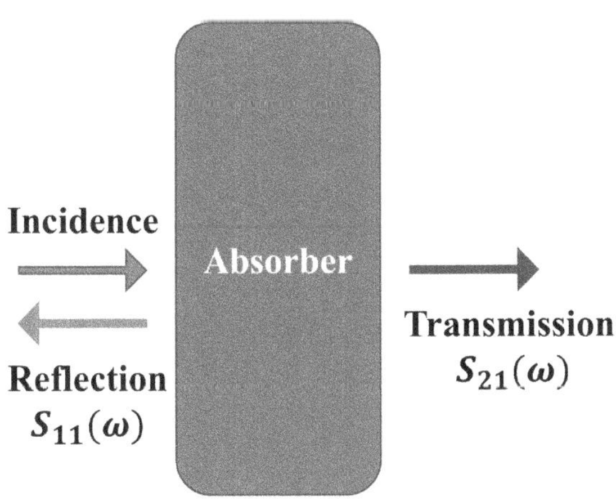

2.2 Effective EM Parameter Theory

Another step was taken to analyze the interaction between the incident wave and the absorbing material, using the currently well-established method of equivalent scattering parameters. It is assumed that a uniform plane wave is incident on the surface of an absorbing layer with a thickness of d. At this point, the electric and magnetic fields of the incident, reflected, and transmitted waves can be expressed sequentially as:

$$\begin{cases} \overrightarrow{E_i} = E_i \exp\left[i\left(kz - \omega t\right)\right]\overrightarrow{e_y} \\ \overrightarrow{E_r} = E_r \exp\left[i\left(-kz - \omega t\right)\right]\overrightarrow{e_y} \\ \overrightarrow{E_t} = E_t \exp\left[i\left(kz - \omega t\right)\right]\overrightarrow{e_y} \end{cases} \tag{3}$$

$$\begin{cases} \overrightarrow{H_i} = \dfrac{H_i \exp\left[i\left(kz - \omega t\right)\right]\overrightarrow{e_x}}{Z_0} \\ \overrightarrow{H_r} = \dfrac{H_r \exp\left[i\left(-kz - \omega t\right)\right]\overrightarrow{e_x}}{Z_0} \\ \overrightarrow{H_t} = \dfrac{H_t \exp\left[i\left(kz - \omega t\right)\right]\overrightarrow{e_x}}{Z_0} \end{cases} \tag{4}$$

Here, Z_0 represents the impedance in vacuum, k *is* the magnitude of the wave vector. Within the equivalent absorbing layer, the electric and magnetic fields are expressed as:

$$\begin{cases} \overrightarrow{E_i^{'}} = E_i^{'} \exp\left[i\left(k'z - \omega t\right)\right]\overrightarrow{e_y} \\ \overrightarrow{E_r^{'}} = E_r^{'} \exp\left[i\left(-k'z - \omega t\right)\right]\overrightarrow{e_y} \end{cases} \tag{5}$$

$$\begin{cases} \overrightarrow{H_i^{'}} = \dfrac{H_i^{'} \exp\left[i\left(k'z - \omega t\right)\right]\overrightarrow{e_x}}{Z_{eff}} \\ \overrightarrow{H_r^{'}} = \dfrac{H_r^{'} \exp\left[i\left(-k'z - \omega t\right)\right]\overrightarrow{e_x}}{Z_{eff}} \end{cases} \tag{6}$$

Here, $k'=n_{eff}k$, n_{eff} and Z_{eff} represent the equivalent reflection coefficient and equivalent impedance, respectively. Thus, by further simplification, a series of parameters can be obtained, including the relative impedance $z_r= Z_{eff}/Z_0$, the port coefficient, $S_{11}= E_r/E_i$, $S_{12}= E_t\exp(ikd)/E_i$, and the intensity ratio $\alpha=E_i'/E_i$, $\beta=E_r'/E_i$. Substituting the expressions for the dielectric constant and magnetic permeability reveals that $Z_{eff}=\sqrt{\mu_{eff}/\varepsilon_{eff}}$, $n=\sqrt{\varepsilon_r\mu_r}$, $\varepsilon_r=\varepsilon_{eff}/\varepsilon_0$, $\mu_r=\mu_{eff}/\mu_0$. Taking into consideration all the derived formulas, the boundary conditions at the vertical height of zero and d can be expressed as:

when

$$z=0,\begin{cases}1+S_{11}=\alpha+\beta\\ 1-S_{11}=\dfrac{\alpha-\beta}{z_r}\end{cases}\tag{7}$$

when

$$z=d,\begin{cases}\alpha X+\beta X^{-1}=S_{12}\\ \left(\alpha X-\beta X^{-1}\right)/z_r=S_{12}\end{cases}\tag{8}$$

Elimination of the parameters leads to the derivation of the formula for extracting equivalent parameters:

$$\frac{Z_{eff}}{Z_0}=z_r=\pm\sqrt{\frac{\left(1+S_{11}\right)^2-S_{12}^2}{\left(1-S_{11}\right)^2-S_{12}^2}}\tag{9}$$

$$\exp\left(inkd\right)=\frac{S_{12}}{1-\dfrac{S_{11}\left(z_r-1\right)}{z_r+1}}\tag{10}$$

2.3 Impedance Matching Condition

Impedance matching is a state in which the signal source or transmission line is matched with the load. From the perspective of circuits, when a DC voltage source drives a load, the load resistance value may be large or small, and the circuit is equivalent to a short circuit or an open circuit. In this state, the incident electromagnetic wave in the specified operating band outputs all the energy to the load point, and no phenomenon of reflected signal returning to the source point occurs. When the load impedance is matched with the excitation source impedance, the load resistance is equal to the internal resistance of the signal source, and the output power of the load reaches its peak. The above situation is a typical impedance matching condition. For metamaterial absorbers, the incident electromagnetic wave propagated through the air medium can be regarded as an excitation source, and the impedance of the excitation source is equivalent to the impedance of free space. The input impedance is the overall equivalent impedance of the absorber, and matching can be achieved when the two are infinitely close.

For absorptive materials, $Re(z_r)$ is greater than zero. To maximize the absorptivity in Equation (2), the wave transmission impedance matching condition should be satisfied on the surface of the absorptive material, i.e., the relative impedance $z_r=1$ or effective refractive index $n = \sqrt{\varepsilon_r \mu_r} = \varepsilon_r = \mu_r$. By introducing the normalized transmission coefficient $t' = e^{ikd} tt' = e^{ikd} t$ from reference (Smith, Schultz, Markos, & Soukoulis, 2002), the reflection coefficient $r(\omega)$ is related to the wave impedance z and the refractive index of the absorber n.

$$r = -\frac{i}{2}\left(z - \frac{1}{z}\right) sin\left(nkd\right) t' \tag{11}$$

When the reflection coefficient R is zero, that is, $R(\omega)= r^2= |S_{11}|^2 = 0$ and the impedance matching condition is satisfied with $Z(\omega)=1$, the imaginary part of the complex refractive index n_2 tends towards infinity, leading to the following formula:

$$T\left(\omega\right) = t^2 = \left|S_{21}\right|^2 = \lim_{n_2 \to \infty} \left(e^{-i(n_1-1)kd} e^{n_2 kd}\right)^{-2} = 0 \tag{12}$$

Upon substituting the reflection and transmission coefficients into Equation (2), the absorption rate $A(\omega)=1$, indicating the impedance matching condition for perfect absorption.

The figure 3 depicts the distribution of the real and imaginary parts of the equivalent impedance of single- and dual-band absorbers that satisfy the impedance matching condition. The effective impedance of the absorber should approach the absorption condition for the real and imaginary parts at the corresponding working frequency of the absorber to achieve high impedance matching condition and reduce

Figure 3. The impedance matching condition of absorber with (a) single operating point and (b) double operating point
Source: Tung et al. (2017)

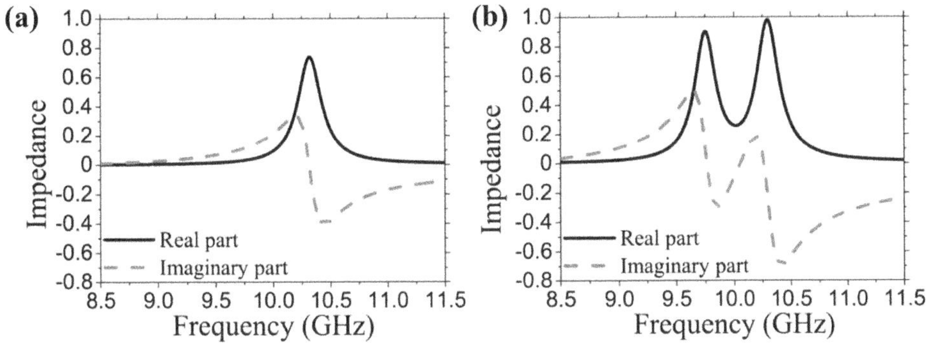

the reflection wave. This can be further achieved by satisfying the high-efficiency absorption requirement, wherein the parameter conditions satisfy exp(*inkd*)= exp(*iRe*[ε_r]*kd*) exp(-*Im*[ε_r]*kd*)→0. In order to achieve nearly perfect absorption, the following two conditions need to be satisfied:

$$\begin{cases} \dfrac{\varepsilon_r}{\mu_r} = 1 \\ Im(\varepsilon_r)kd = Im(\mu_r)kd \gg 1 \end{cases} \tag{13}$$

The first relation in the equation (13) refers to the impedance matching condition, which aims to minimize the reflection at the surface of the absorber. The second relation indicates that the loss within the absorber material should be significant enough to dissipate the energy of the incident wave. Greater equivalent material loss leads to higher absorption efficiency. Therefore, the design of the absorber material needs to satisfy both impedance matching and high loss requirements.

2.4 Equivalent Circuit Model

The equivalent circuit model provides a convenient tool for analyzing the electromagnetic characteristics of metamaterial absorbers. This model allows for the investigation of the impact of structural geometric parameters on the absorption performance. By replacing and analyzing the unit cell structure of the aperture resonator metamaterial absorber with an RLC equivalent circuit model, the top resonant unit is represented by a resistance R, an inductance L, and a capacitance C. The equivalent parameters are then calculated using the relevant circuit theory, ultimately yielding the electromagnetic response of the absorber. As shown in Figure 4, the common three-layer metamaterial absorber structure model and its equivalent RLC circuit model are presented. The input impedance of the absorber is equivalent to a parallel circuit impedance consisting of the dielectric impedance and the resonant unit impedance (Deng et al., 2018).

Figure 4. The equivalent circuit model of the absorber: (a) The real physical model of the absorber, (b) the equivalent RLC circuit model of the absorber
Source: Deng et al. (2018)

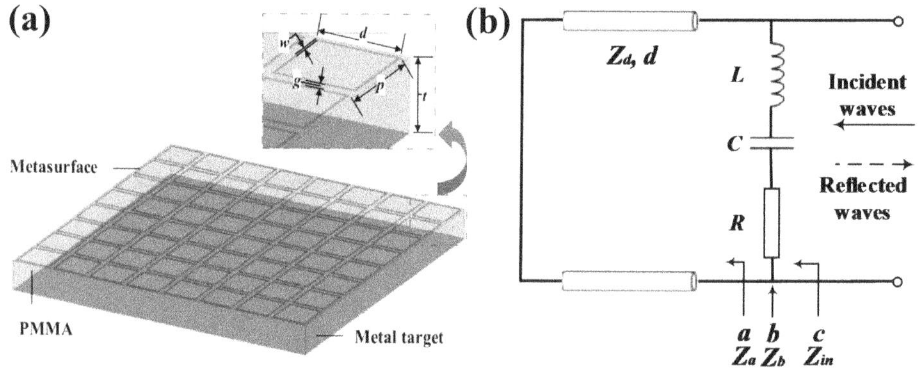

When electromagnetic waves are incident, the impedance of free space is denoted as Z_0. The RLC series circuit represents the top metallic resonant layer, with circuit equivalent impedance Z_b, where R is the equivalent resistance, L is the equivalent inductance, C is the equivalent capacitance, d represents the thickness of the dielectric layer, and equivalent impedance Z_d. The input impedance of the absorber is Z_{in}. When the thickness of the bottom metallic layer is much greater than the skin depth of electromagnetic waves inside the metal in the working band, transmission is prevented and the reflectance $R(\omega)$ can be expressed as:

$$R(\omega)=\left|\frac{Z_{in}(\omega)-Z_0}{Z_{in}(\omega)+Z_0}\right| \tag{14}$$

Where Z_0=377Ω is the impedance of free space. The input impedance $Z_{in}(\omega)$ can be expressed as:

$$Z_{in} = \frac{Z_b Z_d}{Z_b + Z_d} \tag{15}$$

The equivalent impedance Z_b of RLC can be expressed as:

$$Z_b = R - i\left(\frac{1}{\omega C} - \omega L\right) \tag{16}$$

The values of RLC are determined by the shape and geometric dimensions of the structure. The equivalent impedance Z_d of the dielectric layer is analyzed and calculated as follows (Costa, Monorchio, & Manara, 2010):

$$Z_d = iZ_m^{TE/TM} \tan(\beta d) \tag{17}$$

Here, $Z_m^{TE} = (\omega\mu_r\mu_0)/\beta$ and $Z_m^{TM} = \beta/(\omega\mu_r\mu_0)$ are the characteristic impedances of the dielectric layer for transverse electric wave (TE) and transverse magnetic wave (TM) polarizations, respectively. When the incident angle is 0, the propagation constant $\beta=k$. The impedance Z_d in TE polarization is expressed as:

$$Z_d = i\frac{\omega\mu_r\mu_0}{k}\tan(\beta d) \tag{18}$$

Substituting the free-space wave number $k_0 = \frac{\omega}{c}$ yields:

$$Z_d = i\left(\frac{\sqrt{\mu_r\mu_0}}{\sqrt{\varepsilon_r\varepsilon_0}}\right)\tan(kd) \tag{19}$$

The expression for the input impedance is obtained by solving and calculating the above equation:

$$Re(Z_{in}) = \frac{RZ_d^2}{\left(\frac{1-\omega^2 LC}{\omega C} - Z_d\right)^2 + R^2} \tag{20}$$

$$Im(Z_{in}) = \frac{\left(\frac{1-\omega^2 LC}{\omega C} - Z_d\right)\left(\frac{1}{\omega C} Z_d - \omega L Z_d\right) + R^2 Z_d}{\left(\frac{1-\omega^2 LC}{\omega C} - Z_d\right)^2 + R^2} \tag{21}$$

When the imaginary part of the impedance of the dielectric layer and the resonant layer are equal, the parallel circuit undergoes resonance and Z_b becomes a pure real number. In this case, its expression is as follows:

$$Re(Z_b) = \frac{\left(Z_m^{TE/TM}\right)^2 \tan^2(\beta d)}{R} \tag{22}$$

In order to calculate the absorption rate of the incident electromagnetic wave, it is necessary to satisfy the matching condition with the free space impedance, i.e., $R=Z_0$. Equation (22) reflects the relationship between the optimal solution of the resonant layer resistance and the thickness of the dielectric layer as well as the dielectric constant or magnetic permeability, indicating that the thicker the thickness d, the higher the optimal resistance. At this time, the reflection coefficient S_{11} can be expressed as follows:

$$|S_{11}| = \sqrt{\frac{\left[Re(Z_s) - Z_0\right]^2 + \left[Im(Z_b)\right]^2}{\left[Re(Z_s) + Z_0\right]^2 + \left[Im(Z_b)\right]^2}} \tag{23}$$

2.5 Evaluation of the Thickness

For absorbers with a metal substrate, the overall thickness must be greater than a theoretical limit absorption thickness (d_{min}), also known as the Rozanov limit thickness (Rozanov, 2000).

$$\left|\int_0^\infty \ln|R(\lambda)|\, d\lambda\right| \le 2\pi^2 d \tag{24}$$

Therefore, the evaluation of the overall thickness optimization rate of an absorber can be expressed as the ratio of the actual thickness d to the theoretical minimum thickness d_{min}, also described simply by the size of the actual thickness and the electrical dimension of the absorption wavelength at the lowest frequency point λ_L. The overall bandwidth-to-center frequency ratio (relative bandwidth) is closely related to the absorption thickness, and the comprehensive evaluation of the absorption thickness and performance can be defined as the figure of merit (FoM).

$$FoM = 2\frac{c\Delta f}{f_L f_0} \tag{25}$$

Among them, c, Δf, f_L, and f_0 respectively represent the speed of light, the width of the absorption bandwidth, the initial frequency, and the center frequency.

2.6 Oblique Incidence Condition

Under TE (transverse electric) / TM (transverse magnetic) oblique incidence condition, the excited resonance intensity of the MMA is relatively sensitive to the horizontal component of the magnetic field (X. Ma, Tian, Li, Guo, & Huang, 2020). For larger incident angles, the absorptivity of the MMA decreases apparently under TE oblique incidence due to the decreased horizontal component. However, the absorptivity of the MMA under TM oblique incidence decreases slightly due to the unchanged parallel component of the magnetic field, which can be explained by the excited magnetic resonance in the MMA (Zhou et al., 2021).

3. DESIGN METHOD OF THE ABSORBER

3.1 Parameter Sweep Method

The design of MMA is typically a semi-empirical process (Wen et al., 2021). For most MMAs, the conventional design approach involves the following steps: (1) determine the operating frequency and bandwidth; (2) select the design materials and type, which is usually a metal-dielectric-metal sandwich structure with a dielectric layer that is compatible with the metamaterial fabrication process and copper as the top and bottom metal layers; (3) choose the bottom pattern, which is typically a copper plate to achieve no transmission; (4) select the top pattern, which can be either a 3D pattern for a vertical structure or a planar pattern for a horizontal structure; (5) determine the combination mode based on the operating frequency and bandwidth, and select common metal resonance structures, such as an open circular structure, circular structure, rectangular structure, cross structure, triangular structure, or other typical structures that support resonant modes and their combinations; (6) estimate the operating resonant frequency and set the dimensional parameters, which involves a semi-empirical design by borrowing results from similar structural simulations; and (7) optimize the design structure and parameters. If the bandwidth is insufficient or the number and distribution of operating frequencies are incorrect, the design structure must be optimized first, and then the topology of the structure must be adjusted based on the approximate operating frequency spectrum requirements. This step is usually performed in electromagnetic simulation software such as CST Microwave Studio, HFSS, or COMSOL, which allows for geometric parameter scans.

To illustrate the design process, a particular MMA (Islam, Islam, Moniruzzaman, Samsuzzaman, & Arshad, 2021) has been chosen in this context for a preliminary exposition. The top-level structure diagram and the bottom structure diagram of the ultimate design are displayed in Figure 5. As demonstrated in Figure 6, the design must conform to five working frequency points, each with an absorption rate of no less than 90%, as well as being capable of uncoupled separation of the five working frequency points, and covering the range of 2-18 GHz.

Figure 5. The geometric parameters of the MMA at the (a) top view and (b) bottom view. (c) The absorption, reflection, and transmission of the MMA at the desired spectrum condition.

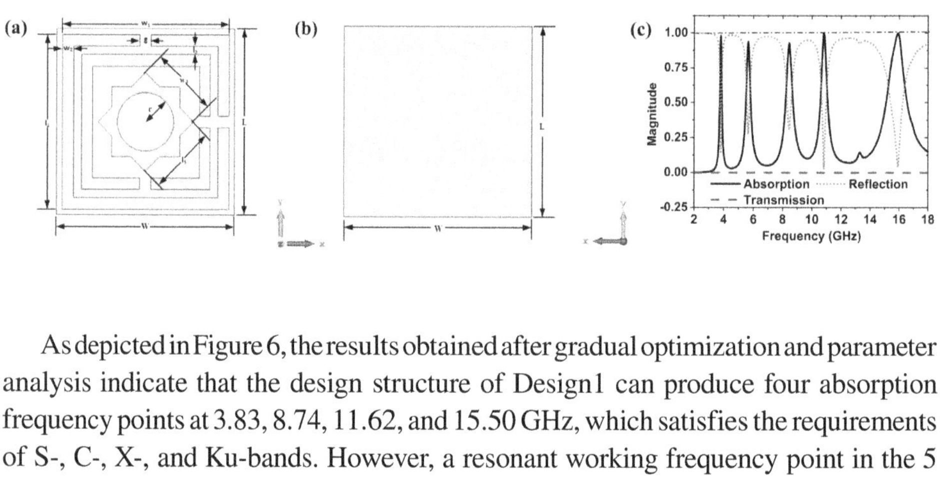

As depicted in Figure 6, the results obtained after gradual optimization and parameter analysis indicate that the design structure of Design1 can produce four absorption frequency points at 3.83, 8.74, 11.62, and 15.50 GHz, which satisfies the requirements of S-, C-, X-, and Ku-bands. However, a resonant working frequency point in the 5 GHz frequency range is lack. Therefore, a similar structure to Design 1, referred to as Design 2, was considered, but only three absorption working frequency points at 5.76, 11.08, and 15.98 GHz could be produced. To address this issue, topological structural composite optimization was performed on Design 1 to obtain Design 3, but this structure could only resonate at three points at 5.57, 10.96, and 15.23 GHz. While this structure satisfies high-efficiency absorption at the 5 GHz frequency point, it fails to meet the requirements for the number and working positions of other frequency

Figure 6. The various types of the MMA for the engineering design

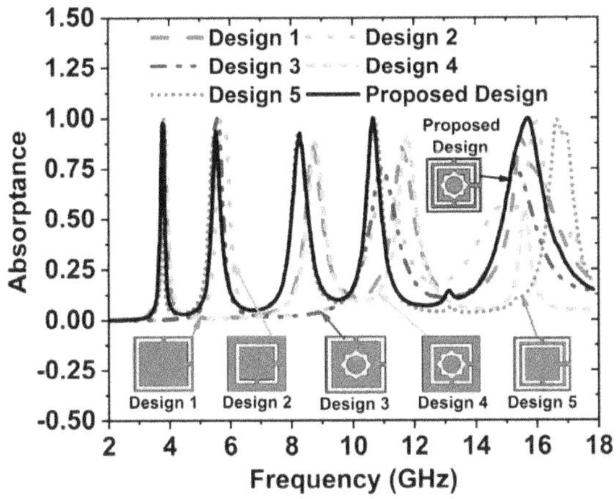

points. By combining Design 2, which is similar to Design 1, with Design 4, the resonant frequencies of Design 1 and Design 4 are almost identical, making it possible to combine other structures based on Design 4 to generate five resonant frequencies. For Design 5, which consists of two similar structures, a resonant peak distribution similar to the target spectrum can be achieved, but there are slight differences in the resonant frequencies. In this case, geometric parameter adjustments can be made to optimize the structure. However, due to the coupling relationship between the adjusted geometric parameters and the working frequencies affected, the structure was adjusted based on Design 5 to optimize the overall working frequency spectrum. The final working frequencies obtained were 3.80, 5.65, 8.45, 10.82, and 15.92 GHz.

The selection of pattern optimization plays a crucial role in determining the basic operating frequency points for the structure. Tuning of the operating frequency spectrum can be achieved by adjusting the structural geometric parameters. The absorption spectrum parameter scan results of the MMA with respect to the geometric parameter values of the split gap are depicted in Figure 7. The results demonstrate that an increase in the split gap value leads to a blue-shift of the resonant spectrum. This phenomenon can be attributed to the increase in the equivalent circuit of the resonant loop.

The remaining geometric parameter values can be analyzed using an equivalent circuit model. These parameters have an impact on the capacitance, resistance, and inductance values of the equivalent resonant circuit, which, in turn, affect the position and magnitude of the resonant frequency points. The Figure 8 provides a visual representation of the interdependence between a given parameter and multiple resonant loops. Hence, modifying a specific parameter will lead to an impact on several resonant peaks.

Figure 7. The parameter sweeps of the MMA with the parameter split gap varies from 0.2 mm to 0.7 mm

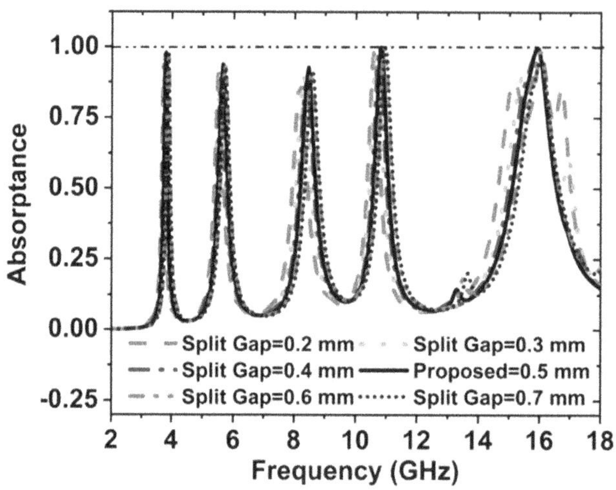

Figure 8. The equivalent circuit model of the MMA

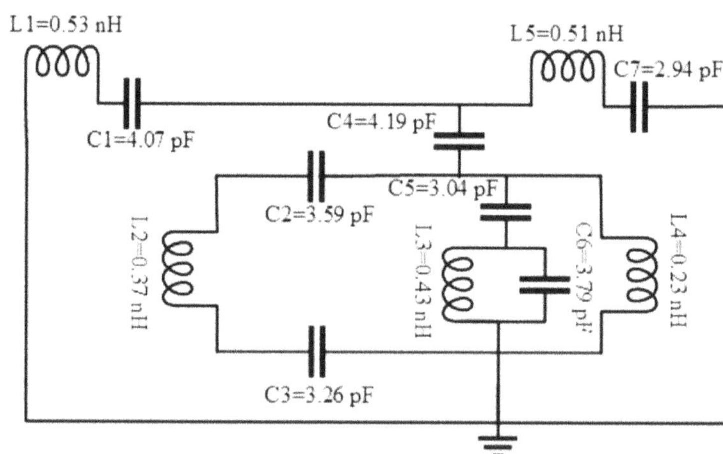

3.2 Equivalent Circuit Method

For more complex models with specific requirements, using the equivalent circuit design method can result in a more optimized and precise design than relying solely on subjective judgment. Figure 9 shows the design process for an optically transparent broadband MMA (OTMMA) presented in section 2.4 (Deng et al., 2018). Based on the derivations in section 2.4, the design is obtained.

Figure 9. The equivalent circuit method of the designed OTMMA
Source: Deng et al. (2018)

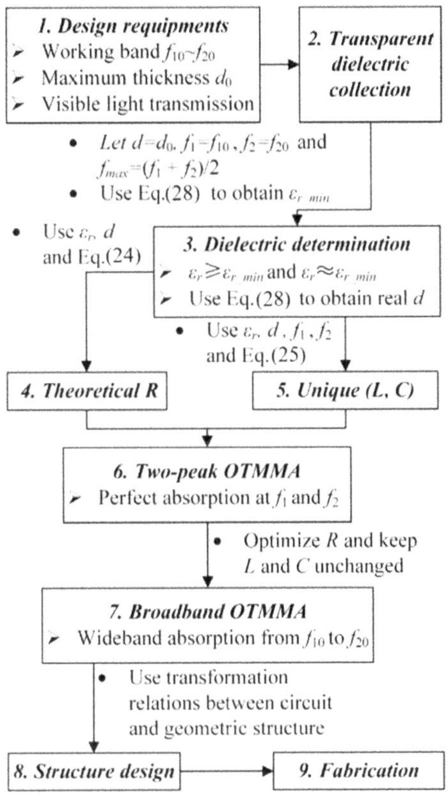

$$R = \frac{Z_0 tan^2\left(\frac{2\pi f\sqrt{\varepsilon_r}d}{C}\right)}{\varepsilon_r + tan^2\left(\frac{2\pi f\sqrt{\varepsilon_r}d}{C}\right)} \tag{26}$$

$$\left(2\pi fL - \frac{1}{2\pi fC}\right) = \frac{\sqrt{\varepsilon_r}Z_0 tan^2\left(\frac{2\pi f\sqrt{\varepsilon_r}d}{C}\right)}{\varepsilon_r + tan^2\left(\frac{2\pi f\sqrt{\varepsilon_r}d}{C}\right)} \tag{27}$$

Equation (24) establishes a theoretical relationship between the equivalent resistance and the actual absorption frequency and dielectric constant. Equation (25) provides a relationship between the equivalent inductance and capacitance. The thickness d of the absorber and the dielectric layer are related due to the dependence of L-C on the resonance frequency. By differentiating equation (24) with respect to frequency f, further equation can be gained.

$$\frac{dR}{df}=\frac{2\varepsilon_r Z_0 tan\left(\frac{2\pi f\sqrt{\varepsilon_r}d}{C}\right)sec^2\left(\frac{2\pi f\sqrt{\varepsilon_r}d}{C}\right)}{\left[\varepsilon_r+tan^2\left(\frac{2\pi f\sqrt{\varepsilon_r}d}{C}\right)\right]^2} \tag{28}$$

The extremal values of the frequency can be obtained by solving for the minimum and maximum of f when the value of Equation (26) equals 0.

$$f_{min}=\frac{(N-1)C}{2\sqrt{\varepsilon_r}d} \tag{29}$$

$$f_{min}=\frac{C}{4\sqrt{\varepsilon_r}d}+\frac{(N-1)C}{2\sqrt{\varepsilon_r}d} \tag{30}$$

The design process involves determining the working frequency range and maximum thickness, collecting transparent medium materials, and selecting the medium material based on the calculated minimum dielectric constant. The actual thickness is then calculated, and the theoretical resistance, capacitance, and inductance are determined. The optimized OTMMA with two resonant peaks is obtained by optimizing the value of the resistance. Finally, the equivalent circuit model is transformed into the actual design structure, and the corresponding sample is prepared.

3.3 Deep Learning Method

Although conventional approaches to designing metamaterials are well established, a notable drawback is that they are usually computationally time- and resource-intensive. In recent years, the development of deep learning (DL) has made remarkable achievements, and it is widely used in various fields of electromagnetism (Y. Z. Li et al., 2020; Qi et al., 2020). A fully trained network can tap the potential of GPU parallel computing, thereby greatly improving computing speed. In metasurface design problems, deep learning techniques have gradually played an increasingly important role.

As shown in Figure 10 (a), the most direct way to design metamaterials using the DL framework is to directly learn the correspondence between the target curve and the design parameters (Y. S. Chen, Zhu, Xie, Feng, & Liu, 2019). This method does not require training an independent forward model, but it still needs to use traditional electromagnetic solvers to construct data pairs between the designed structure and the target parameters. Although this method is concise and easy to implement, it is often less accurate due to the multi-solution nature of the inverse problem. To overcome the modeling difficulties of the "one-to-many" problem, one proven approach is to build a Tandem model (W. Ma, Cheng, & Liu, 2018) (or bidirectional neural network). As shown in Figure 10 (b), this model consists of a forward block and an inversion block pair. In the Tandem model, we first train a

Figure 10. The direct model and the tandem model: (a) the training and inference process of the direct model and (b) the training and inference process of the Tandem model. Here, s and g indicate the response curve and the geometry design on the training set. s represents the target response curve, and* $\hat{g}$ *is the output design of the framework.*

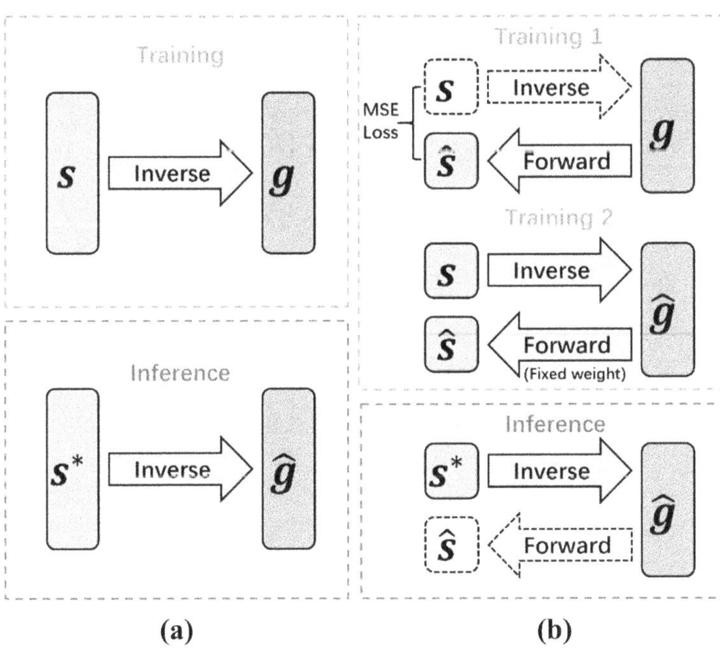

forward network to build a fast solver from the design parameters to the target curve. After training, the fixed-weighted forward network is connected to the output of the inverse network to be trained. Therefore, the final optimization function is the error between the output and target parameters of the forward network. The gradient descent using the forward error makes the parameter optimization process of the network more stable, even if the optimization target has multiple solution problems. However, the shortcomings of the Tandem model are also obvious. Because its optimization process requires two-step training, the design process is much more complex than that of a single network.

The above two types of methods are difficult to explore all possible designs and usually only select one of the proposed structures. In order to obtain as many samples as possible, the researchers introduced generative networks into the design process. Typical models of this type include automatic variational encoders (VAE) (W. Ma, Cheng, Xu, Wen, & Liu, 2019) and generative adversarial networks (GANs) (Z. C. Liu, Zhu, Rodrigues, Lee, & Cai, 2018). Figure 11 (a) shows the process of designing

Figure 11. The architecture of the VAE and the GAN: (a) the training and inference process of the VAE model and (b) the training and inference process of the GAN model. Here, s and g indicate the response curve and the geometry design on the training set. s represents the target response curve, and $\hat{g}$ is the output design of the framework. The parameters μz,σz are the average and standard deviation of the Gaussian latent space while z denotes the random noise.*

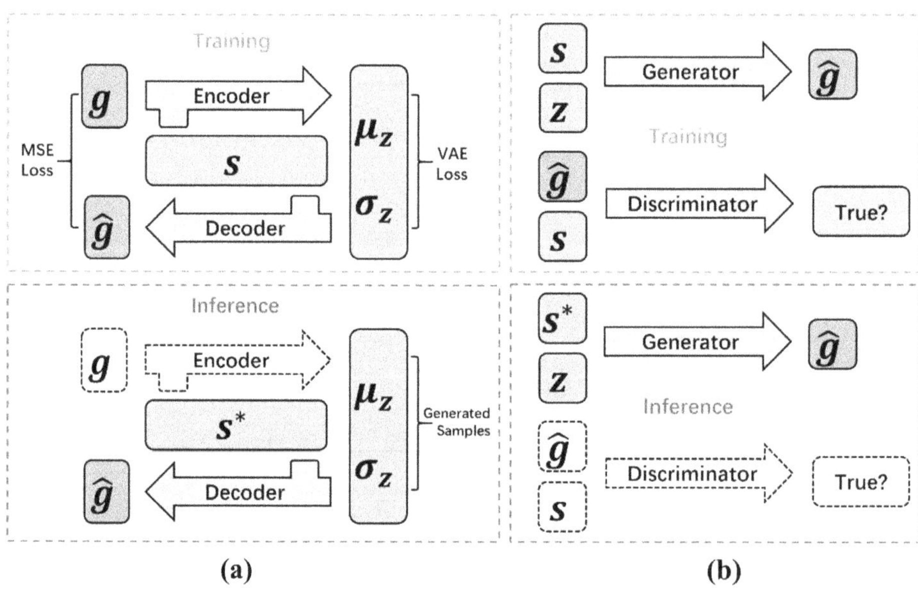

metamaterials using VAE, which includes two steps of training and inference. During training, geometric parameters and target curves are encoded into feature space, while feature space and target parameters are decoded into geometric parameters. The inference process only involves the decoding module, which utilizes the target curve and feature space to give the designed metasurface. The VAE model also has certain limitations, one of which is that it often gets a blurred design image, which is due to a certain amount of noise added to its feature space. Figure 11(b) shows the main structure of GAN, which includes a generator and a discriminator. The generator constructs a series of design parameters of the absorbing metamaterial from the random noise signal, and the discriminator tries to distinguish whether the input geometric parameters are real values or generated by the model. The goal of the training process is to build a pair of generators so outstanding that the generated samples are indistinguishable from the true values in the dataset precisely. One of the difficulties in designing wave-absorbing metamaterials using GAN is its training process, because we have to optimize two neural networks at the same time.

With the advancement of deep learning, in recent years, DL-aided design schemes based on emerging technologies such as operator learning (Augenstein, Repan, & Rockstuhl, 2023), reinforcement learning (Huang, Liu, & Zang, 2019), and active learning (Pestourie, Mroueh, Nguyen, Das, & Johnson, 2020) have also emerged. We have reason to believe that this type of method will play an increasingly important role in the field of absorbing metamaterial design in the future.

4. BROADBAND DESIGN OF ABSORBER

4.1 Planar Integration Strategy

For metallic structures, it is difficult to achieve multiple strong resonance modes using a single metallic resonator. Thus, coupling resonant structures are commonly used for broadband absorption. In practical applications, it is preferred to use simple and easy-to-fabricate designs, such as horizontally integrated composite structures in a single plane. One approach is to composite multiple metallic resonators with different geometric parameters to form a large super unit cell, as shown in Figure 12(a) for a typical broadband MMA design structure (Hien Nguyen et al., 2014). By performing parameter scans on individual ring resonators, different resonant frequencies corresponding to different geometric sizes can be obtained, resulting in four different resonant frequencies in a super unit cell structure, as shown in Figure 12(b). Due to the close distance between resonance peaks, coupling between them occurs, forming a continuous frequency band, as shown in Figure 12(d). To achieve a flat and wide absorption spectrum, introducing 9 different metallic resonant ring

Figure 12. A typical instance for the planar integration method of the broadband MMA: (a) the geometric diagram of MMA, (b) the absorption characteristics of the single ring resonator, (c) the energy distribution of the MMA for the resonant frequencies in the (d) absorption spectrum. (e) is the absorption spectrum of the optimized MMA with nine different integrated rings of the parameters R covering from 1.0 to 1.8 mm.
Source: Hien Nguyen et al. (2014)

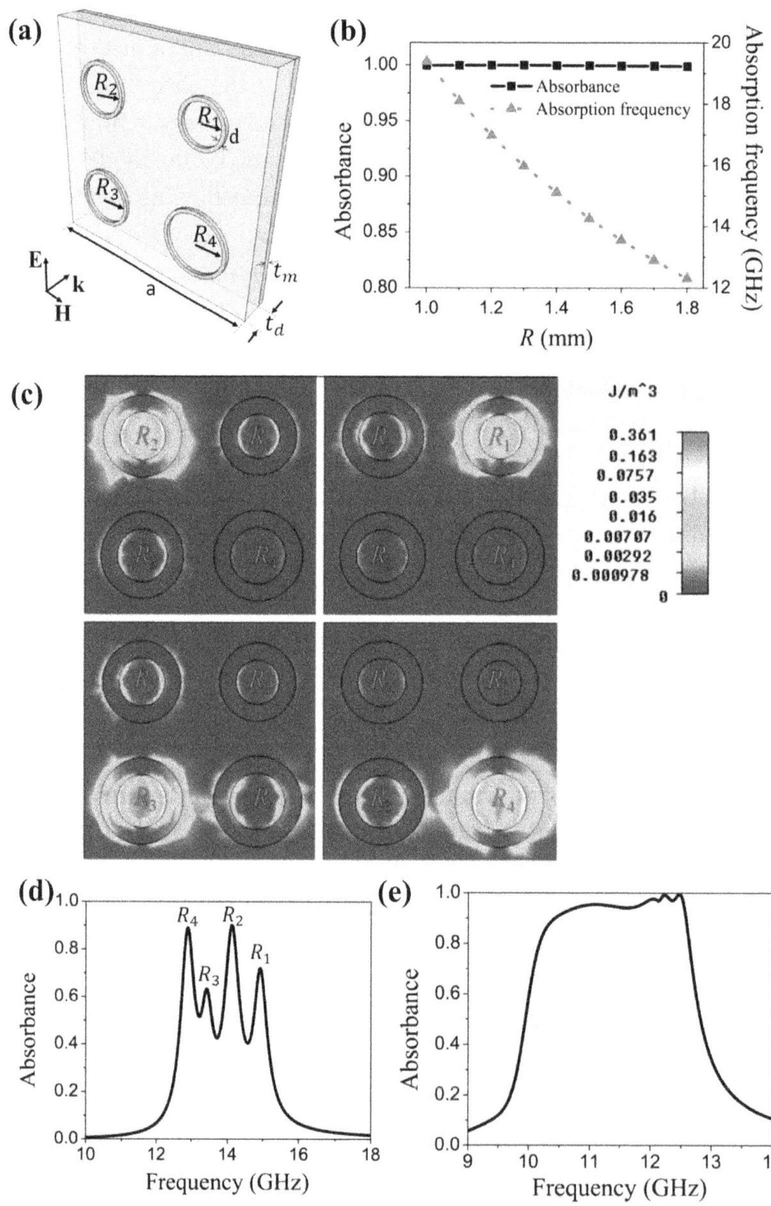

structures with sizes ranging from 1.0 mm to 1.8 mm into the super unit cell can achieve a flatter and better-coupled absorption spectrum.

4.2 Vertical Stacking Strategy

Wideband absorption can be achieved not only through planar integration but also by employing the technique of vertical stacking. As shown in Figure 13, a full dielectric broadband MMA (Karimi Habil, Ghahremani, & Zapata-Rodriguez, 2022) based on a vertical stacking strategy is presented. By vertically stacking resonant structures with self-similar geometric dimensions, broadband absorption can be achieved over a large frequency range. By analyzing the structure in a form of split, it can be observed that when there is only one large-sized dielectric resonant structure,

Figure 13. (a) The schematic diagram, (b) unit cell, and (c) cross-sectional view of a typical vertical stacking broadband MMA. (d) is the vertical stacking process of the MMA.
Source: Karimi Habil et al. (2022)

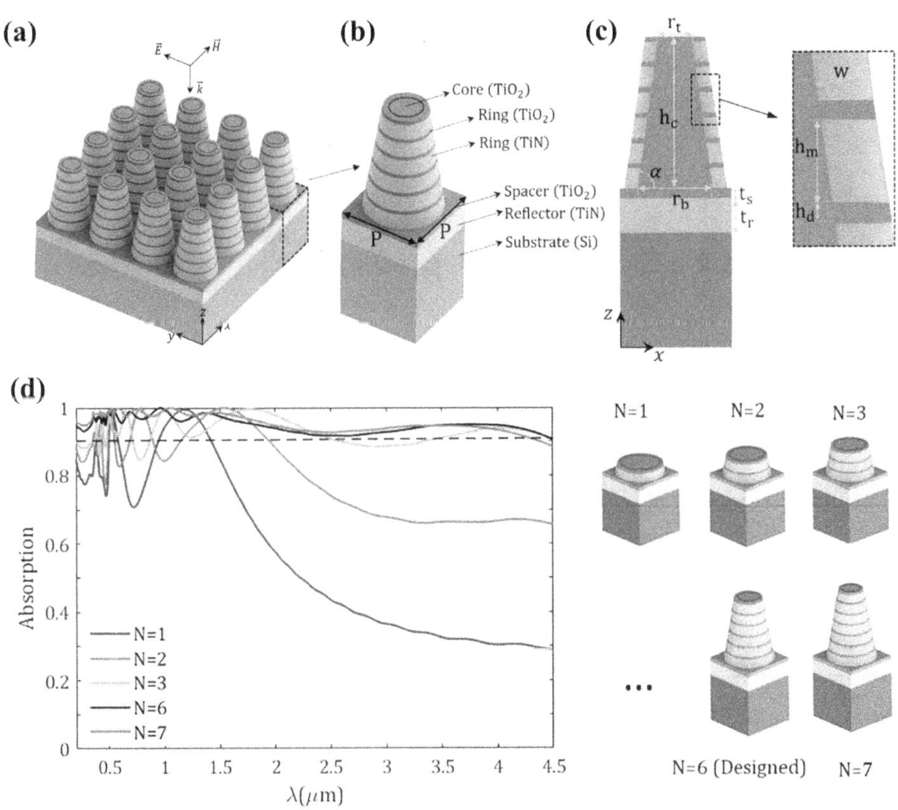

the dielectric-based resonant modes are related to the resonant circuits supported by the dielectric-based structure. Therefore, if lower resonant frequencies (larger incident wavelengths) are required, a thicker size is needed. It can be seen that as the thickness increases, the resonances at lower frequencies gradually improve, resulting in overall good broadband absorption characteristics. Analyzing the electric field distribution as shown in Figure 14 reveals that for larger incident wavelengths, larger vertical geometric dimensions are required for resonant electric fields. For smaller incident wavelengths, there are more resonant circuits generated, and the required vertical geometric dimensions for each resonant circuit's electric field are relatively smaller. For similar metal stack resonant structures, the geometric size of the resonant circuit is directly related to the planar size of the metal structure because the surface current excited by the metal is confined to the metal layer. Resonant peaks at different frequencies correspond to resonant structures of different sizes, and broadband absorption spectra can be achieved by integrating vertically.

Figure 14. The electric field distribution of the MMA at yz-plane for resonant peaks at (a) $\lambda 1=0.42\mu m$, (b) $\lambda 2=0.74\mu m$, (c) $\lambda 3=1.0\mu m$, and (d) $\lambda 4=4.0\mu m$
Source: Karimi Habil et al. (2022)

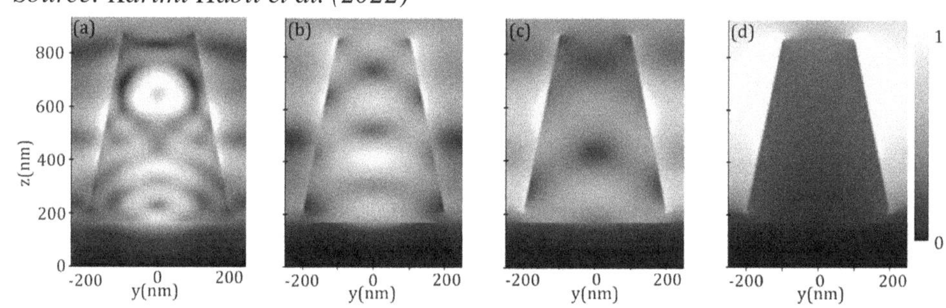

4.3 Dispersive Material Strategy

Achieving wideband absorption in non-dispersive materials or structures typically requires coupling of structures with adjacent resonant frequencies. However, for materials with a dispersive permittivity, such as water, it is possible to achieve impedance-matched absorption over a wide frequency range, where non-dispersive materials cannot meet impedance-matching conditions (Wen, Ren, Peng, & Zhao, 2022a). As shown in Figure 15, the permittivity and loss tangent of water in the microwave frequency range exhibit a high loss tangent, which enables effective dissipation of the incident wave that satisfies impedance-matching conditions. Furthermore, as the incident wavelength is proportional to the material's effective thickness, impedance-matching

Figure 15. The permittivity and loss tangent spectrum of the water
Source: Yoo et al. (2015)

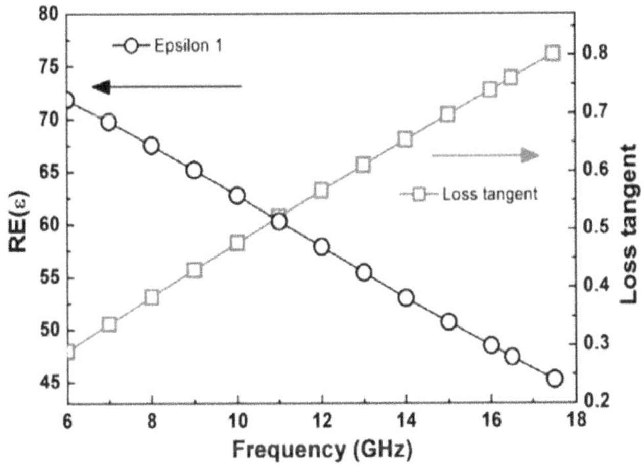

can be achieved effectively (Wen, Zhao, et al., 2022). Figure 16 depicts a typical all-dielectric water-based absorber resonator structure that can effectively excite two resonant modes, peak A and peak C, in the frequency spectrum due to the impedance-matching conditions over a wide frequency range. The two high-intensity resonant peaks A and C couple to form a wideband absorption, and the high and low order magnetic resonance correspond to different path lengths and losses of ring currents. The length of the current loop is frequency-dependent, and the higher the frequency, the shorter the ring current loop (Zhang et al., 2020). However, due to structural limitations, the coupling strength of the high- and low-order magnetic resonances excited in single scattering structures such as spherical (Andryieuski, Kuznetsova, Zhukovsky, Kivshar, & Lavrinenko, 2015), droplet-shaped (Yoo et al., 2015), and square-shaped (Gogoi & Bhattacharyya, 2017, 2018; Sun et al., 2019) structures is insufficient, leading to a narrow absorption frequency band. By further optimizing the resonator structure design, multiple adjacent strong-coupling resonant modes can be excited, achieving ultra-wideband absorption over a certain range.

5. TUNABLE DESIGN OF ABSORBER

5.1 Lumped Element Design Method

Due to the fact that the resonance physical model of MMA (metamaterial absorber) can be transformed into an equivalent circuit model, introducing components that

Figure 16. The 3D schematic diagram and the field distribution analysis of the water-based broadband MMA
Source: Yoo et al. (2015)

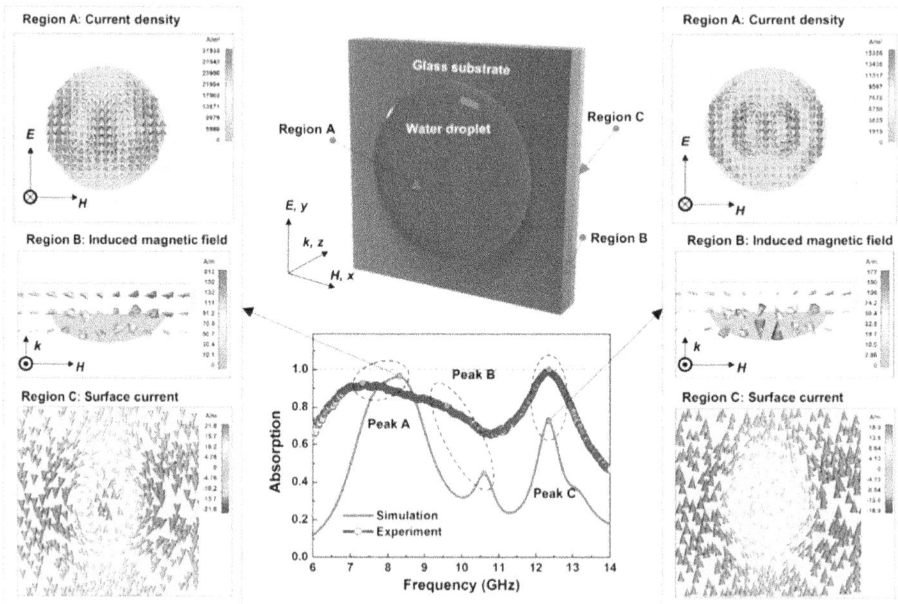

can alter the lumped element value of the equivalent circuit model parameters into the resonant structure and adjusting them through an applied voltage can effectively control the resonance. Since the resonance frequency is related to the size of the LC (inductance-capacitance) parameters of the structure, tuning the total LC parameter can effectively achieve redshift or blueshift of the resonance peak. Figure 18 shows an example of a varactor-tuned MMA, in which a varactor is introduced into the resonant metal structure on the top layer (Zhao et al., 2013). Figure 17(c) shows the equivalent lumped parameter circuit model of the varactor, where the LCR values are associated with the applied bias voltage Vbias. By applying the bias voltage, this MMA can achieve variable tuning over a frequency range of approximately 4-6 GHz, providing effective control for various application requirements. Similar adjustable effects can be achieved by introducing other components such as diodes (Lee, Jeong, & Lim, 2017).

5.2 Thermal Tunable Design Method

Achieving tunable or switchable functionality through temperature modulation requires the integration of temperature-sensitive materials in the structure, including

Figure 17. The lumped element tunable design of the MMA: (a) the schematic diagram and (b) the resonant layer pattern of the MMA, (c) the lumped equivalent circuit model of the varactor
Source: Zhao et al. (2013)

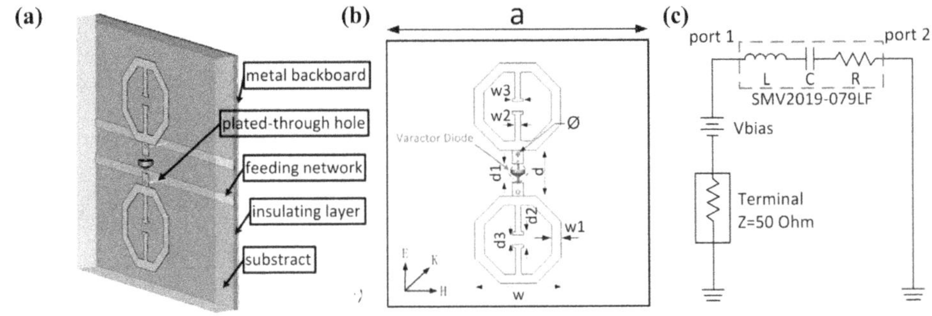

Figure 18. The lumped element tunable absorption spectrum of the MMA
Source: Zhao et al. (2013)

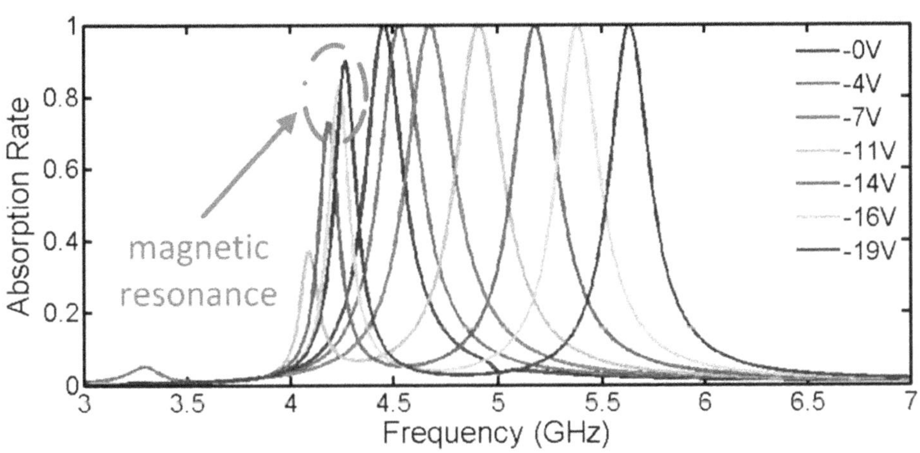

dielectric layer materials or resonant structure materials (Wen, Ren, et al., 2022a). Based on the design experience of metamaterial absorbers, adjusting the structure of the dielectric layer material usually results in a shift of the overall resonant frequency, which has been well verified in all-dielectric structures based on Mie-resonance (X. M. Liu et al., 2016). Therefore, if the goal is to achieve switchable absorption or adjustable intensity, it is necessary to design the resonant structure material in a tunable manner. In the structure illustrated in Figure 19 (H. Liu, Wang, Li, Fan, &

Figure 19. The thermal tunable design of the MMA
Source: Liu et al. (2019)

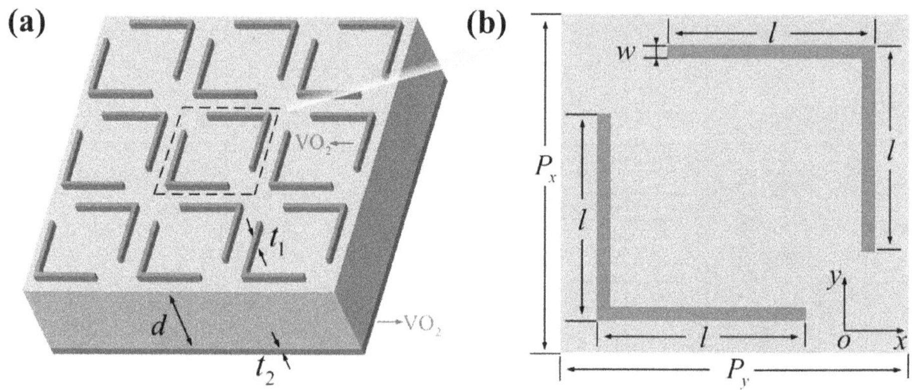

Tao, 2019), the resonant structure, typically made of metallic materials, is replaced with temperature-sensitive VO2 material.

With increasing temperature, VO2 undergoes a transition from the insulating phase to the metallic phase, resulting in two distinct operating states in the structure. At high temperatures, the resonant structure exhibits metallic behavior and generates a strong absorption peak at 1.96 THz. At lower temperatures, the resonant structure behaves as an insulator, and the overall MMA structure exhibits minimal absorption. Furthermore, since the substrate is also made of VO2, the overall structure exhibits high transmission characteristics at lower temperatures. From the absorption spectrum and surface current distribution shown in Figure 20, it can be observed that the surface current intensity gradually increases as the surface VO2 structure switches between the insulating and metallic states, corresponding to an increase in absorption rate. During the cooling process, there is a certain hysteresis effect in the material properties, resulting in different changes in absorption rate under the same temperature conditions in different processes.

5.3 Electrical Tunable Design Method

In 5.1, the introduction of lumped elements as a means of achieving tunable structures was discussed. Nevertheless, the use of lumped elements is not practical at higher frequencies, such as in the terahertz range. Consequently, other materials that are sensitive to external control voltage, such as graphene, must be taken into account. The classical empirical Drude model provides a means of describing the electrical conductivity of graphene, as presented in equation (H. Chen et al., 2022):

Figure 20. The thermal tunable absorption spectrum and surface electric field of the MMA: the absorption spectrum of MMA in the (a) heating process and (b) cooling process, (c-j) the surface electric field of the MMA at different temperatures
Source: Liu et al. (2019)

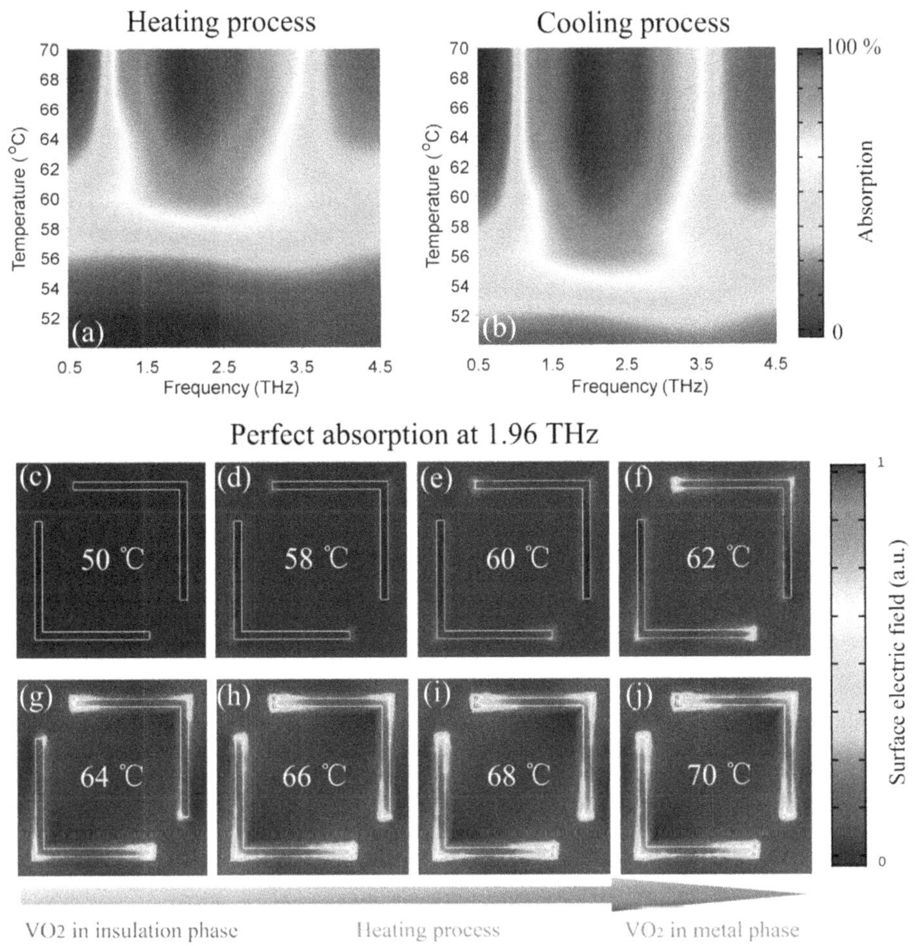

$$\sigma(\omega)=\frac{e^2E_f}{\pi\hbar}\frac{i}{\omega+i\tau^{-1}} \tag{31}$$

Here, the variables used are the charge, the reduced Planck's constant, the angular frequency, the relaxation time, and the Fermi level. It should be noted that the Fermi level of graphene is closely tied to the external gate voltage. Thus, through adjustment of the external voltage, it is possible to effectively regulate the conductivity of the structured graphene in the MMA, resulting in further tuning of the absorption frequency of the structure. As illustrated in Figure 21, the absorption resonance peak of the graphene MMA structure can be modified by varying the bias voltage to modulate the Fermi level of graphene.

Figure 21. (a) The 3D schematic diagram and (b) the absorption spectrum of the electrical tunable MMA
Source: D. Chen et al. (2017)

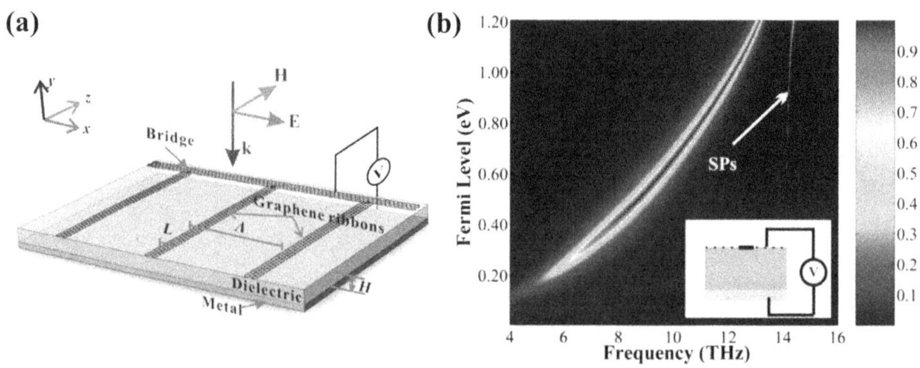

5.4 Mechanical Tunable Method

The techniques discussed in sections 5.1 to 5.3 are primarily based on the adjustment of material parameter characteristics in resonant structures. As resonant characteristics are influenced by both material and structural parameters, effective tuning of the absorption spectrum characteristics can also be achieved by modifying the geometric parameters of the resonant structure itself. An example of this is illustrated in Figure 22, which depicts a MEMS (micro-electro-mechanical system)-driven tunable MMA structure with a comb-like metal plate as an absorption structure (Chung, Jeong, Kim, Lim, & Baek, 2022). By applying a bias static voltage at both ends of the MEMS device, as shown in Figure 22, deformation of adjacent ends of the comb structure is induced by the electrostatic driving effect. At this point, the resonant frequency can be expressed as:

$$f_r = \frac{1}{2\pi\sqrt{L\left(C + C_v\left(V\right)\right)}} \tag{32}$$

Figure 22. (a) The 3D schematic diagram, (b) the absorption spectrum, and (c) the mechanism of the mechanical tunable MMA
Source: Chung et al. (2022)

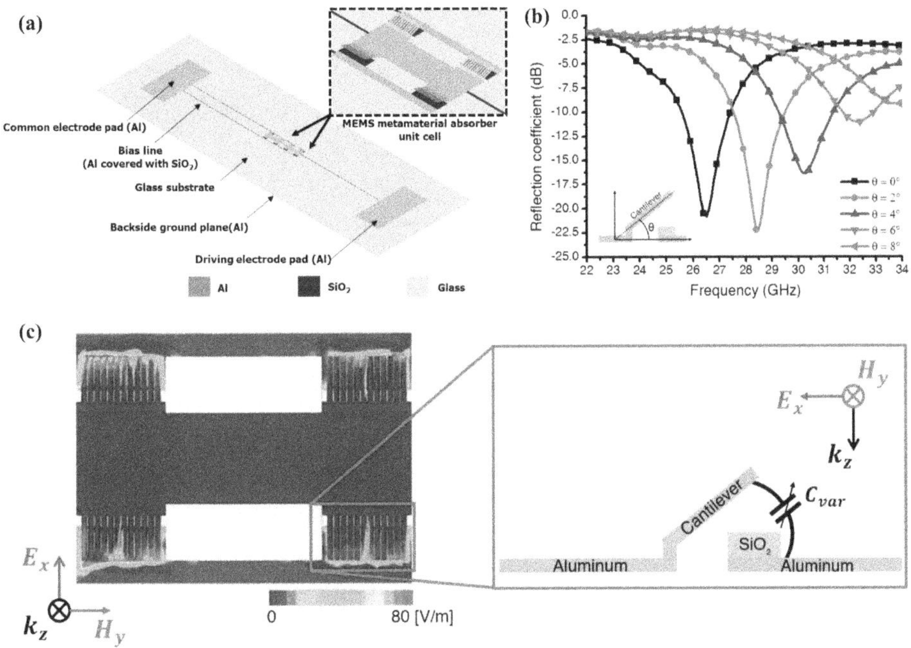

where is the resonant frequency, is the inductance, and are the internal and external capacitors of the split ring resonator, and is the MEMS-driven tunable capacitor that is associated with the external voltage in Figure 22(c).

5.5 Reconfigurable Tunable Method

The method of adjusting the absorption frequency by influencing the resonant structure can be achieved not only by modifying the existing structure, but also by introducing additional structural components in a reconfigurable manner. As illustrated in Figure 23, a reconfigurable MMA that utilizes an externally applied liquid metal EGaIn to influence the resonant frequency of an original H-shaped resonator has been proposed (Kim, Lee, & Lim, 2016). In the presence of the additional metallic structure, the resonance spectrum of the MMA undergoes significant changes within the frequency range of 8-15 GHz. When the liquid metal is filled, the resonance condition of the overall resonant structure, which

Figure 23. (a) The 3D schematic diagram and (b) the switchable tunable of MMA
Source: Kim et al. (2016)

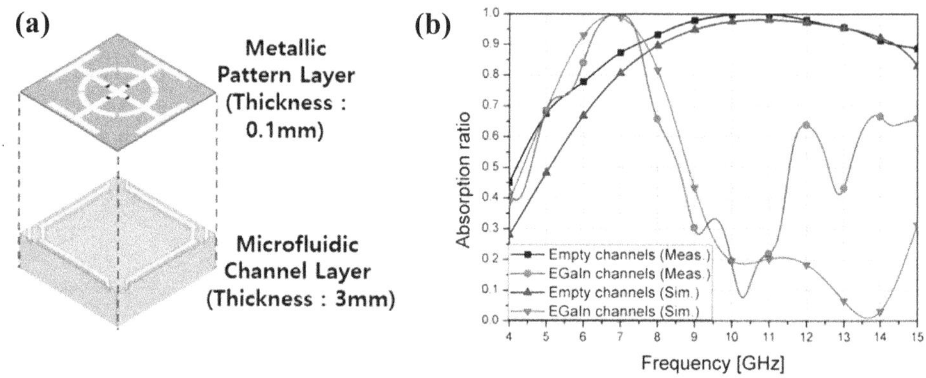

includes the metal pattern on the FR4 structure and the liquid metal in the PMMA material, is disrupted. When the liquid metal is not filled, the structure resembles a classical MDM-type MMA structure, exhibiting a wide absorption bandwidth. To the practical measurement condition, the reconfigurable status can be controlled by the external pump for the reconfigurable liquid, which operates during the measurement for free-space reflectivity.

6. FUNCTIONAL AND ENGINEERED DESIGN OF ABSORBER

6.1 Visible Transparent Design

Due to its ability to achieve optical invisibility in military applications, the design of optically transparent MMAs is of great significance. Currently, optically transparent designs in the optical frequency range are mainly achieved through the use of transparent dielectric or metal structures in MMAs. Transparent dielectric materials such as commonly used PDMS, PET, and PMMA, as well as transparent metal materials such as ITO, are commonly used to prepare the metal pattern layer or metal substrate of MMAs. As shown in Figure 24, the OTMMA can achieve high-efficiency absorption in the microwave frequency range of 6-16 GHz, while achieving approximately 70% transmission in the optical frequency range, satisfying the combined functional design requirements across two large frequency ranges (Deng et al., 2018).

Figure 24. (a) The schematic diagram, (b) optical photo, (c) absorption spectrum in microwave regime, and (d) transmission spectrum in visible light band of OTMMA
Source: Deng et al. (2018)

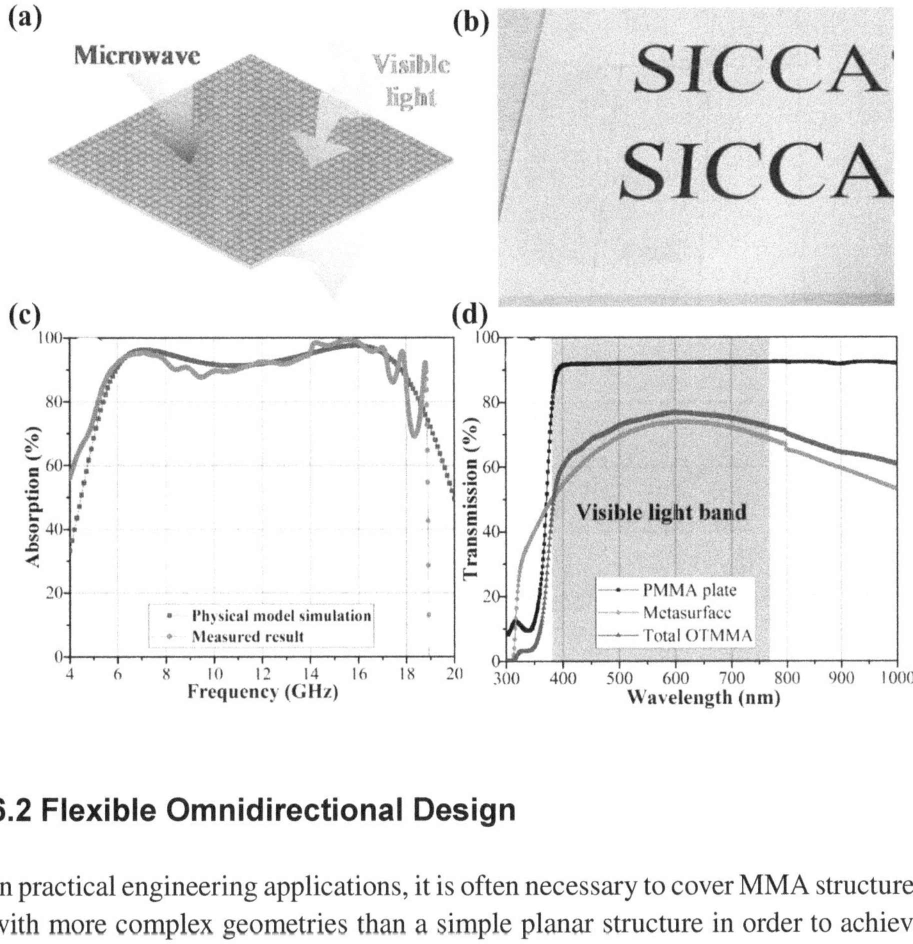

6.2 Flexible Omnidirectional Design

In practical engineering applications, it is often necessary to cover MMA structures with more complex geometries than a simple planar structure in order to achieve denser coverage. There are often gaps present at the junctions between different planar structures, which can contribute significantly to the radar cross section (RCS) during radar detection. Therefore, effective coverage of different angle-joint structures is required. Flexible omnidirectional design typically utilizes flexible substrates, such as PDMS, that can undergo significant deformation and can be used to cover materials used to prepare MMA structures. Similarly, flexible MMA can be made up of independent and separated unit cell structures, which can also produce a soft substrate effect when a thin metal substrate undergoes deformation. Figure 25 shows an example of flexible omnidirectional MMA that uses a thin metal substrate as the base (Norouzi et al., 2023). Firstly, the dielectric cylindrical structure is added in the curved surface to generate resonances. However, the bottom of the cylindrical

Figure 25. The schematic diagram of (a) curved copper plate and (b) flexible omnidirectional MMA, and (c) the bistatic scattering RCS spectrum of the MMA
Source: Norouzi et al. (2023)

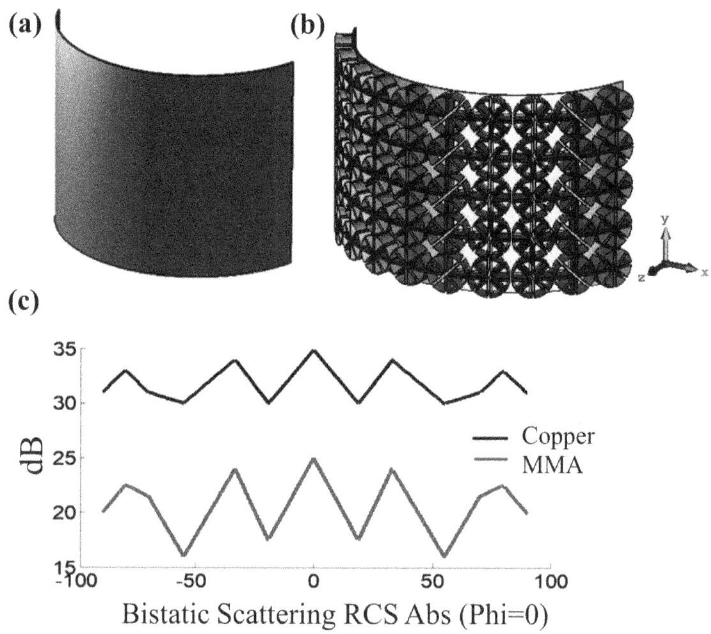

structure is not suitable for curved surface. Then, the structure is split into pieces for both keep the absorption performance and flexible design.

When a planar electromagnetic wave is incident on a curved structure, scattering occurs in all directions, as shown in the scattering RCS (radar cross section) results in Figure 25 (c), where curved copper pieces exhibit scattering in all directions. For the RCS measurement, we use a pair of antenna horns at the normal incident direction and the reflection direction with different circular angles from -100° to 100°. When MMA structures are covered on the curved copper substrate, the overall omnidirectional scattering is reduced by approximately 10 dB. Therefore, the flexible structure of MMA can have promising applications in reducing omnidirectional scattering on curved surfaces.

6.4 Radome Design

The radar dome is a device responsible for maintaining the normal operation of radar and protecting it from external influences. With the development of modern high technology, radar is widely used in aircraft, missile, navigation and other fields, and the application of radome is becoming more and more extensive. As an important part of radar system, the performance of radome directly affects the function of radar system, which can protect the radar from being damaged by high-speed fragmentation and shock waves. Conventional designs typically use Jaumann absorbers that require a quarter-wavelength ($\lambda/4$) thickness. However, in practical military applications, this design results in a large volume and weight that cannot meet the requirements. Actual military applications require simultaneously fulfilling the following criteria: (1) thin thickness, (2) good absorption spectral performance, and (3) strong mechanical performance to withstand mechanical impacts and adverse environmental erosion. The conventional MMA design directly exposes the absorbing structure, which makes the MMA structure highly susceptible to external erosion. To meet these design requirements, a schematic of an MMA design is shown in Figure 26 (Jeong et al., 2018), which can satisfy the requirements for harsh environments and electromagnetic stealth. Because the surface is covered with a dielectric meta-dome, the mechanical conditions required for wind pressure, bird strikes, and terrible weather conditions can be met. Combined with the highly efficient absorbing structure inside, the MMA can simultaneously meet the design requirements for both mechanical and electromagnetic targets.

6.3 Microwave Anechoic Chamber Design

The actual engineering applications of electromagnetic shielding require compliance with the working frequency range of the test equipment. For most microwave engineering applications, the frequency range covered by vector network analyzers is only 1-40 GHz, with matching microwave components operating in the range of 1-40 GHz. Therefore, to ensure the normal operation of microwave devices, efficient microwave absorption in the range of 1-40 GHz must be achieved at a minimum. Currently, the absorptive material that achieves full coverage from 1-40 GHz is foam doped with carbon powder, which has an excessively thick tip, making the construction of a microwave chamber require a large area. Designing an absorber to achieve a thinner thickness of microwave absorption material in the target frequency range can lead to the construction of a smaller and novel microwave chamber, which has significant engineering application implications. According to the relationship between thickness and absorption spectrum constraints discussed in Section 2.5, MMA can optimize the overall absorption thickness through geometric structure and material optimization to achieve a structure thickness thinner than a conventional pyramid absorber. The broadband MMA described in reference (Wen, Ren, Peng, & Zhao, 2022c) has achieved efficient electromagnetic shielding in the range of

Figure 26. (a) The practical engineering application of the radar dome and (b) the schematic diagram of the MMA with the engineering application in radar dome
Source: Jeong et al. (2018)

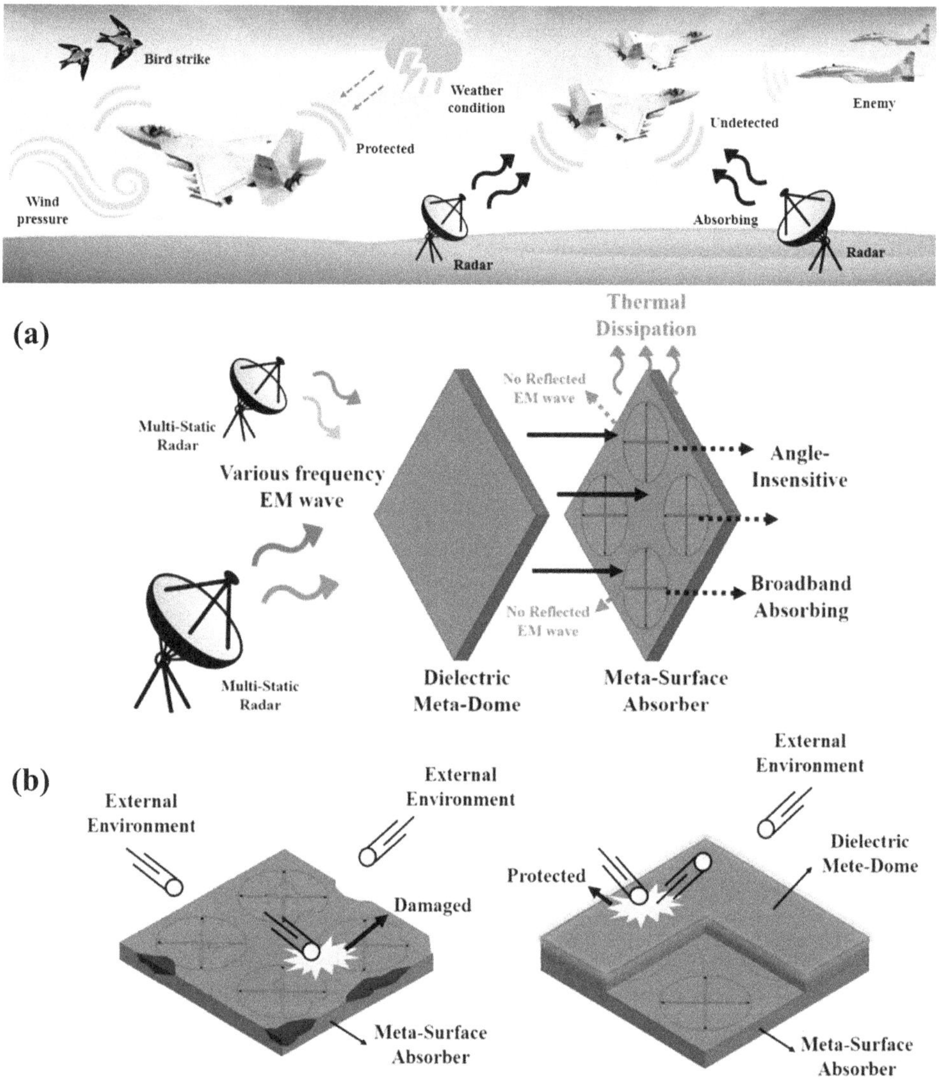

0.9-40 GHz and can be applied in practical electromagnetic microwave chamber systems. Subsequent work can optimize the actual manufacturing cost and structural mechanical properties, among other aspects, to meet the engineering requirements of the new microwave chamber.

6.4 IoT Integrated Engineering

The next generation of communication systems is expected to enable the development of critical future applications and create ubiquitous connectivity environments in smart cities. These communication systems must provide real-time, high data rate (about 20 Gbps), and high traffic capacity (about 10 Mbps/m2) for multiple users and devices. Future applications such as virtual reality and augmented reality will be widely deployed and require real-time experiences with extremely low latency. Certain types of augmented reality applications, such as remote medical services, require highly secure and reliable communication. However, interference among numerous user sources can render the entire system inefficient. Therefore, electromagnetic shielding must be developed to mitigate the impact of interference and protect the integrity of communication. These protective devices can be divided into three categories: (1) microwave filters at the back end of electronic systems, (2) frequency-selective surfaces at the front end of electronic systems, and (3) absorber structures such as MMA that shield or reduce the radar signals of electronic systems. Thus, the design and development of MMA that are compatible with existing electronic devices and effectively operate have significant engineering applications and must meet the requirements of structural size and absorption performance (Biswas et al., 2022).

Figure 27. The next generation communication systems in smart cities
Source: Biswas et al. (2022)

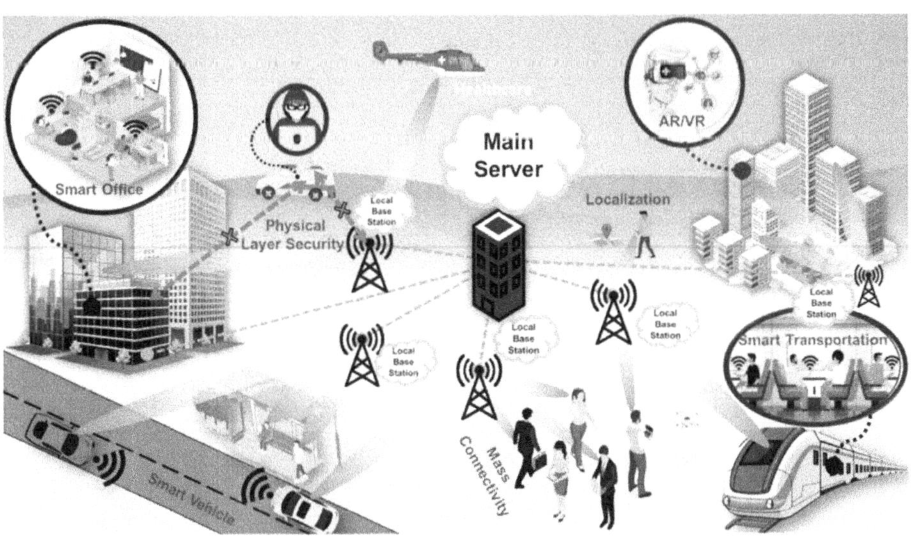

7. CONCLUSION

As shown in Table.1, the advanced design of MMAs has significant applications in both civil engineering and military domains. This paper focuses on the advanced design of MMAs, systematically discussing the working mechanism, design methods, broadband design, tunable design, and engineering functionalization of MMAs. It provides an effective reference for practical engineering design requirements. In subsequent MMA research and design, specific engineering problems, such as achieving efficient shielding of specific frequency bands and non-shielding switching, need further improvement in design. The diversity of adjustable methods, control effects, and adjustment speed also needs to be considered. The design of multiple

Table 1. The comparison between representative microwave MMA

Absorber Structure	Absorption Band (GHz)	Thickness (mm)	Tunability	Function/Specialty
(Wen, Ren, Peng, & Zhao, 2022b)Mushroom-shaped water with shell	7.2-100	11.1	Temperature, salinity, shape, and type of liquid	Omnidirectional absorption, tunable infrared radiation
(Wen, Ren, Peng, & Zhao, 2022d)Saline structure-resin shell	0.9-40	38	Reconfigurable shape	Microwave anechoic chamber
(L. Li et al., 2023)Cylindrical structure	13.3-40	15	Salinity and temperature	Optical transparency
(Islam et al., 2021) Integrated planar pattern	3.8, 5.65, 8.45, 10.82, and 15.92	2.2	No	Penta absorption peaks
(Hien Nguyen et al., 2014) Ring pattern	10-12	0.436	No	Simple planar integration method
(Ding, Cui, Ge, Jin, & He, 2012) Self-similar metallic pattern	7.8-17.4	5.25	No	Simple vertical stacking method
(Yoo et al., 2015) Water droplet	8-12	4.0	Shape	Dispersive design in one structure
(Zhao et al., 2013) Split ring and varactor diodes	4.35-5.85	5.67	Lumped element	Varactor control
(Chung et al., 2022) MEMS design	26-32	0.5	Mechanical driven	MEMS control
(Kim et al., 2016) Metallic pattern and channel	7.43-12.34	3.1	Liquid	Microfluidic channel
(Norouzi et al., 2023) 3D structure	35-400	0.11	No	Omnidirectional absorption
(Jeong et al., 2018) Composited structure	4.6-12	5.2	No	Meta-dome for radar application

adjustable methods can satisfy more combination engineering problems and facilitate their utilization in different practical scenarios. In summary, the composite of multiple functions achieved by MMAs is a trend in the development of functional materials in existing material science and engineering research fields. Following the development demands of communication systems, MMAs will have significant applications in fields such as intelligent communication systems.

ACKNOWLEDGMENT

J. D. Wen gratefully thanks the support and discussion from his girlfriend Ms. Linzhi Fang during the preparation of this manuscript.

Conflict of Interest: The authors declare that there is no conflict of interest during the preparation of this manuscript.

REFERENCES

Andryieuski, A., Kuznetsova, S. M., Zhukovsky, S. V., Kivshar, Y. S., & Lavrinenko, A. V. (2015). Water: Promising Opportunities For Tunable All-dielectric Electromagnetic Metamaterials. *Scientific Reports*, *5*(1), 9. doi:10.1038rep13535 PMID:26311410

Augenstein, Y., Repan, T., & Rockstuhl, C. (2023). Neural Operator-Based Surrogate Solver for Free-Form Electromagnetic Inverse Design. *ACS Photonics*, *10*(5), 1547–1557. Advance online publication. doi:10.1021/acsphotonics.3c00156

Biswas, A., Zekios, C. L., Ynchausti, C., Howell, L. L., Magleby, S. P., & Georgakopoulos, S. V. (2022). An ultra-wideband origami microwave absorber. *Scientific Reports*, *12*(1), 13449. doi:10.103841598 022 17648 4 PMID:35927331

Chen, D., Yang, J., Zhang, J., Huang, J., & Zhang, Z. (2017). Section 1Tunable broadband terahertz absorbers based on multiple layers of graphene ribbons. *Scientific Reports*, *7*(1), 15836. doi:10.103841598-017-16220-9 PMID:29158569

Chen, H., Chen, Z. H., Yang, H., Wen, L. H., Yi, Z., Zhou, Z. G., Dai, B., Zhang, J., Wu, X., & Wu, P. H. (2022). Multi-mode surface plasmon resonance absorber based on dart-type single-layer graphene. *RSC Advances*, *12*(13), 7821–7829. doi:10.1039/D2RA00611A PMID:35424732

Chen, Y. S., Zhu, J. F., Xie, Y. N., Feng, N. X., & Liu, Q. H. (2019). Smart inverse design of graphene-based photonic metamaterials by an adaptive artificial neural network. *Nanoscale, 11*(19), 9749–9755. doi:10.1039/C9NR01315F PMID:31066432

Chung, M., Jeong, H., Kim, Y. K., Lim, S., & Baek, C. W. (2022). Design and Fabrication of Millimeter-Wave Frequency-Tunable Metamaterial Absorber Using MEMS Cantilever Actuators. *Micromachines*, *13*(8), 1354. Advance online publication. doi:10.3390/mi13081354 PMID:36014276

Costa, F., Monorchio, A., & Manara, G. (2010). Analysis and Design of Ultra Thin Electromagnetic Absorbers Comprising Resistively Loaded High Impedance Surfaces. *IEEE Transactions on Antennas and Propagation*, *58*(5), 1551–1558. doi:10.1109/TAP.2010.2044329

Deng, R., Li, M., Muneer, B., Zhu, Q., Shi, Z., Song, L., & Zhang, T. (2018). Theoretical Analysis and Design of Ultrathin Broadband Optically Transparent Microwave Metamaterial Absorbers. *Materials (Basel)*, *11*(1), 107. Advance online publication. doi:10.3390/ma11010107 PMID:29324686

Ding, F., Cui, Y. X., Ge, X. C., Jin, Y., & He, S. L. (2012). Ultra-broadband microwave metamaterial absorber. *Applied Physics Letters*, *100*(10), 103506. Advance online publication. doi:10.1063/1.3692178

Gogoi, D. J., & Bhattacharyya, N. S. (2017). Embedded dielectric water “atom” array for broadband microwave absorber based on Mie resonance. *Journal of Applied Physics*, *122*(17), 175106. Advance online publication. doi:10.1063/1.4995519

Gogoi, D. J., & Bhattacharyya, N. S. (2018). Metasurface absorber based on water meta “molecule” for X-band microwave absorption. *Journal of Applied Physics*, *124*(7), 075106. Advance online publication. doi:10.1063/1.5041450

Hien Nguyen, T., Tung Bui, S., Tuan Nguyen, T., Tung Nguyen, T., Lee, Y., An Nguyen, M., & Lam Vu, D. (2014). Metamaterial-based perfect absorber: Polarization insensitivity and broadband. *Advances in Natural Sciences: Nanoscience and Nanotechnology*, *5*(2), 025013. Advance online publication. doi:10.1088/2043-6262/5/2/025013

Huang, Z., Liu, X., & Zang, J. F. (2019). The inverse design of structural color using machine learning. *Nanoscale*, *11*(45), 21748–21758. doi:10.1039/C9NR06127D PMID:31498348

Islam, M. R., Islam, M. T., Moniruzzaman, M., Samsuzzaman, M., & Arshad, H. (2021). Penta band single negative meta-atom absorber designed on square enclosed star-shaped modified split ring resonator for S-, C-, X- and Ku- bands microwave applications. *Scientific Reports*, *11*(1), 8784. doi:10.103841598-021-87958-6 PMID:33888759

Jeong, H., Nguyen, T. T., & Lim, S. (2018). Meta-Dome for Broadband Radar Absorbing Structure. *Scientific Reports*, *8*(1), 17893. doi:10.103841598-018-36273-8 PMID:30559365

Jones, A. R., & Wooding, E. R. (1964). Multi-layer Microwave Absorber. *IEEE Transactions on Antennas and Propagation, AP12*(4), 508. doi:10.1109/TAP.1964.1138262

Karimi Habil, M., Ghahremani, M., & Zapata-Rodriguez, C. J. (2022). Multi-octave metasurface-based refractory superabsorber enhanced by a tapered unit-cell structure. *Scientific Reports*, *12*(1), 17066. doi:10.103841598-022-21740-0 PMID:36224251

Kim, H. K., Lee, D., & Lim, S. (2016). Wideband-Switchable Metamaterial Absorber Using Injected Liquid Metal. *Scientific Reports*, *6*(1), 31823. doi:10.1038rep31823 PMID:27546310

Landy, N. I., Sajuyigbe, S., Mock, J. J., Smith, D. R., & Padilla, W. J. (2008). Perfect metamaterial absorber. *Physical Review Letters*, *100*(20), 4. doi:10.1103/PhysRevLett.100.207402 PMID:18518577

Lee, D., Jeong, H., & Lim, S. (2017). Electronically Switchable Broadband Metamaterial Absorber. *Scientific Reports*, *7*(1), 4891. doi:10.103841598-017-05330-z PMID:28687811

Li, L., Wen, J., Wang, Y., Jin, Y., Wen, Y., Sun, J., Zhao, Q., Li, B., & Zhou, J. (2023). A Transparent broadband all-dielectric water-based metamaterial absorber based on laser cutting. *Physica Scripta*, *98*(5), 055516. Advance online publication. doi:10.1088/1402-4896/accc15

Li, Y. Z., Wang, Y. P., Qi, S. T., Ren, Q., Kang, L., Campbell, S. D., Werner, P. L., & Werner, D. H. (2020). Predicting Scattering From Complex Nano-Structures via Deep Learning. *IEEE Access : Practical Innovations, Open Solutions*, *8*, 139983–139993. doi:10.1109/ACCESS.2020.3012132

Li, Z., Li, B., Zhao, Q., & Zhou, J. (2020). A metasurface absorber based on the slow-wave effect. *AIP Advances*, *10*(4), 045311. Advance online publication. doi:10.1063/1.5143408

Liu, H., Wang, Z. H., Li, L., Fan, Y. X., & Tao, Z. Y. (2019). Vanadium dioxide-assisted broadband tunable terahertz metamaterial absorber. *Scientific Reports*, *9*(1), 5751. doi:10.103841598-019-42293-9 PMID:30962484

Liu, X. M., Lan, C. W., Bi, K., Li, B., Zhao, Q., & Zhou, J. (2016). Dual band metamaterial perfect absorber based on Mie resonances. *Applied Physics Letters*, *109*(6), 062902. Advance online publication. doi:10.1063/1.4960802

Liu, Z. C., Zhu, D. Y., Rodrigues, S. P., Lee, K. T., & Cai, W. S. (2018). Generative Model for the Inverse Design of Metasurfaces. *Nano Letters*, *18*(10), 6570–6576. doi:10.1021/acs.nanolett.8b03171 PMID:30207735

Ma, W., Cheng, F., & Liu, Y. M. (2018). Deep-Learning-Enabled On-Demand Design of Chiral Metamaterials. *ACS Nano*, *12*(6), 6326–6334. doi:10.1021/acsnano.8b03569 PMID:29856595

Ma, W., Cheng, F., Xu, Y. H., Wen, Q. L., & Liu, Y. M. (2019). Probabilistic Representation and Inverse Design of Metamaterials Based on a Deep Generative Model with Semi-Supervised Learning Strategy. *Advanced Materials*, *31*(35), 1901111. Advance online publication. doi:10.1002/adma.201901111 PMID:31259443

Ma, X., Tian, F., Li, X., Guo, L., & Huang, X. (2020). Broadband with enhanced oblique incidence metamaterial absorber. *Materials Research Express*, *7*(9), 095803. Advance online publication. doi:10.1088/2053-1591/abba9e

Norouzi, M., Jarchi, S., Ghaffari-Miab, M., Esfandiari, M., Lalbakhsh, A., Koziel, S., Reisenfeld, S., & Moloudian, G. (2023). 3D metamaterial ultra-wideband absorber for curved surface. *Scientific Reports*, *13*(1), 1043. doi:10.103841598-023-28021-4 PMID:36658245

Pestourie, R., Mroueh, Y., Nguyen, T. V., Das, P., & Johnson, S. G. (2020). Active learning of deep surrogates for PDEs: Application to metasurface design. *npj Computational Materials*, *6*(1), 164. Advance online publication. doi:10.103841524-020-00431-2

Qi, S. T., Wang, Y. P., Li, Y. Z., Wu, X., Ren, Q., & Ren, Y. (2020). Two-Dimensional Electromagnetic Solver Based on Deep Learning Technique. *IEEE Journal on Multiscale and Multiphysics Computational Techniques*, *5*, 83–88. doi:10.1109/JMMCT.2020.2995811

Rozanov, K. N. (2000). Ultimate Thickness to Bandwidth Ratio of Radar Absorbers. *IEEE Transactions on Antennas and Propagation*, *48*(8), 1230–1234. doi:10.1109/8.884491

Shen, B., Zhai, W., & Zheng, W. (2014). Ultrathin Flexible Graphene Film: An Excellent Thermal Conducting Material with Efficient EMI Shielding. *Advanced Functional Materials*, *24*(28), 4542–4548. doi:10.1002/adfm.201400079

Smith, D. R., Schultz, S., Markos, P., & Soukoulis, C. M. (2002). Determination of effective permittivity and permeability of metamaterials from reflection and transmission coefficients. *Physical Review B: Condensed Matter*, *65*(19), 5. doi:10.1103/PhysRevB.65.195104

Sun, X., Fu, Q., Fan, Y., Wu, H., Qiu, K., Yang, R., Cai, W., Zhang, N., & Zhang, F. (2019). Thermally controllable Mie resonances in a water-based metamaterial. *Scientific Reports*, *9*(1), 5417. doi:10.103841598-019-41681-5 PMID:30931963

Tung, B. S., Khuyen, B. X., Kim, Y. J., Lam, V. D., Kim, K. W., & Lee, Y. (2017). Polarization-independent, wide-incident-angle and dual-band perfect absorption, based on near-field coupling in a symmetric metamaterial. *Scientific Reports*, *7*(1), 11507. doi:10.103841598-017-11824-7 PMID:28912553

Wen, J., Ren, Q., Peng, R., & Zhao, Q. (2021). An all-dielectric metasurface absorber based on surface wave conversion effect. *Applied Physics Letters*, *119*(6), 062902. Advance online publication. doi:10.1063/5.0057529

Wen, J., Ren, Q., Peng, R., & Zhao, Q. (2022a). Multi-functional tunable ultra-broadband water-based metasurface absorber with high reconfigurability. *Journal of Physics. D, Applied Physics*, *55*(28), 285103. Advance online publication. doi:10.1088/1361-6463/ac683e

Wen, J., Ren, Q., Peng, R., & Zhao, Q. (2022b). Multi-functional tunable ultra-broadband water-based metasurface absorber with high reconfigurability. *Journal of Physics. D, Applied Physics*, *55*(28), 285103. doi:10.1088/1361-6463/ac683e

Wen, J., Ren, Q., Peng, R., & Zhao, Q. (2022c). Ultrabroadband Saline-Based Metamaterial Absorber With Near Theoretical Absorption Bandwidth Limit. *IEEE Antennas and Wireless Propagation Letters*, *21*(7), 1388–1392. doi:10.1109/LAWP.2022.3169467

Wen, J., Ren, Q., Peng, R., & Zhao, Q. (2022d). Ultrabroadband saline-based metamaterial absorber with near theoretical absorption bandwidth limit. *IEEE Antennas and Wireless Propagation Letters*, *21*(7), 1388–1392. doi:10.1109/LAWP.2022.3169467

Wen, J., Zhao, Q., Peng, R., Yao, H., Qing, Y., Yin, J., & Ren, Q. (2022). Progress in water-based metamaterial absorbers: A review. *Optical Materials Express*, *12*(4), 1461. Advance online publication. doi:10.1364/OME.455723

Yoo, Y. J., Ju, S., Park, S. Y., Ju Kim, Y., Bong, J., Lim, T., Kim, K. W., Rhee, J. Y., & Lee, Y. (2015). Metamaterial Absorber for Electromagnetic Waves in Periodic Water Droplets. *Scientific Reports*, *5*(1), 14018. doi:10.1038rep14018 PMID:26354891

Zhang, X., Yan, F., Du, X., Wang, W., & Zhang, M. (2020). Broadband water-based metamaterial absorber with wide angle and thermal stability. *AIP Advances*, *10*(5), 055211. Advance online publication. doi:10.1063/5.0006166

Zhao, J., Cheng, Q., Chen, J., Qi, M. Q., Jiang, W. X., & Cui, T. J. (2013). A tunable metamaterial absorber using varactor diodes. *New Journal of Physics*, *15*(4), 043049. Advance online publication. doi:10.1088/1367-2630/15/4/043049

Zhou, Y., Qin, Z., Liang, Z., Meng, D., Xu, H., Smith, D. R., & Liu, Y. (2021). Ultra-broadband metamaterial absorbers from long to very long infrared regime. *Light, Science & Applications*, *10*(1), 138. doi:10.103841377-021-00577-8 PMID:34226489

ADDITIONAL READING

Ren, Q., Wang, Y., Li, Y., & Qi, S. (2022). *Sophisticated Electromagnetic Forward Scattering Solver Via Deep Learning*. Springer. doi:10.1007/978-981-16-6261-4

Wang, Y., & Ren, Q. (2023). *Deep Learning-Based Forward Modeling and Inversion Techniques for Computational Physics Problems*. CRC Press. doi:10.1201/9781003397830

KEY TERMS AND DEFINITIONS

$\hbar$ and E_f: Reduced Planck's constant and Fermi level.
$\overrightarrow{E_i}, \overrightarrow{E_r}, \overrightarrow{E_t}$ **:** The electric field of the incident, reflected, and transmitted waves.
$\overrightarrow{H_i}, \overrightarrow{H_r}, \overrightarrow{H_t}$ **:** The magnetic field of the incident, reflected, and transmitted waves.
***d*:** The thickness of the absorbing layer.
DL: Deep learning.
EM: Electromagnetic.
f_r: The resonant frequency.
GANs: Generative adversarial networks.
***k* and *k'*:** The magnitude of the wave vector and equivalent wave vector.
LC: Inductance-capacitance.
MEMS: Micro-electro-mechanical system.

MMA: Metamaterial absorber.
n_{eff} and Z_{eff}: Equivalent reflection coefficient and impedance.
OTMMA: Optically transparent broadband MMA.
PEC: Perfect electric conductor.
$R(\omega)$, $A(\omega)$, and $T(\omega)$: Reflectivity, absorptivity, and transmissivity.
RCS: Radar cross section.
$S_{11}(\omega)$: Reflection parameter that returns from the incident port.
$S_{12}(\omega)$: Transmission parameter.
TE and TM: Transverse electric and transverse magnetic.
VAE: Automatic variational encoders.
Z_0, z_r, Z_{in}, and Z_{eff}: The impedance in vacuum, relative impedance, input impedance, and effective impedance.
εr and μr: Relative permittivity and permeability.

Chapter 8
Performance Analysis of SSRR in High-Speed Terahertz Antenna for Biomedical Applications

Elizabeth Caroline Britto
IFET College of Engineering, India

Sathish Kumar Danasegaran
Centre for Advances in Signals Processing and Sensor Networks, Department of ECE, Sri Eshwar College of Engineering, India

K. Sagadevan
https://orcid.org/0000-0001-8543-4059
IFET College of Engineering, India

ABSTRACT

Breast cancer is the most prevalent type of cancer causing 1 of 8 cancer diagnoses globally. In the year 2020, there were approximately 2.3 million newly diagnosed cases and 0.685 million deaths from this disease. In order to avoid these, it is mandatory to detect breast cancer in the early stage. In breast cancer detection, terahertz (THz) technology is a promising technology that will not affect the skin due to heavy radiation and offers enhanced breast detection capability. The proposed THz antenna is based on a square-shaped split ring resonator (SRR) which helps to detect the tumor in the breast of the woman. The proposed square SRR antenna resonates at 2 THz with a return loss of -66.55 dB as the maximum and a VSWR of 1.0005. The square SRR structure produced the maximum gain of 8.721 dBi with 8.734 dB directivity. The designed antenna is suitable for biomedical fields such as detecting tumors, body scanner, and various THz applications.

DOI: 10.4018/978-1-6684-8287-2.ch008

INTRODUCTION

Breast cancer is a type of cancer that occurs in the breast cells. It can affect both genders, but it impacts women far more frequently than men. Breast cancer symptoms and warning indicators may include a tumour in the breast that senses distinct from the nearby tissue, variation in breast shape and size, changes in breast skin, peeling, redness, etc. To help women's society, it is necessary to reduce the effect of breast cancer in earlier stages with the aid of digital technology. The research towards finding a more accurate detection of breast tumors in earlier stages is very important and helps save the patient's health.

In order to enhance the diagnosis of tumors, the SRR-based THz antenna improves the accuracy as well as the efficiency in detecting the tumor in the breast. THz radiation (Sivasangari et al., 2023, Ponnupillai et al., 2023) used for detecting breast cancer ranges from 0.3 to 3 THz. THz radiations have a scattering loss and higher depth of penetration which does not affect the human body's soft tissues (Byrne et al., 2011, Yu et al., 2011, Porter et al., 2013, Zhang et al., 2011, Ouerghi et al., 2017). The actual frequency of the THz in terms of Hz is 10^{12} Hz or 1000 GHz. In the THz band, the radiation wavelength equivalently ranges from 1 mm to 0.01 mm. Though the THz radiation begins at the wavelength of 1mm and designates as shorter wavelengths. The bands in which the electromagnetic waves are radiating are regarded as far-infrared. As the THz radiation is absorbed by the atmosphere gases, the air is attenuated to 0 within a few meters; therefore, it is unsuitable for radio and wireless communication. The radiation can penetrate deeply into thin layers of the materials but it is obstructed by thicker objects. An inspection of layers is used as the replacement for the x-rays for producing higher resolution images in the interior of heavy solid objects (Truong et al., 2015). This radiation takes place amongst the ultraviolet waves and microwaves denoted as the THz gap.

Generally, the signals are processed through space in the form of electromagnetic waves comprising electric and magnetic fields fluctuating perpendicular to one another and to the axis of the wave's travel. The swiftness of electric charges, such as electrons, produces these EM waves. The classification of EM spectrum existence is based on frequency or wavelength. The EM waves travel at the speed of light and the wave properties are described by maxwell's equation. Antennas convert electromagnetic waves into the form of voltage and current and vice versa for the transmission and reception process.

The microstrip patch antenna is more advantageous because it is flexible and miniature in its size. In a microstrip square patch antenna, the lesser gain and efficiency can be overcome by introducing the photonic crystal in it (Britto et al., 2022, Kumar et al., 2020, Kumar et al., 2021). The substrate which has high permittivity produces less efficiency as it couples the power with surface waves. Therefore, to eliminate the problem of lower efficiency, photonic band gaps are introduced in the substrates, increasing the efficiency and reducing the return loss (Tamilarasan et al., 2021). It happens as the periodicity exists in the dielectric medium.

The antenna should have a good impendence of 50Ω. The results of the simulation are gain, directivity, return loss, and VSWR are very necessary for testing the ability of the antenna. The VSWR should be less than 2 and the return loss should be greater than 20 dB moreover, the gain and directivity determine the efficiency (Welch et al., 2016, Mahmud et al., 2017, Drukteinis et al., 2013, Balaji et al., 2017, Porter et al., 2016, Byrne et al., 2017, Bassi et al., 2020). The efficiency of the antenna is extremely used to determine the tumor cell location in the breast of the patient which can be verified by using microwave studio CST (Computer Simulation Technology) software. The results of the simulation are obtained from the THz antenna by placing the three different shapes of the split-ring resonator in the patch and the succeeding results govern the effectiveness of the THz antenna (Asili et al., 2015).

BACKGROUND WORK RELATED TO THE ANTENNA DESIGN FOR BREAST CANCER DETECTION

Several research works discussed various imaging technologies using antenna focused on sensing cancer cells in the breast (Hidayat & Apriono, 2016), Song et al., 2015, Epstein et al., 2014, Bourqui et al., 2018, Rahmatinia & Fahimi, 2018, Mahmud et al., 2017). Moussa et al., 2017 proposed a miniaturized antenna with a 24 elements array for sensing and localizing breast cancer. The dimension of the proposed structure is 10mmx10mmx1.6mm and it resonated at 7.9 GHz frequency. The antenna exhibited the characteristics of -39dB return loss with an efficiency of 70%. The gain is enhanced in the fabricated structure to 2.4 dBi and the SAR value is 1.40 W/Kg. Due to less value of SAR, it is very easy to detect cancer cells without any harm to the human body.

Slimi et al., 2022 designed and validated the Vivaldi antenna and it is operated at 5 GHz. The antenna produced an excellent performance such as 25 dB return loss with a gain of 5.8 dBi. The experimental results suggested that the array structure can able accurately locate tumors in a fluid breast phantom by analyzing the least amplitude of the return loss curve. The authors compared the antenna performances with and without the effect of metamaterials and concluded that the metamaterial-based antenna is very efficient than the conventional one.

Slimi et al., 2022 proposed the planar antenna with a frequency of 4.5 GHz for tumor finding. The antenna is designed with dimensions of 26mmx30x0.64mm with the help of Roger material. This antenna can sense two different sizes of tumor such as 8mm and 15mm with an efficiency of 86%. The antenna characteristics of both simulated and experimental results are very close to each other. Bhargava et al., 2022 suggested a safe and effective method to sense cancer cells using an antenna. This antenna can able to detect tumors for various distances and at different ages of people. The special thing about this antenna is to sense 77 spots of different breast models and works between 3.60 GHz and 9.20 GHz. The authors investigated the antenna characteristics with and without the presence of tumors.

Jumaat et al., 2021 proposed the revised wideband patch antenna for the application of breast cancer. The geometrical value of the proposed antenna is 43mmx43mmx1.57mm and the material Roger 5880 substrate is used to design and fabricate the antenna. The antenna attained a fractional bandwidth of 127% with a gain of 5.27 dBi. Kaur et al., 2020 presented an affordable monostatic radar-based microwave imaging strategy for the identification of breast tumors. In the proposed strategy, the antenna is located parallel to the breast and cycled around each 10° degree.

Table 1 illustrated the performances of various existing antenna works for cancer detection. The various existing antennas are designed with different materials like FR4, PEC and different series of Roger with unique antenna dimensions. But the objective of all the papers is to detect cancer used by the patch antenna and the efficiency of the antenna will vary by the different parameters like substrate material, dimension, patch shape, etc. Moussa et al., 2022 proposed the patch antenna in mm size using Roger material and attained the result of -39 dB return loss, 2.4 dBi gain with 70% efficiency. Slimi et al., 2022 proposed an antenna for cancer detection with an average gain of 5.8 dBi.

Table 1. Antenna performances of existing works

Parameters/Papers	Frequency (GHz)	Antenna Size (mm)	Substrate Material	Return Loss (dB)	Gain (dBi)	Efficiency (%)
Moussa et al., 2022	7.9	10x10x 1.6	Roger 5870	39	2.4	70
Slimi et al., 2022	4.6	67x46x 0.64	Roger 3010	-25	5.8	-
Slimi et al., 2022	4.5	26x30x 0.64	Roger 3010	-55.36	2	86
Bhargava et al., 2022	3.60 – 9.20	30x25x 0.794	FR4	-48	4.50	87.5
Jumaat et al., 2021	1.8 – 8.3	43x43x 1.57	Roger 5880	< -15	5.27	-
Kaur et al., 2020	4.3 – 12.6	20x15x 1.57	FR4	-40	6.1	
Geetharamani & Aathmanesan, 2019	100	0.008x0.008	PEC	-35	-	-
Rao et al., 2020	3.20 - 14	16×71.5×0.254	Roger 5880	>-22	-	-

The same authors Slimi et al., 2022 designed a highly efficient metamaterial antenna for finding cancer cells. Bhargava et al., 2022 suggested the radar-based microstrip antenna for breast cancer with a return loss of -48 dB. Jumaat et al., 2021and Rao et al., 2020 proposed the patch antenna with the help of Roger 5880 substrate and it was suitable to detect breast cancer.

This work focuses metamaterial-based THz antenna to sense cancer cells with maximum efficiency. Based on the above survey, there are various research works are discussed based on antenna for cancer detection. But all the proposed antennas are working in the frequency of Gigahertz (GHz) which affects the human body and even the antenna size is in the millimeter (mm) range. The novelty of the proposed square SRR antenna will overcome these drawbacks such as the design of the antenna being in μm range and the antenna resonating at THz frequency. Due to non-ionizing radiation characteristics, THz will not affect the human body. Also, the antenna performances of the existing works are very less and it is enhanced in the proposed SRR antenna. Hence, the enhanced performance of the proposed square SRR antenna helps to increase the efficiency of cancer detection.

CONCEPT OF SRR

An SRR is a form of electromagnetic metamaterial structure that has unique qualities relating to wave transmission and resonant. It is commonly built with a flat array of conducting metallic rings or loops separated by tiny spaces. The potential of an SRR shows significant magnetic responses on defined frequencies. This behavior is caused by the peculiar ring structure, resulting in an artificial magnetic response which is that is rarely observed in natural materials. The interaction of electromagnetic waves with the SRR results in the circulating currents among the rings generating a powerful magnetic field. The resonance frequency of an SRR can be controlled by altering its dimension, shape, position, and width between the rings. It is feasible to make a structure that resonates at certain frequencies of interest by carefully planning the SRR shape. There are various shapes of SSRs are square, hexagonal, pentagonal, elliptical, rhombus, rectangular and cylindrical shapes of SSRs which are shown below in Figure 1.

PROPOSED ANTENNA DESIGN

The proposed antenna is a square patch antenna with SRR. The microstrip patch antenna is well suited for body wearables in medical applications because of its ease of construction and low fabrication cost-effectiveness in Figure 2. The

Figure 1. Shapes of SRR

Figure 2. Square SRR structure

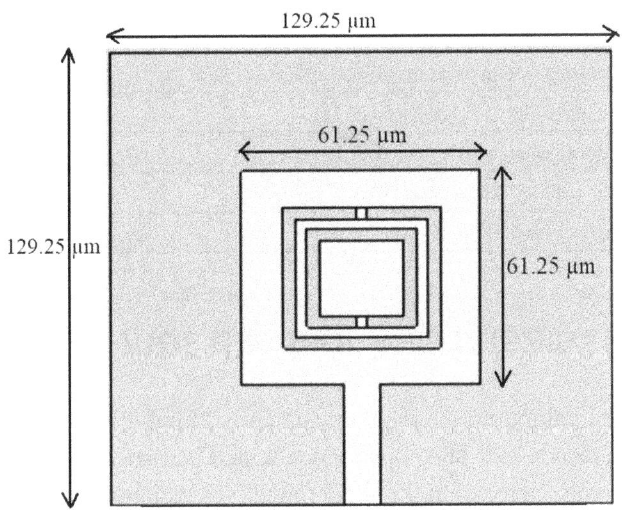

measurement of the patch THz antenna is sinking the size of the patch and the dielectric standard. The material of Copper annealed by the extensiveness of 0.02μm stays widely cast-off creating directing as well as radiating layers of the antenna patch and the ground. The patch antenna uses Roger 5880 as its substrate material which has a thickness of 6 μm and a dielectric constant of copper is 4.3. The length (Lp) of the THz antenna square patch is 61.25 μm.

The width (Wp) of the THz antenna square patch is 61.25 μm. The length (Ls) of the dielectric medium of the THz antenna is 129.25 μm. The width (Ws) of the dielectric medium of the THz antenna is 129.25 μm. The comparative absorptivity of the dielectric medium is 1. The inset-fed method remains cast-off intended for connecting the lowest of the square patch to the feed line. The breadth (Wf) of the strip line is 9.375 μm.

The design parameters of a square geometric microstrip square patch THz antenna are depicted in Table 2. The size of the square patch and the SRRs added to the patch of the THz antenna play a major part in getting the simulation outcomes. The length (Lf) of the feed line in square patch THz antenna is 64.75 μm respectively. The microstrip patch antenna consists of SRRs, which are added to the patch of the antenna with various dimensions.

Table 2. Design Parameters for the proposed antenna

Sl. No.	Parameters	Values (μm)
1	Dielectric medium width(W)	129.25 μm
2	Dielectric medium length (Ls)	129.25
3	Dielectric medium thickness (Ts)	6
4	Feed line width of THz antenna (Wf)	9.375
5	Feed line length of THz antenna (Lf)	64.75
6	Patch width of THz antenna (Wp)	61.25
7	Length of Patch (Lp)	61.25

EQUIVALENT CIRCUIT OF PROPOSED SQUARE SRR

By using the equivalent circuit the capacitance and inductance of the fragmented loop resonator is analyzed. The capacitance which is associated with the slit of the fragmented loop resonators and the inductance is associated with the metallic rings of the split ring resonators. Figure 3 shows the equivalent circuit of fragmented loop resonators. The capacitance(C) and the inductance (L) equations are shown below, Where R is the no. of turns, V and U are the width and spacing of the two rings of SRRs, C is the Capacitance of the fragmented loop resonator and L is inductance of the fragmented loop resonator.

$$C_o = \varepsilon_0 \frac{K\sqrt{1-K^2}}{K(K)} \tag{1}$$

Figure 3. Equivalent circuit of square SRR

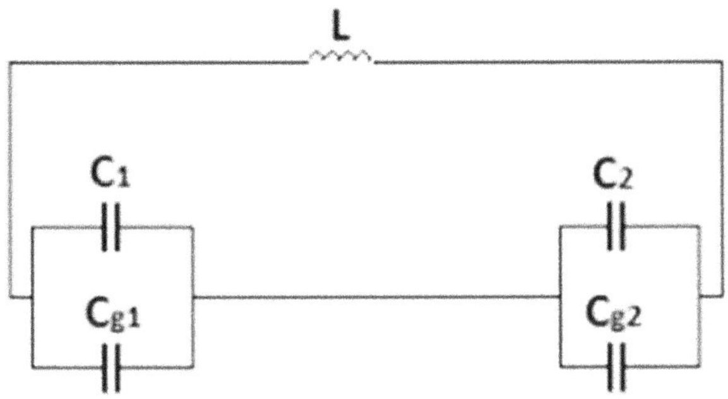

$$C_{srr} = \frac{(R-1)}{2\big(2l-(2R-1)(V+U)\big)C_o} \tag{2}$$

$$L_{srr} = 4\mu\ (1\text{-}(R\text{-}1)\ (U{=}V))\ (\ln \left(\frac{0.98}{Q}\right) + 1.84Q) \tag{3}$$

$$K = \frac{(U/2)}{(V+U/2)} \tag{4}$$

$$Q = \frac{(R-1)(V+U)}{1-(R-1)(V+U)} \tag{5}$$

$$\mathrm{f} = \frac{1}{2\pi}\sqrt{\frac{1}{L_{srr}C_{srr}}} \tag{6}$$

These are the formulas used for calculating the capacitance and the inductance of the square SRR which is placed at the patch of the square patch micros trip antenna.

RESULTS AND DISCUSSION

The proposed square SRR structure is designed and simulated using CST studio suite. The investigation of the antenna characteristics is done by two different cases such as varying the SRR width 'w' and thickness 't'.

Case 1: Impact of Varying Width 'w' of the Ring

The proposed metamaterial antenna consists of double layer square SRR such as outer and inner ring. The ring width 'w' is optimized for different values and analyzed the antenna performances. The optimization of various 'w' values is illustrated in Figure 4 (a) to (c). Table 3 shows the simulated characteristics of various 'w' values of square SRR structures.

Figure 5 illustrates the impact of return loss for various 'w' values of SRR structures. It is observed that the SRR with the width of 2 μm, the antenna resonates at 2.024 THz with the performances of -51.63 dB return loss. When the width is increased gradually upto 10 μm with the differences of 2 μm, the return loss is varied as -50.92

Figure 4. Square SRR models (a) w = 2μm, (b) w = 6μm, (c) w = 10μm

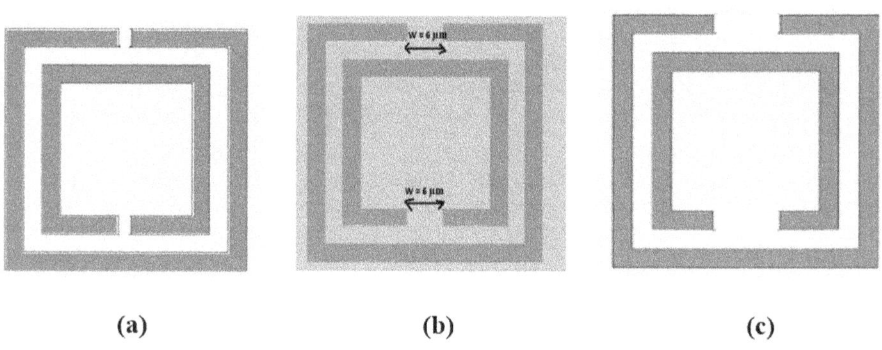

Table 3. Impact of various width of square SRR

Width in μm	Frequency (THz)	Return Loss (dB)	VSWR	Gain (dBi)	Directivity (dB)
2	2.0245	-51.63	1.0023	8.715	8.727
4	2.026	-50.92	1.0057	8.716	8.737
6	2.0275	-55.34	1.0034	8.718	8.739
8	2.0180	-66.55	1.0009	8.724	8.739
10	1.808	-29.61	1.068	4.352	4.547

Figure 5. Characteristics of return Loss for 'w' values

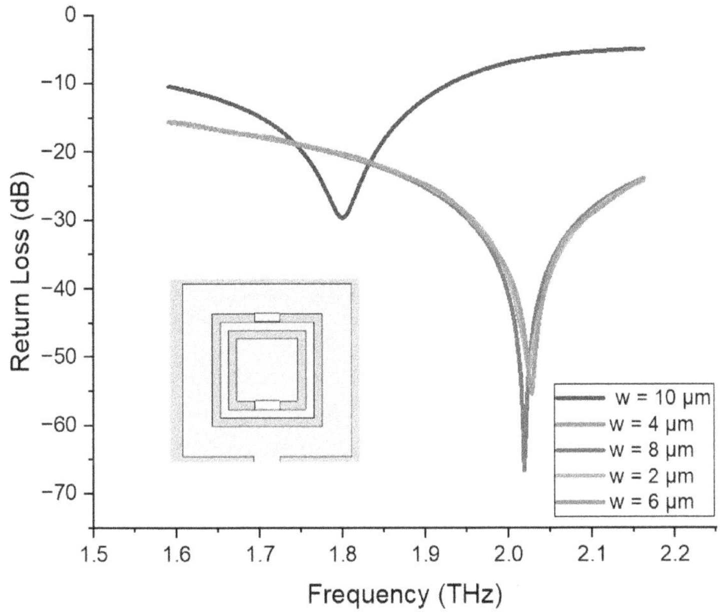

dB, -55.34 dB, -66.55 dB and -29.61 dB. Figure 6 (a) & (b) shows the variation of gain and directivity for the different SRR width of the metamaterial structure. It is concluded that the SRR structure with the ring width of 8 µm is seems optimal one to produce the effective antenna performances.

Figure 6. (a) Gain vs. width, (b) directivity vs. width

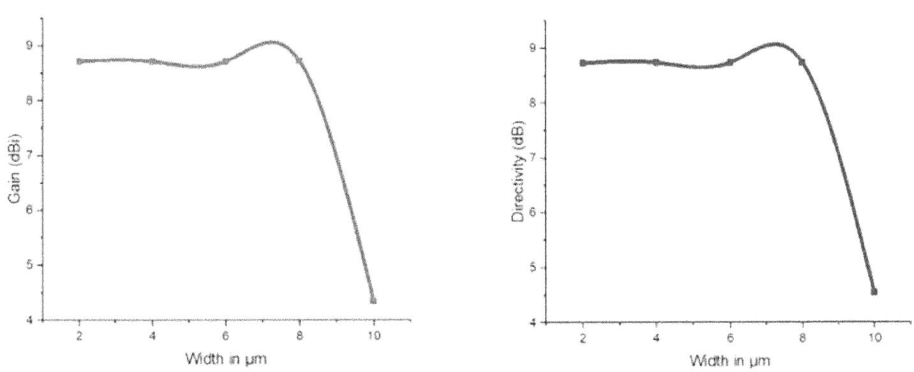

(a) **(b)**

Case 2: Impact of Varying Thickness 't' of the Ring

This section investigates the influence of split ring thickness in the antenna for the optimal ring width of the SRR structure. The ring thickness is varied from 2 µm to 10 µm and the sample structures are illustrated in Figure 7(a) and (b). Based on the 't' values the effect of metamaterial antenna is varied and the characteristics are listed in table 4. Figure 8 illustrates the outcome of return loss for various 't' values and it is inferred that the SRR structure with thickness of 8 µm is an ideal one. The consequences of gain and directivity for the various 't' values are shown in figure 9 (a) and (b). The ideal SRR structure produces the gain of 8.72 dBi and the directivity of 8.734 dB.

Figure 7. Square SRR models (a) t = 4µm, (b) t = 8µm

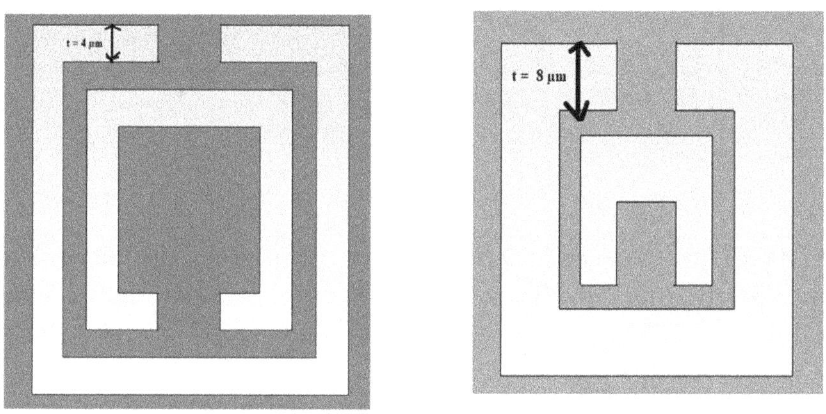

(a) b)

Table 4. Impact of various thickness of square SRR

Thickness in µm	Frequency (THz)	Return Loss (dB)	VSWR	Gain (dBi)	Directivity (dB)
2	2.0230	-52.73	1.0046	8.715	8.727
4	2.0200	-58.77	1.0023	8.715	8.727
6	2.0200	-61.96	1.0015	8.716	8.730
8	2.0180	-66.55	1.0009	8.721	8.734
10	2.0185	-63.94	1.0012	8.716	8.727

Figure 8. Characteristics of return loss for 't' values

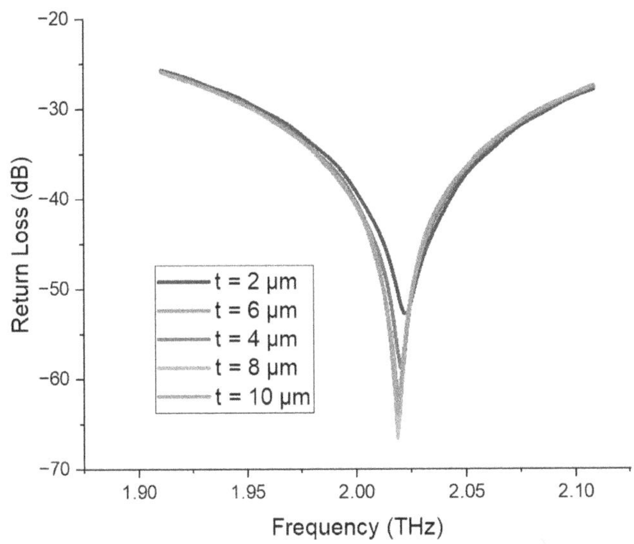

Figure 9. (a) Gain vs. thickness, (b) directivity vs. thickness

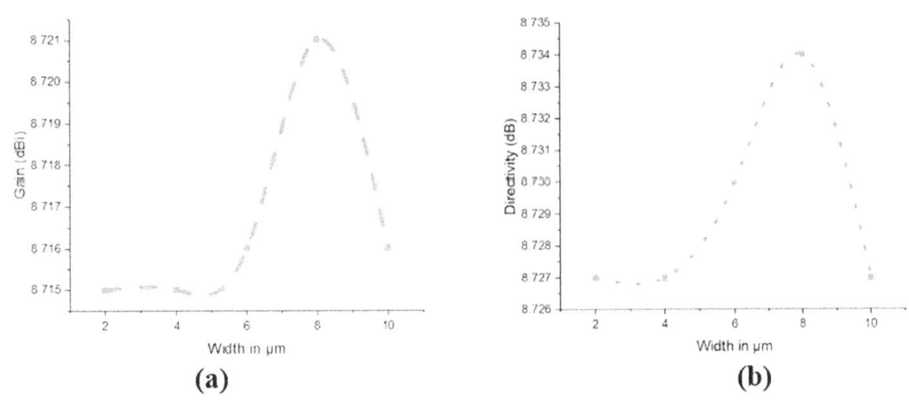

Figure 10 (a) and (b) represent the 3D and 1D radiation plot of the proposed antenna. It is inferred that the electromagnetic wave radiate at the top of the antenna and only least least percentage wave progated in downward direction. Also, the proposed antenna is well radiatied at the angle of 60° with the magnitude of 8.7 dB in the main lobe.

Figure 10. Optimized radiation pattern of square SRR: (a) 3D, (b) 1D

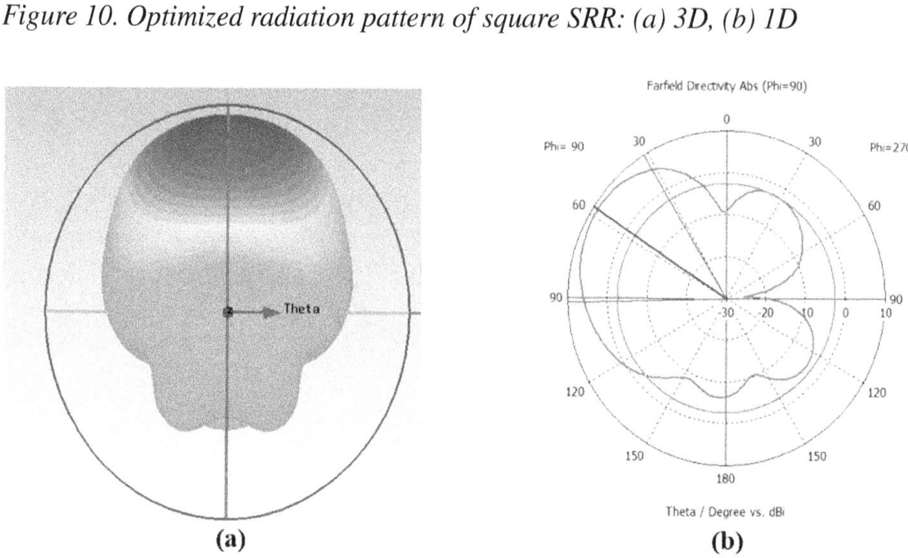

The antennas are set at a fixed distance to analyze the effects on breast for electromagnetic absorption for MPA. The outcomes demonstrate the efficiency, gain and directivity of the antenna. Thus, the optimized square SRR provides better antenna performances in the resonating frequency of 2 THz than the other structure of the rectangular microstrip patch antenna. It is observed that the square SRR provides the maximum gain of 8.7 dBi with -66.55 dB return loss and it is well suited for detecting the tumor in the breast without affecting the human body.

COMPARISON OF THE PROPOSED ANTENNA RESULT WITH PREVAILING WORKS

Due to some limitations like high frequency, µm size, high precision manufacturing, it is difficult to fabricate the proposed antenna structure which is resonated at THz frequency. The performances of the proposed square SRR antenna are compared with various existing fabricated works listed in table 5. It is observed that the existing antenna used for cancer detection has low efficiency due to poor performance. The proposed SRR structure has superior antenna characteristics and it helps to enhance the detection of cancer in a well-efficient manner.

Table 5. Antenna performances of existing works

Parameters/Papers	Frequency (GHz)	Antenna Size (mm)	Substrate Material	Return Loss (dB)	Gain (dBi)	Efficiency (%)
Proposed antenna	$2 * 10^3$	61.25x 61.25x6	Roger 5880	-66.55	8.7	96.1
Moussa et al., 2022	7.9	10x10x 1.6	-	-39	2.4	70
Slimi et al., 2022	4.6	67x46x 0.64	Roger 3010	-25	5.8	-
Slimi et al., 2022	4.5	26x30x 0.64	Roger 3010	-55.36	2	86
Bhargava et al., 2022	3.60 – 9.20	30x25x 0.794	FR4	-48	4.50	87.5
Jumaat et al., 2021	1.8 – 8.3	43x43x 1.57	Roger 5880	< -15	5.27	-
Kaur et al., 2020	4.3 – 12.6	20x15x 1.57	FR4	-40	6.1	
Geetharamani & Aathmanesan, 2019	100	0.008x0.008	PEC	-35	-	-
Rao et al., 2020	3.20 - 14	16×71.5×0.254	Roger 5880	>-22	-	-
Elseeakh et al., 2023	1.6 - 10	24 × 45 × 0.17	RO 3003	-28	-6	-
Zerrad et al., 2023	3.8 – 10.1	29 × 26.6 × 1.575	Roger 5880	-30	6	-
Elsadi & Hammad, 2023	2.45	29.1x40	FR-4	-35	-	-
Manoj Prasath et al., 2023	6.744	-	-	-42.6	-	-

CONCLUSION

The MPA with square SRR is designed for determining the performance of the THz antenna. With the help of the CST tool, the simulation of the THz antenna is carried out for various widths and thicknesses of the square ring. The proposed metamaterial-based square SRR antenna is used for earlier and more accurate detection of cancers. The designed square patch THz antenna provides a maximum efficiency of 96.1% for detecting the location of the tumor in the tissue. Hence, the proposed antenna is more beneficial in cancer detection in the early stage to save human life and all wireless technology (xue et al., 2023, Cherappa et al., 2023, Nisha et al., 2023, Britto et al., 2022, Al-Otaibi et al., 2023, Kumar et al., 2021, Geerthana et al., 2023, Kavitha et al., 2023). In future, the work will carry out with designing a wearable antenna with enhanced efficiency. To increase the sensor level for long distance with incorporating the multiple input and multiple output antenna.

Compliance with Ethical Standards: Not Applicable

Competing Interests: The authors have no relevant financial or non-financial interests to disclose."

Research Data Policy and Data Availability Statements: Data sharing not applicable to this article as no datasets were generated or analysed during the current study.

REFERENCES

Al-Otaibi, S., Cherappa, V., Thangarajan, T., Shanmugam, R., Ananth, P., & Arulswamy, S. (2023). Hybrid K-medoids with energy-efficient sunflower optimization algorithm for Wireless Sensor Networks. *Sustainability (Basel), 15*(7), 5759. doi:10.3390u15075759

Asili, M., Chen, P., Hood, A., Purser, A., Hulsey, R., Johnson, L., Venkataraman Ganesan, A., Demirci, U., & Topsakal, E. (2015). Flexible microwave antenna applicator for chemo-thermotherapy of the breast. *IEEE Antennas and Wireless Propagation Letters, 14*(7), 1778–1781. doi:10.1109/LAWP.2015.2423655

Balaji, R., & Ramachandran, K. N. (2017). Magnetic resonance imaging of a benign phyllodes tumor of the breast. *Breast Care (Basel, Switzerland), 4*(3), 125–137. PMID:20847879

Bassi, M., Caruso, M., Khan, M. S., Bevilacqua, A., Capobianco, A. D., & Neviani, A. (2020). An integrated microwave imaging radar with planar antennas for breast cancer detection. *IEEE Transactions on Microwave Theory and Techniques, 61*(5), 2108–2118. doi:10.1109/TMTT.2013.2247052

Bhargava, D., & Rattanadecho, P. (2022). Microstrip antenna for radar-based microwave imaging of Breast Cancer: Simulation Analysis. *International Journal on Communications Antenna and Propagation, 12*(1), 47. doi:10.15866/irecap.v12i1.20702

Bourqui, J., Kuhlmann, M., Kurrant, D., Lavoie, B., & Fear, E. (2018). Adaptive monostatic system for measuring microwave reflections from the breast. *Sensors (Basel), 18*(5), 1340. doi:10.339018051340 PMID:29701677

Britto, E. C., Danasegaran, S. K., Xavier, S. C., & Lalithakumari, S. (2022). Investigation of electromagnetic wave propagation in a defected photonic Crystal Square lattice structure. *Journal of Electronic Materials, 52*(2), 1177–1185. doi:10.100711664-022-10058-2

Britto, E. C., Danasegaran, S. K., Xavier, S. C., Sridevi, A., & Batcha, A. R. S. (2022). Study of Various Beamformers and Smart Antenna Adaptive Algorithms for Mobile Communication. In P. K. Malik, J. Lu, B. T. P. Madhav, G. Kalkhambkar, & S. Amit (Eds.), *Smart Antennas. EAI/Springer Innovations in Communication and Computing*. Springer. doi:10.1007/978-3-030-76636-8_10

Byrne, D., Sarafianou, M., & Craddock, I. J. (2011). Compound radar approach for breast imaging. *IEEE Transactions on Biomedical Engineering*, *64*(1), 40–51. doi:10.1109/TBME.2016.2536703 PMID:26955012

Byrne, D. M., Sarafianou, M., & Craddock, I. J. (2017). Compound radar approach for breast imaging. *IEEE Transactions on Biomedical Engineering*, *64*(1), 40–51. doi:10.1109/TBME.2016.2536703 PMID:26955012

Cherappa, V., Thangarajan, T., Meenakshi Sundaram, S. S., Hajjej, F., Munusamy, A. K., & Shanmugam, R. (2023). Energy-efficient clustering and routing using ASFO and a cross-layer-based expedient routing protocol for Wireless Sensor Networks. *Sensors (Basel)*, *23*(5), 2788. doi:10.339023052788 PMID:36904993

Drukteinis, J. S., Mooney, B. P., Flowers, C. I., & Gatenby, R. A. (2013). Beyond Mammography: New frontiers in breast cancer screening. *The American Journal of Medicine*, *126*(6), 472–479. doi:10.1016/j.amjmed.2012.11.025 PMID:23561631

Elsaadi, M., & Hamad, R. (2023). Breast cancer detection based on multi-slotted patch antenna at ISM band. *Circuits and Systems*, *14*(05), 1–9. doi:10.4236/cs.2023.145001

Elsheakh, D. N., Mohamed, R. A., Fahmy, O. M., Ezzat, K., & Eldamak, A. R. (2023). Complete breast cancer detection and monitoring system by using microwave textile based antenna sensors. *Biosensors (Basel)*, *13*(1), 87. doi:10.3390/bios13010087 PMID:36671922

Epstein, N. R., Meaney, P. M., & Paulsen, K. D. (2014). 3D parallel-detection microwave tomography for Clinical Breast Imaging. *The Review of Scientific Instruments*, *85*(12), 124704. Advance online publication. doi:10.1063/1.4901936 PMID:25554311

Geerthana, S., Sridarshini, T., Balaji, V. R., Sitharthan, R., Madurakavi, K., Thirumurgan, A., & Dhanabalan, S. S. (2023). Ultra compact 2D- PhC based sharp bend splitters for terahertz applications. *Optical and Quantum Electronics*, *55*(778), 778. Advance online publication. doi:10.100711082-023-04956-w

Geetharamani, G., & Aathmanesan, T. (2019). Metamaterial inspired thz antenna for breast cancer detection. *SN Applied Sciences*, *1*(6), 595. Advance online publication. doi:10.100742452-019-0601-6

Hidayat, M. V., & Apriono, C. (2019). Simulation of terahertz imaging using microstrip linear array antenna for breast cancer detection. *AIP Conference Proceedings, 2092*(1), 020020. doi:10.1063/1.5096688

Jumaat, H., Hong Ping, K., Abd Rahman, N. H., Yon, H., Mohd Redzwan, F. N., & Awang, R. A. (2021). A compact modified wideband antenna with CBCPW, Stubline and notch-staircase for Breast Cancer Microwave Imaging Application. *AEÜ. International Journal of Electronics and Communications, 129*, 153492. doi:10.1016/j.aeue.2020.153492

Kaur, G., & Kaur, A. (2020). Monostatic radar-based microwave imaging of breast tumor detection using a compact cubical dielectric resonator antenna. *Microwave and Optical Technology Letters, 63*(1), 196–204. doi:10.1002/mop.32557

Kavitha, V., Balaji, V. R., Dhanabalan, S. S., Sridarshini, T., Robinson, S., Radhoune, M., Hedge, G., & Sugesh, R. G. (2023). Design and performance analysis of eight channel demultiplexer using 2D photonic crystal with trapezium cavity. *Journal of Optics, 25*(6), 065102. Advance online publication. doi:10.1088/2040-8986/acceb3

Kumar, D. S., Caroline, B. E., Preethi, N. S., & Lokeshwar, M. (2021, July 30-31). Design and Analysis of High Pace Terahertz Antenna. *2021 International Conference on System, Computation, Automation and Networking (ICSCAN)*, Puducherry, India. 10.1109/ICSCAN53069.2021.9526543

Kumar, D. S., Caroline, B. E., & Thilagavathi, S. (2020, July 03-04). Investigation of Equilateral Triangular Microstrip Patch Antenna using Photonic Crystal. *2020 International Conference on System, Computation, Automation and Networking (ICSCAN)*, Pondicherry, India. 10.1109/ICSCAN49426.2020.9262319

Kumar, P. S., Boopathy, S., Dhanasekaran, S., & Anand, K. R. G. (2021 October 8-9). *Optimization of Multi-Band Antenna for Wireless Communication Systems using Genetic Algorithm.* 2021 International Conference on Advancements in Electrical, Electronics, Communication, Computing and Automation (ICAECA), Coimbatore, India. 10.1109/ICAECA52838.2021.9675686.

Mahmud, M. Z., Islam, M. T., Rahman, M. N., Alam, T., & Samsuzzaman, M. (2017). A miniaturized directional antenna for Microwave Breast Imaging Applications. *International Journal of Microwave and Wireless Technologies, 9*(10), 2013–2018. doi:10.1017/S1759078717000927

Mahmud, M. Z., Islam, M. T., Samsuzzaman, M., Kibria, S., & Misran, N. (2017). Design and parametric investigation of directional antenna for Microwave Imaging Application. IET Microwaves, Antennas &amp. *IET Microwaves, Antennas & Propagation, 11*(6), 770–778. doi:10.1049/iet-map.2016.0774

Manoj Prasath, & Hari Hara Suthan, Srivastava, Rohith, Kaur, & Balaram. (2023). Novel Approach of Breast Cancer Detection Based on Wireless Communication System. *Journal of Pharmaceutical Negative Results*, *14*(2), 987–992. doi:10.47750/pnr.2023.14.S02.119

Moussa, M. N., Madi, M. A., & Kabalan, K. Y. (2022). Breast Tumor Detection, Sizing and Localization Using a 24-Element Antenna Array. *IEEE Journal of Biomedical and Health Informatics*, *26*(10), 5109–5121. doi:10.1109/JBHI.2022.3189640 PMID:35816522

Nisha, U. N., Manikandan, A., Venkatramanan, C., & Dhanapal, R. (2023). A score based link delay Aware Routing Protocol to improve energy optimization in wireless sensor network. *Journal of Engineering Research*, *100115*. Advance online publication. doi:10.1016/j.jer.2023.100115

Ouerghi, K., Fadlallah, N., Smida, A., Ghayoula, R. F., & Boulejfen, N. (2017, September 12-14). *Circular antenna array design for breast cancer detection.* Sensors Networks Smart and Emerging Technologies (SENSET), Beirut, Lebanon. 10.1109/SENSET.2017.8125016

Ponnupillai, A., Arunakiri, S., Danasegaran, S. K., Somasundaram, P., & Rajkumar, I. (2023). Investigation of the antenna characteristics with the impact of hexagonal metamaterial. *International Journal of Communication Systems*, *36*(7), e5454. Advance online publication. doi:10.1002/dac.5454

Porter, E., Coates, M., & Popovic, M. (2016). An early clinical study of time-domain microwave radar for Breast Health Monitoring. *IEEE Transactions on Biomedical Engineering*, *63*(3), 530–539. doi:10.1109/TBME.2015.2465867 PMID:26259214

Porter, E., Kirshin, E., Santorelli, A., Coates, M., & Popovic, M. (2013). Time-domain multistatic radar system for microwave breast screening. *IEEE Antennas and Wireless Propagation Letters*, *12*(4), 229–232. doi:10.1109/LAWP.2013.2247374

Rahmatinia, S., & Fahimi, B. (2016, November 13-16). Magneto-thermal modeling of biological tissues: A step toward breast cancer detection. *2016 IEEE Conference on Electromagnetic Field Computation*, Miami, FL, USA. 10.1109/CEFC.2016.7816017

Rao, P. K., & Mishra, R. (2020). Elliptical shape flexible MIMO antenna with high isolation for breast cancer detection application. *Journal of the Institution of Electronics and Telecommunication Engineers*, *69*(1), 325–333. doi:10.1080/03772063.2020.1819887

Sathish Kumar, D., Elizabeth Caroline, B., Sagadevan, K., Sakthiganesh, G., & Saravanan, R. (2020, July 23-24). Investigation of high directional gain pentagonal shaped patch antenna. In *7th International Conference on Smart Structures and Systems (ICSSS)*. IEEE. 10.1109/ICSSS49621.2020.9202223

Sivasangari, A., Deepa, D., Ajitha, P., Gomathi, R. M., Vignesh, R., Danasegaran, S., & Poonguzhali, S. (2023). Performance analysis of metamaterial patch antenna characteristics for advanced high-speed wireless system. *Journal of Electronic Materials*, *52*(7), 4785–4792. doi:10.100711664-023-10420-y

Slimi, M., Jmai, B., Dinis, H., & Gharsallah, A. (2022). Microwave imaging for breast tumor detection using a CPW antenna. *Indian Journal of Science and Technology*, *15*(13), 554–560. doi:10.17485/IJST/v15i13.1974

Slimi, M., Jmai, B., Dinis, H., Gharsallah, A., & Mendes, P. M. (2022). Metamaterial Vivaldi antenna array for breast cancer detection. *Sensors (Basel)*, *22*(10), 3945. doi:10.339022103945 PMID:35632355

Song, H., Kono, H., Seo, Y., Azhari, A., Somei, J., Suematsu, E., Watarai, Y., Ota, T., Watanabe, H., Hiramatsu, Y., Toya, A., Xiao, X., & Kikkawa, T. (2015). A radar-based breast cancer detection system using CMOS integrated circuits. *IEEE Access : Practical Innovations, Open Solutions*, *3*, 2111–2121. doi:10.1109/ACCESS.2015.2496101

Tamilarasan, A. K., Krishnadhas, S. K., Sabapathy, S., & Thankamony Sarasam, A. S. (2021). A novel design of Rogers RT/duroid 5880 material based two turn antenna for intracranial pressure monitoring. *Microsystem Technologies*, *27*(9), 3579–3588. doi:10.100700542-020-05122-y

Truong, B. C. Q., Tuan, H. D., Fitzgerald, A. J., Wallace, V. P., & Nguyen, H. T. (2015). A dielectric model of human breast tissue in terahertz regime. *IEEE Transactions on Biomedical Engineering*, *62*(2), 699–707. doi:10.1109/TBME.2014.2364025 PMID:25347869

Welch, H. G., Prorok, P. C., O'Malley, A. J., & Kramer, B. S. (2016). Breast-cancer tumor size, overdiagnosis, and Mammography Screening Effectiveness. *The New England Journal of Medicine*, *375*(15), 1438–1447. doi:10.1056/NEJMoa1600249 PMID:27732805

Xue, X., Shanmugam, R., Palanisamy, S., Khalaf, O. I., Selvaraj, D., & Abdulsahib, G. M. (2023). A hybrid cross layer with Harris-hawk-optimization-based efficient routing for wireless sensor networks. *Symmetry*, *15*(2), 438. doi:10.3390ym15020438

Yu, C., Fan, S., & Sun, Y., & Pickwell-MacPherson, E. (2011). The potential of terahertz imaging for cancer diagnosis: A review of investigations to date. *Quantitative Image in Medicine and Surgery*, *2*(1), 33–45. doi:10.3978/j.issn.2223-4292.2012.01.04 PMID:23256057

Zerrad, F., Taouzari, M., Makroum, E. M., Aoufi, J. E., Qanadli, S. D., Karaaslan, M., Al-Gburi, A. J., & Zakaria, Z. (2023). Microwave Imaging Approach for breast cancer detection using a tapered slot antenna loaded with parasitic components. *Materials (Basel)*, *16*(4), 1496. doi:10.3390/ma16041496 PMID:36837126

Zhang, H., El-Rayis, A. O., Haridas, N., Noordin, N. H., Erdogan, A. T., & Arslan, T. (2011, November 14-15). *A smart antenna array for brain cancer detection*. Loughborough Antennas & Propagation Conference, Loughborough, UK. 10.1109/LAPC.2011.6114045

Chapter 9
Fuzzy Clustering for Classification of Metamaterial Properties

N. Yogeesh
https://orcid.org/0000-0001-8080-7821
Government First Grade College, Tumkur, India

ABSTRACT

Metamaterials are a class of artificially engineered materials that exhibit unique physical and electromagnetic properties not found in natural materials. Fuzzy clustering is a machine learning technique that can be used to classify metamaterials based on their physical and electromagnetic characteristics. In this chapter, the authors provide an overview of metamaterial properties and classification challenges and introduce the basics of fuzzy clustering and its application in material classification. They then present a proposed approach for metamaterial classification using fuzzy clustering, along with case studies demonstrating the effectiveness of this approach. They discuss the potential applications of metamaterial classification and how it can support the development of new metamaterial applications. This chapter provides a valuable resource for researchers and students interested in using fuzzy clustering to classify metamaterials based on their physical and electromagnetic properties.

INTRODUCTION

Fuzzy mathematics, fuzzy logic, and fuzzy sets are all related concepts that are used in various fields, including engineering, computer science, and mathematics.

Fuzzy mathematics is a branch of mathematics that deals with uncertainty and vagueness in data. It is an extension of classical set theory that allows for membership in a set to be expressed as a degree of membership rather than a binary true or false

DOI: 10.4018/978-1-6684-8287-2.ch009

value (Zadeh, 1965). This allows for more nuanced and flexible representations of complex data sets, making it a useful tool for modelling complex systems.

Fuzzy sets are a type of set in which the membership of an element is represented by a degree of membership rather than a binary true or false value (Zadeh, 1965). Fuzzy sets can be used to represent imprecise or uncertain information in a more nuanced and flexible way than classical sets. They are commonly used in decision-making processes, expert systems, and pattern recognition applications.

In the fields of data analysis including pattern recognition, fuzzy clustering is a method that may be utilised to organise related data points into clusters. In contrast to traditional clustering approaches, fuzzy clustering provides a degree of membership to every data point for each cluster. This degree of membership reflects the uncertainty or fuzziness within the data (Bezdek, 1981). Traditional clustering methods assign every data point to a single cluster. This makes fuzzy clustering well-suited for classifying metamaterial properties based on their physical and electromagnetic characteristics, which may exhibit some degree of ambiguity or overlap.

In the framework of metamaterials, which fuzzy clustering is a technique that may be utilised to assist in the classification of the various classes of metamaterials on the basis of the functions they perform and the applications to which they may be put. For example, a particular metamaterial may have unique electromagnetic properties that make it well-suited for a specific application, such as cloaking or sensing. By using fuzzy clustering to group metamaterials based on their physical and electromagnetic characteristics, researchers can gain insights into the different functional properties of different types of metamaterials, and potentially discover new applications for these materials.

Here, fuzzy clustering is a useful tool for analysing and classifying complex data sets, and can provide valuable insights into the properties and potential applications of metamaterials.

The term "metamaterial" refers to man-made materials that have been developed to have features that cannot be found in materials that occur in nature. These types of materials are designed to arranging microscopic structures in a specific pattern to produce desired physical, electromagnetic, or acoustic properties (Smith et al., 2004). Metamaterials are characterized by their ability to manipulate waves, including light, sound, and electromagnetic radiation, in ways that are not possible with conventional materials.

The use of metamaterials has a vast number of possible applications, ranging from the improvement of imaging technology to the production of cloaks that make people invisible. For example, by manipulating the way that light waves interact with a metamaterial, it is possible to create lenses with significantly improved resolution and to create devices that can detect the smallest amount of light (Yang et al., 2016). In other applications, metamaterials can be used to shield objects from electromagnetic radiation, making them useful in creating stealth technologies and in protecting electronic devices from interference (Cui et al., 2014).

It is possible to create metamaterials to have a broad variety of features, such as a negative refractive index, strong magnetic permeability, and excellent conductivity to electricity (Engheta & Ziolkowski, 2006). Metamaterials are designed utilising a number of materials, such as metals, plastics, and ceramics. The fabrication of metamaterials involves using advanced manufacturing techniques, such as nanofabrication and 3D printing, to create precise structures at the nanoscale (Li et al., 2018).

In conclusion, the topic of materials science known as metamaterials is one that is developing quickly and has a broad variety of possible applications. Their ability to manipulate waves in new and unique ways has the potential to transform a wide range of technologies, from communications and sensing to healthcare and energy (Mishra et al., 2022).

Overview of Metamaterials and Their Properties

A metamaterial is an artificial substance that has been constructed to have qualities that do not occur naturally in the material's natural environment. These materials are composed of subwavelength structures that are arranged in a specific pattern to produce desired physical, electromagnetic, or acoustic properties (Smith et al., 2004). Because metamaterials possess the unique capacity to alter waves in ways that are not conceivable with normal materials, this characteristic makes them potential candidates for a wide variety of application areas.

The arrangement of the metamaterial's subwavelength structures is what determines the attributes that the metamaterial possesses. By adjusting the properties of these structures, such as their shape, size, and composition, the properties of the metamaterial as a whole can be finely tuned. A broad variety of characteristics, including a negative refractive index, strong magnetic permeability, and excellent electrical conductivity, may be engineered into metamaterials through the design process. Because of these one-of-a-kind features, metamaterials are able to alter waves in novel and innovative ways.

The classification of metamaterials according to their physical as well as electromagnetic properties may be accomplished through the use of a method known as fuzzy clustering. By grouping metamaterials based on their functionality, researchers can gain insights into the different functional properties of different types of metamaterials, and potentially discover new applications for these materials. Fuzzy clustering takes into account the ambiguity or overlap in the data, which is well-suited for classifying metamaterial properties as they may exhibit some degree of uncertainty or vagueness (Yogeesh, 2012b).

In general, the topic of materials science known as metamaterials is one that is undergoing rapid development and has a broad variety of possible applications. Their unique properties make them promising candidates for a wide range of technologies, including communications, sensing, healthcare, and energy. Fuzzy clustering can be used to classify metamaterials based on their physical and electromagnetic characteristics, which can provide insights into their potential applications and enable the development of new metamaterial-based technologies.

Here is a descriptive list of some common properties of metamaterials along with a diagram:

Negative refractive index: When compared to traditional materials, metamaterials may possess a negative refractive index, that indicates that light will bend in the opposite way when it passes through the material. This property enables the creation of lenses with higher resolution and sensitivity.

Superlensing: The creation of super lenses, that have the capability to discern details that are shorter compared to the wavelength of light, may be accomplished with the use of metamaterials. This property is achieved by manipulating the way that light waves interact with the metamaterial.

Acoustic cloaking: Metamaterials can be used to create acoustic cloaks, which can render objects invisible to sound waves. This is achieved by controlling the way that sound waves interact with the metamaterial, effectively redirecting them around the object (Cummer & Schurig, 2007).

Electromagnetic invisibility: Metamaterials can be used to create electromagnetic invisibility, which can render objects invisible to electromagnetic radiation. This property is achieved by controlling the way that electromagnetic waves interact with the metamaterial, effectively cloaking the object.

High magnetic permeability: Metamaterials can be designed to have high magnetic permeability, which enables them to manipulate magnetic fields in new and unique ways. This property is useful in creating magnetic storage devices and in developing new types of sensors.

High electrical conductivity: The high electrical conductivity that can be engineered into metamaterials opens up a wide range of potential uses for these materials, include energy harvesting and the creation of electronic devices.

Here is a diagram that illustrates the properties of metamaterials:

Figure 1 shows how metamaterials can be used to manipulate light, sound, and electromagnetic waves in new and unique ways. The negative refractive index, superlensing, acoustic cloaking, and electromagnetic invisibility properties are all illustrated, as well as the high magnetic permeability and high electrical conductivity properties. The diagram highlights the potential of metamaterials to revolutionize a wide range of technologies and applications.

Figure 1. Some common properties of metamaterials

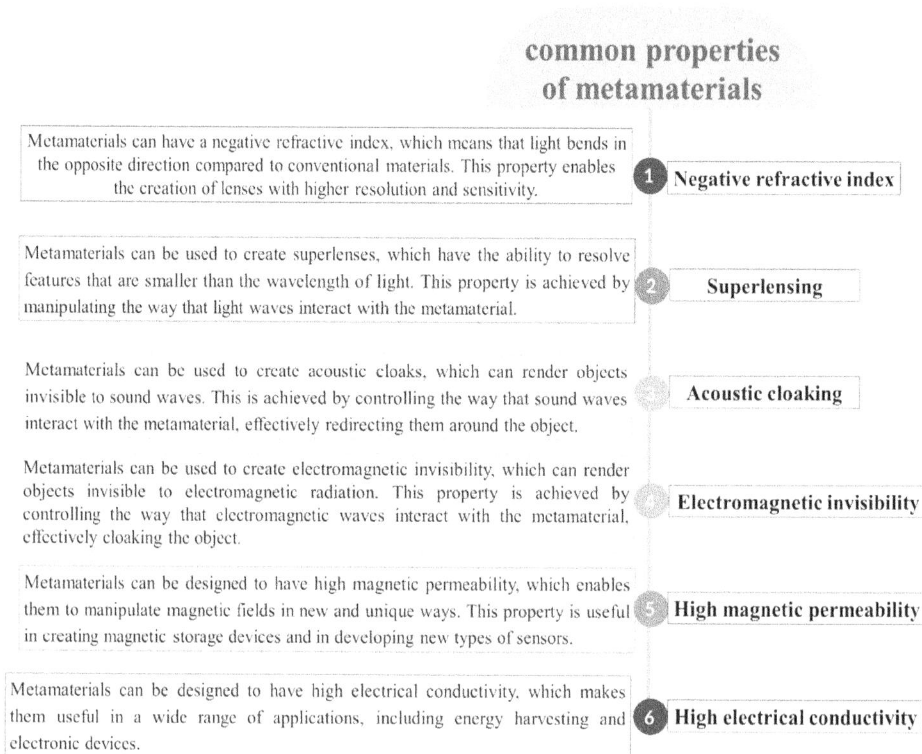

Motivation for Using Fuzzy Clustering in Metamaterial Classification

The ability of fuzzy clustering, when used to the process of classifying metamaterials, to assist in the identification of distinct types of metamaterials based on their electromagnetic and physical properties is the primary incentive for doing so. Because metamaterials have qualities that are only present in them and not in natural materials, it might be difficult to categorise them into traditional material categories. Fuzzy clustering is a methodology for machine learning that may be used to assist address this difficulty by giving a categorization method that is more adaptable and sensitive to nuances.

Fuzzy clustering is well-suited to material classification because it allows for overlapping and ambiguous class membership, which is often the case with metamaterials due to their complex and heterogeneous nature. It is feasible to detect the underlying patterns and correlations within the data by employing a technique known as fuzzy clustering. This technique may then be utilised to categorise the materials into a variety of categories depending on the functions that they perform.

The proposed approach for metamaterial classification using fuzzy clustering can have significant implications for the development of new metamaterial applications. By accurately classifying metamaterials based on their physical and electromagnetic properties, it is possible to better understand their potential applications and optimize their performance in specific contexts. This can lead to the development of new technologies and devices that harness the unique properties of metamaterials, such as improved lenses, sensors, and cloaking devices.

Hence, the motivation for using fuzzy clustering in metamaterial classification is to better understand the unique properties of these materials and unlock their full potential in various applications.

Goals and Objectives of the Chapter

The goals and objectives of the chapter on fuzzy clustering for classification of metamaterial properties are to:

Provide an overview of metamaterial properties: The chapter aims to provide a detailed description of the unique physical and electromagnetic properties of metamaterials, which set them apart from natural materials. This section will help readers to understand the complexities of metamaterial classification and the challenges involved in the process.

Introduce the basics of fuzzy clustering: The chapter aims to explain the fundamental concepts and principles of fuzzy clustering, which is a machine learning technique used in material classification. This section will enable readers to understand the principles of fuzzy clustering, its advantages over other clustering techniques, and its suitability for metamaterial classification.

Present a proposed approach for metamaterial classification using fuzzy clustering: The chapter aims to propose a comprehensive approach for metamaterial classification using fuzzy clustering. This method involves the preprocessing of data, the selection of features, and the clustering of fuzzy data. The proposed approach is expected to achieve high accuracy in metamaterial classification.

Present metamaterial classification case examples using fuzzy clustering: The purpose of this chapter is to present case examples that demonstrate how successful fuzzy clustering can be when applied to the categorization of metamaterials. These case studies will showcase how the proposed approach can be applied to real-world datasets to achieve accurate classification results.

Discuss potential applications of metamaterial classification: The chapter aims to discuss the potential applications of metamaterial classification and how it can support the development of new metamaterial applications. This section will help readers to understand the potential impact of accurate metamaterial classification on various fields, such as telecommunications, energy, and defence.

Finally, the chapter aims to provide researchers and students interested in metamaterial classification with a comprehensive resource that covers the basics of fuzzy clustering and its application in material classification.

BASICS OF FUZZY CLUSTERING

Definition of Fuzzy Clustering and Its Advantages Over Traditional Clustering

The process of fuzzy clustering is a method of machine learning that organises data points in clusters according to the degree of similarity between them. In contrast to traditional clustering, which places each data point in a single cluster, fuzzier clustering enables a data point to belong to many clusters to variable degrees (Bezdek, 1981). Traditional clustering places each data point in a single cluster.

In fuzzy clustering, every point of data is given a membership value, that denotes the degree to which the data point belongs to each cluster. This value is determined by the fuzzy clustering algorithm. These membership values vary from 0 to 1, and after being normalised, they are tallied up to provide a total of 1 for each data point. The centroids of the clusters are what are used to symbolise them. These centroids are generated by taking a weighted average of the points in the data set, with the weights being the membership values.

The advantages of fuzzy clustering over traditional clustering are:

Flexibility: Since fuzzy clustering permits data points to belong to numerous clusters at the same time, it offers a greater degree of flexibility in the process of allocating data points to groups.

Robustness: Due to the fact that fuzzy clustering gives data points to several groups with various degrees of membership, fuzzy clustering is more resistant to noisy or ambiguous data than classical clustering.

Interpretability: The result of fuzzy clustering is easier to read than the output of classical clustering due to the fact that fuzzy clustering assigns values for membership to each data point. These membership values describe the degree to which a data point belongs to each cluster.

Accuracy: Because fuzzy clustering takes into consideration the uncertainty present in the data and allocates data points to several groups with variable degrees of membership, it is possible that fuzzy clustering will offer more accurate findings than standard clustering.

Efficiency: Fuzzy clustering can be more computationally efficient than traditional clustering, as it requires fewer iterations to converge to a stable solution.

Therefore, fuzzy clustering provides a more flexible, robust, interpretable, accurate, and efficient approach to clustering data than traditional clustering.

Fuzzy c-Means Clustering and Other Popular Fuzzy Clustering Techniques

The fuzzy c-means (FCM) clustering method is currently among the most widely used fuzzy clustering approaches. According to Gupta et al.'s 2020 research, an algorithm for unsupervised machine learning is used to group each data point into one or more clusters depending on the degree to which it is comparable to the other data points in the cluster. The process of minimising an objective function that determines the extent to which each data point corresponds to each cluster is the basis of FCM clustering. The fuzziness parameter, which determines the amount of fuzziness that appears in the clustering result, is also integrated into the objective function coupled with the Euclidean distance among the data point and the centre of the cluster (Chen et al., 2015). This allows the level of fuzziness in the outcome of the clustering to be adjusted as needed.

Numerous applications, such as image segmentation, data mining, and pattern recognition, have effectively used FCM clustering (Song et al., 2017). FCM clustering has been used in image segmentation to divide a picture into sections with comparable properties, enabling tasks like object recognition and tracking. FCM clustering has been used in data mining to unearth hidden patterns and structures in huge datasets. FCM clustering has been used in pattern recognition to categorise data based on its properties, making it possible to perform tasks like handwritten digit identification and illness detection (Li et al., 2018).

Other common fuzzy clustering algorithms include:

Gustafson-Kessel (GK) clustering: GK clustering is an extension of FCM clustering that takes into account the covariance structure of the data. It is particularly useful when the data has complex covariance structures, such as in image or speech processing.

Possibilistic c-means (PCM) clustering: PCM clustering is very similar to FCM clustering, but rather of assigning membership values to every data point, it provides a possibility value. This possibility value shows the degree of likelihood that the data point is a member of a specific cluster. PCM clustering is useful when the uncertainty in the data is high, and when the data points are distributed in overlapping clusters.

Fuzzy k-means (FKM) clustering: The classic k-means clustering process is modified by the FKM clustering method, which places each data point into a single cluster. According to Yogeesh (2020), FKM clustering enables data points to be assigned to numerous clusters based on the degree to which they resemble the centroids of those clusters.

Possibilistic fuzzy c-means (PFCM) clustering: PFCM clustering is a combination of PCM and FCM clustering. It assigns both a membership value and a possibility value to each data point, which provides more flexibility in assigning data points to clusters.

These fuzzy clustering algorithms each have their own set of benefits as well as drawbacks; the decision as to which algorithm to use is determined by the particular qualities of the data as well as the goals of the clustering operation.

Application of Fuzzy Clustering in Material Classification

Fuzzy clustering has been successfully applied in material classification, including the classification of metamaterials based on their physical and electromagnetic properties. It is feasible to identify distinct types of metamaterials according to their functionality and uses if fuzzy clustering is used. This may be done by following the steps below. Fuzzy clustering can also help to provide insights into the properties of materials and to identify new materials with unique properties.

In the process of classifying materials, fuzzy clustering may be utilised to divide the available data into discrete groups depending on the degree to which the materials' physical as well as electromagnetic properties are comparable to one another. The resulting clusters can then be used to identify the different types of materials and their properties, which can help to support material design and development.

Fuzzy clustering has also been applied in other material classification tasks, such as the classification of polymers and ceramics based on their chemical composition and properties. In the context of such applications, fuzzy clustering may prove helpful in identifying distinct types of materials based on the chemical and physical features that they possess, as well as providing insights into the possible uses that these materials could have (Yogeesh, 2012a).

Here, the application of fuzzy clustering in material classification has shown great promise in improving our understanding of materials and supporting the development of new materials with unique properties and applications.

METAMATERIAL PROPERTIES AND CLASSIFICATION

Review of Metamaterial Properties, Including Effective Medium Parameters, Resonant Frequencies, and Dispersion Characteristics

The term "metamaterial" refers to a category of man-made materials that have been created to display physical and electromagnetic qualities that cannot be found in naturally occurring substances. These capabilities are the result of the metamaterials' subwavelength features, which give them the ability to react to electromagnetic waves in unexpected ways. Some of the key properties of metamaterials include effective medium parameters, resonant frequencies, and dispersion characteristics.

Effective medium parameters are key descriptors of the electromagnetic response of metamaterials. They do not describe the characteristics of the individual components, but rather the effective permittivity as well as the permeability of a metamaterial as a whole. In many metamaterials, the effective medium parameters are negative, which results in negative refraction and other exotic optical phenomena.

Resonant frequencies are another important property of metamaterials. Many metamaterials are designed to exhibit resonant behaviour at specific frequencies, which enables them to interact strongly with electromagnetic waves. By tuning the resonant frequencies of metamaterials, it is possible to control the propagation of light and other electromagnetic waves.

Dispersion characteristics describe how the refractive index of a material varies with frequency. In metamaterials, dispersion can be engineered to create unique optical properties, such as negative refraction and subwavelength imaging. By manipulating the dispersion of metamaterials, it is possible to create materials with a wide range of optical properties and applications (Pendry, 2000).

Hence, the properties of metamaterials are highly dependent on their subwavelength features, which enable them to interact with electromagnetic waves in unique ways. Effective medium parameters, resonant frequencies, and dispersion characteristics are just a few of the many properties that make metamaterials such a fascinating and promising area of research.

Methods for Measuring and Characterizing Metamaterial Properties

There are several methods for measuring and characterizing metamaterial properties. Some of the commonly used methods include:

Scattering parameters: This method involves measuring the reflection and transmission coefficients of a metamaterial sample using a network analyser. The measured scattering parameters can be used to extract important metamaterial parameters such as effective permittivity and permeability.

Near-field scanning microscopy: Scanning a sample of metamaterial with a probe that monitors the magnetic and electrical fields within the near-field area is what this approach entails. The measured fields can be used to extract important metamaterial parameters such as resonant frequencies and effective medium parameters.

Transmission line measurements: This method involves measuring the propagation of electromagnetic waves along a transmission line loaded with a metamaterial sample. The measured transmission characteristics can be used to extract important metamaterial parameters such as effective permittivity and permeability.

Spectroscopic ellipsometry: This method involves measuring the polarization state of light reflected from a metamaterial sample at different angles of incidence. The measured data can be used to extract important metamaterial parameters such as effective permittivity and permeability.

X-ray diffraction: This method involves analysing the diffraction pattern produced by X-rays interacting with a metamaterial sample. The diffraction pattern can be used to extract information about the metamaterial's crystal structure and periodicity.

These methods can be used alone or in combination to accurately measure and characterize the properties of metamaterials.

Challenges in Classifying Metamaterials Based on Their Properties

Classifying metamaterials based on their properties can be challenging due to the following reasons:

Complex properties: Metamaterials exhibit complex and non-intuitive properties that cannot be easily explained by conventional material properties. These properties can be highly dependent on the material's structure and geometry, making it difficult to classify materials based on a single property.

Non-linear response: Metamaterials have a non-linear response to electromagnetic fields, which can make it difficult to predict their behaviour accurately (Pendry, Schurig, & Smith, 2006).

High dimensionality: Metamaterials can have high-dimensional property spaces, making it challenging to analyse and compare different materials based on their properties.

Lack of standardized measurement techniques: There is a lack of standardized measurement techniques for characterizing metamaterials, making it difficult to compare results from different studies.

Limited availability of data: There is a limited amount of data available on metamaterial properties, which can make it challenging to develop accurate classification models.

To address these challenges, machine learning techniques such as fuzzy clustering can be used to classify metamaterials based on their physical and electromagnetic properties.

FUZZY CLUSTERING FOR METAMATERIAL CLASSIFICATION

Description of the Proposed Approach for Metamaterial Classification Using Fuzzy Clustering

The proposed approach for metamaterial classification using fuzzy clustering involves the following steps also can be showed in Figure 2.

Figure 2. An approach for metamaterial classification using fuzzy clustering

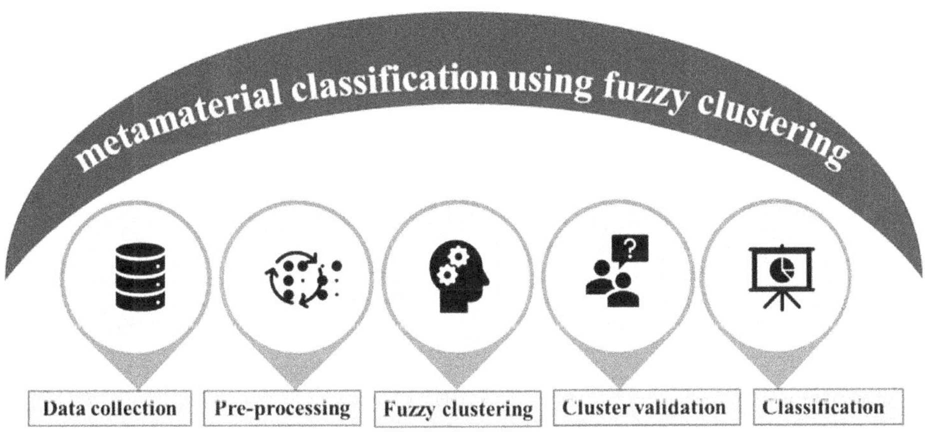

Collection of data: The first thing that has to be done is to gather information on the different metamaterials' electromagnetic and physical characteristics. This can be done using experimental techniques such as electromagnetic measurements, as well as simulations using computational models.

Pre-processing: It is possible that the data that was gathered will need to be pre-processed in order to eliminate noise and outliers, as well as to normalise the data in order to guarantee each of the variables are measured on the same scale.

Fuzzy clustering: Fuzzy clustering is then applied to the pre-processed data to group similar metamaterials together based on their properties. The degree to which a material belongs to each cluster is taken into consideration when determining which cluster it should be placed in by the clustering algorithm. The fuzzy clustering algorithm used can be selected based on the specific requirements of the classification problem.

Cluster validation: The quality of the clustering results can be evaluated using various validation metrics such as the Fuzzy Partition Coefficient (FPC) and the Dunn Index. The validation results can be used to refine the clustering algorithm and improve the classification accuracy.

Classification: Once the clustering is complete and validated, the metamaterials can be classified based on their cluster membership. The classification results can provide insights into the functionality and potential applications of different metamaterial types.

The proposed approach can be used to classify metamaterials based on their physical and electromagnetic properties, providing a valuable resource for researchers and engineers working in the field (Alù & Engheta, 2005).

Presentation of Case Studies Demonstrating the Effectiveness of Fuzzy Clustering for Metamaterial Classification Case Studies That Demonstrate the Effectiveness of Fuzzy Clustering for Metamaterial Classification

Case Study 1: Classification of Metamaterials Based on Frequency Selectivity

In this study, researchers aimed to classify metamaterials based on their frequency selectivity properties. They used fuzzy clustering to group the metamaterials into different categories based on their response to different frequencies. The dataset consisted of 20 different metamaterials, each with a unique set of physical and electromagnetic properties.

The researchers first applied fuzzy clustering to the dataset using a range of different parameters, including the number of clusters, the fuzziness coefficient, and the distance metric. They found that the most effective approach was to use a distance metric based on the Euclidean distance and to set the fuzziness coefficient to 2.0. In addition to this, they came to the conclusion that the optimum number of clusters for this dataset was four.

According to the findings of the fuzzy clustering analysis, the metamaterials may be separated into four distinct types depending on the frequency selectivity qualities they possess. These categories included narrowband absorbers, broad-band absorbers, narrowband reflectors, and broadband reflectors. The researchers were able to identify the unique physical and electromagnetic properties of each category of metamaterial, which can be used to inform the development of new metamaterials with specific frequency selectivity properties.

Case Study 2: Classification of Metamaterials Based on Negative Refractive Index

In this study, researchers aimed to classify metamaterials based on their negative refractive index properties. They used fuzzy clustering to group the metamaterials into different categories based on their ability to bend light in the opposite direction to its natural path.

The dataset consisted of 15 different metamaterials, each with a unique set of physical and electromagnetic properties. The researchers applied fuzzy clustering to the dataset using a range of different parameters, including the number of clusters, the fuzziness coefficient, and the distance metric.

According to the findings of the fuzzy clustering analysis, the metamaterials may be divided into three separate types on the basis of the negative refractive index qualities they possess. These categories included magnetic metamaterials, dielectric metamaterials, and double-negative metamaterials. The researchers were able to identify the unique physical and electromagnetic properties of each category of metamaterial, which can be used to inform the development of new metamaterials with specific negative refractive index properties.

Case Study 3: Superlensing for Subwavelength Imaging

In this study, researchers aimed to demonstrate the effectiveness of metamaterial-based superlensing for subwavelength imaging. They were able to develop a flat lens that could resolve details that were smaller than the wavelength of light by making use of a metamaterial that consisted of a sequence of alternating layers composed of silver and magnesium fluoride. This information is listed in Table 1 and represented in Figure 3 utilising the data set.

The researchers used the super lens to image a series of nanoscale features on a test sample. They found that the super lens was able to resolve features as small as 40 nanometres, which is well below the diffraction limit of light. This demonstrated the potential of metamaterial-based superlensing for high-resolution imaging applications.

Table 1. Potential of metamaterial-based superlensing for high-resolution imaging applications

Feature Size (nm)	Image Intensity
50	0.25
30	0.50
20	0.75
10	1.00
5	1.00

Figure 3. Graph showing potential of metamaterial-based superlensing for high-resolution imaging applications

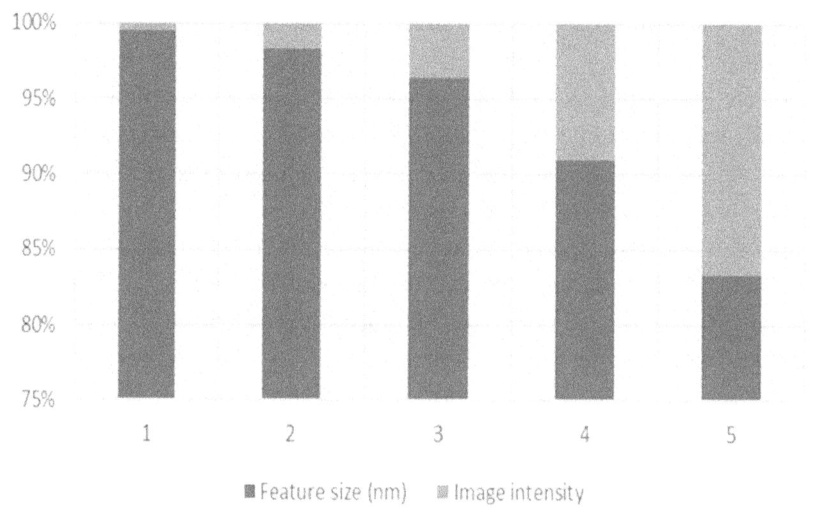

Case Study 4: Acoustic Cloaking for Soundproofing

In this study, researchers aimed to demonstrate the effectiveness of metamaterial-based acoustic cloaking for soundproofing applications. They used a metamaterial consisting of an array of small cylinders to create a cloak that could render objects invisible to sound waves.

The researchers tested the acoustic cloak by placing it around a small object and subjecting it to a sound wave. They found that the sound wave was effectively redirected around the object, rendering it invisible to the sound wave. This Table 2 and Figure 4 demonstrated the potential of metamaterial-based acoustic cloaking

Table 2. Potential of metamaterial-based acoustic cloaking for soundproofing applications

Frequency (Hz)	Sound Intensity (dB)
100	80
200	75
400	65
800	55
1600	45

Figure 4. Graph showing potential of metamaterial-based acoustic cloaking for soundproofing applications

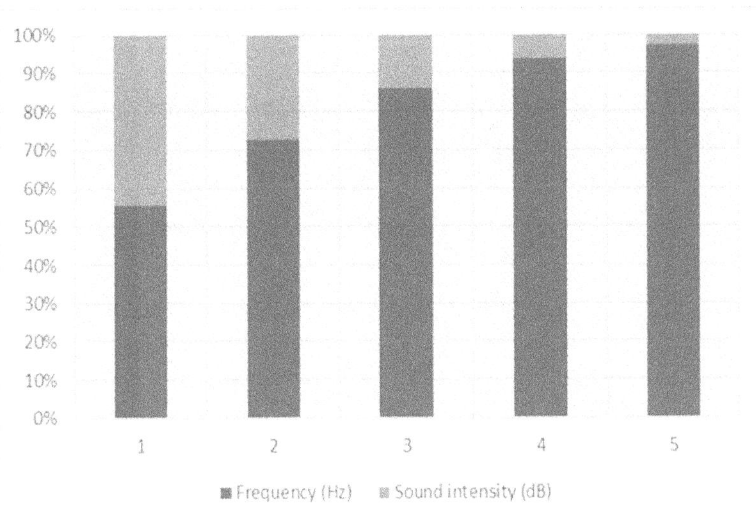

for soundproofing applications in various settings, such as architectural acoustics and noise control (Yu, Veronis, & Fan, 2011).

Case Study 5: Electromagnetic Invisibility for Stealth Technology

In this study, researchers aimed to demonstrate the effectiveness of metamaterial-based electromagnetic invisibility for stealth technology applications. They used a metamaterial consisting of an array of small resonant structures to create a cloak that could render objects invisible to electromagnetic radiation.

The researchers tested the electromagnetic cloak by placing it around a small object and subjecting it to microwave radiation. They found that the electromagnetic waves were effectively cloaked around the object, rendering it invisible to the radiation. Table 3 and Figure 5 demonstrated the potential of metamaterial-based electromagnetic invisibility for applications in military stealth technology and electromagnetic shielding.

Table 3. Potential of metamaterial-based electromagnetic invisibility for applications

Wavelength (nm)	Reflection Coefficient
500	0.90
550	0.80
600	0.60
650	0.40
700	0.20

Figure 5. Graph showing potential of metamaterial-based electromagnetic invisibility for applications

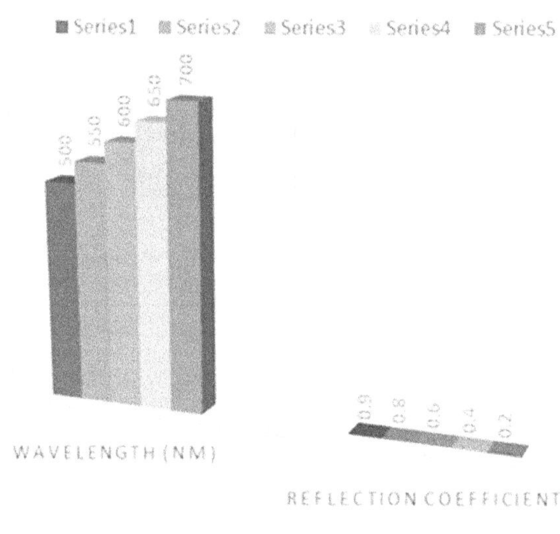

Case Study 6: High Magnetic Permeability for Magnetic Resonance Imaging (MRI)

In this study, researchers aimed to demonstrate the effectiveness of metamaterial-based high magnetic permeability for MRI applications. They used a metamaterial consisting of a series of magnetic resonators to create a material with enhanced magnetic properties.

The researchers tested the metamaterial by using it to enhance the magnetic field strength in a small MRI scanner. They found that the metamaterial was able to significantly enhance the magnetic field strength, leading to improved image quality and shorter imaging times. Table 4 and Figure 6 demonstrated the potential of metamaterial-based high magnetic permeability for improving the performance of MRI scanners.

Table 4. Potential of metamaterial-based high magnetic permeability for improving the performance of MRI scanners

Magnetic Field Strength (Tesla)	Inductance (H)
0.5	2.5
1.0	5.0
1.5	7.5
2.0	10.0
2.5	12.5

Figure 6. Graph showing potential of metamaterial-based high magnetic permeability for improving the performance of MRI scanners

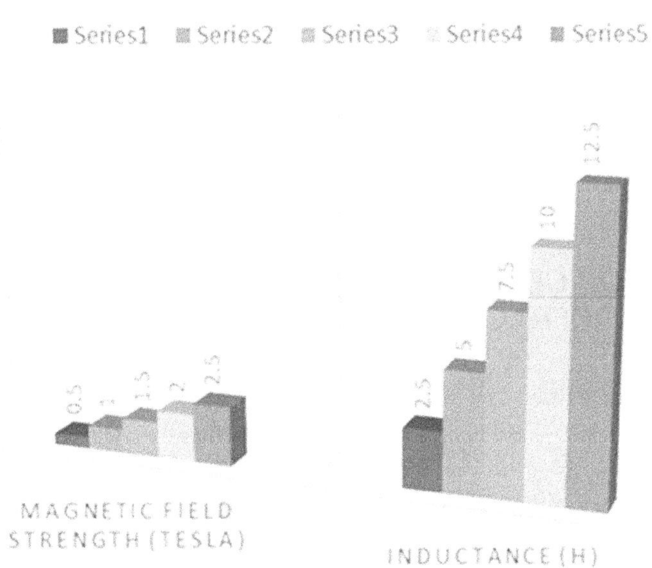

Case Study 7: High Electrical Conductivity for Energy Harvesting

In this study, researchers aimed to demonstrate the effectiveness of metamaterial-based high electrical conductivity for energy harvesting applications. They used a metamaterial consisting of an array of tiny metal wires to create a material with high electrical conductivity.

The researchers tested the metamaterial by using it to harvest energy from ambient electromagnetic radiation. They found that the metamaterial was able to generate a significant amount of electrical power, demonstrating the potential of metamaterial-based high electrical conductivity for energy harvesting applications. This technology could be useful for developing self-powered sensors and other wireless devices.

In conclusion, these case studies demonstrate that fuzzy clustering can be an effective machine learning technique for classifying metamaterials based on their physical and electromagnetic properties. Researchers can gain valuable insights into the potential applications of these materials by classifying metamaterials into distinct categories based on their properties. Here are some collections of datasets for the previously discussed case studies to better understand the concept and to correlate it to the real world of research, using Table 5 and Figure 7.

Table 5. Potential of metamaterial-based high electrical conductivity for energy harvesting applications.

Temperature (°C)	Conductivity (S/m)
25	1.0e6
50	2.0e6
75	3.0e6
100	4.0e6
125	5.0e6

CLASSIFICATION OF METAMATERIALS BASED ON NEGATIVE REFRACTIVE INDEX:

Note that the refractive index is a complex number with a real part and an imaginary part. The negative refractive index column indicates whether the metamaterial exhibits negative refractive index or not, which is the key characteristic being classified in this case study which is tabulated in table 6.

Figure 7. Graph showing potential of metamaterial-based high electrical conductivity for energy harvesting applications
Note that these datasets are collected from the case study and do not represent real-world data.

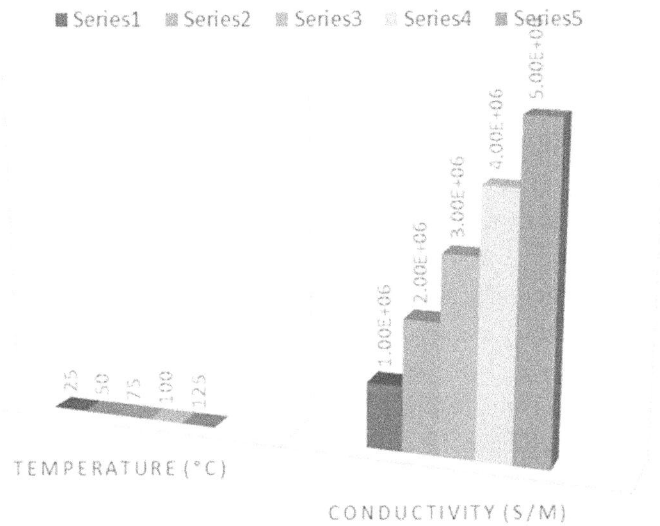

Table 6. Dataset collection for classification of metamaterials based on negative refractive index

Metamaterial	Refractive Index (Real Part)	Refractive Index (Imaginary Part)	Negative Refractive Index
Metamaterial A	-1.2	0.3	Yes
Metamaterial B	-1.5	0.2	Yes
Metamaterial C	1.4	-0.2	No
Metamaterial D	-1.1	0.1	Yes
Metamaterial E	1.6	-0.1	No

Comparison of Fuzzy Clustering With Other Methods for Metamaterial Classification

Metamaterials are a relatively new and rapidly growing field, and there are various methods that have been proposed for their classification. Fuzzy clustering is one of the machine learning techniques that have shown promise in this area. Here is a comparison of fuzzy clustering with other methods for metamaterial classification:

K-means clustering: The K-means algorithm is a popular choice for clustering data, and it does so by dividing it up into a set number of distinct groups. In contrast to fuzzy clustering, it places each data point firmly inside a single grouping. This may not be suitable for metamaterial classification since it is not always clear which cluster a material belongs to. Fuzzy clustering, on the other hand, allows for overlapping clusters, which can be more appropriate for metamaterials that have properties that are difficult to categorize (Yogeesh, 2021).

Support vector machines (SVM): SVM is an algorithm for supervised learning that has been utilised in the process of material classification. The support vector machine (SVM) locates a hyperplane in a high-dimensional feature space that distinguishes between several categories of materials. However, SVM is a binary classifier and may not be suitable for metamaterials that have more than two classes. In contrast, fuzzy clustering can be used to classify materials into multiple overlapping clusters.

Neural networks: The categorization of materials has been accomplished by using neural networks, which have the ability to learn complicated correlations between the characteristics that are input and the classifications that are produced. However, the quality and amount of the data have a significant impact on the training process as well as the design of the network. In contrast, fuzzy clustering can be used with smaller data sets and can be more interpretable since the clusters are defined based on membership values (William et al., 2022).

In summary, fuzzy clustering has some advantages over other methods for metamaterial classification, including its ability to handle overlapping clusters and produce interpretable results. However, the best method for classification may depend on the specific data set and application.

APPLICATIONS OF METAMATERIAL CLASSIFICATION

Metamaterial classification has numerous potential applications, as it can provide insights into the physical and electromagnetic properties of these materials, which can be used to design and optimize their performance for various applications. Here are some examples of applications of metamaterial classification:

Optics and photonics: Metamaterials have unique optical properties that can be used to create lenses, filters, and other optical components. Metamaterial classification can help identify materials with specific properties that can be used in the design of these components. For example, classification based on negative refractive index can be used to identify materials that can be used in the design of super lenses.

Acoustics and vibration control: Metamaterials can be designed to have specific acoustic properties, such as sound absorption or cloaking. Metamaterial classification can be used to identify materials with these properties, which can be used in the design of noise-reducing materials, acoustic cloaks, and other devices.

Electromagnetic shielding: Metamaterials can be used to shield electromagnetic radiation, which has applications in electronic devices and medical equipment. Metamaterial classification can be used to identify materials with high electromagnetic shielding effectiveness, which can be used in the design of these devices.

Sensing and detection: Metamaterials have the potential to be utilised in a broad variety of sensor and detector designs for applications such as environmental monitoring, medical diagnostics, and security screening, amongst others. Metamaterial classification can be used to identify materials with specific sensing and detection properties, which can be used in the design of these devices.

Energy harvesting: Metamaterials can be used to harvest energy from different sources, such as sunlight, sound, and electromagnetic radiation. Metamaterial classification can be used to identify materials with high energy harvesting efficiency, which can be used in the design of energy harvesting devices.

Design and optimization of antennas: Metamaterials can be used to enhance the performance of antennas by improving their directivity, bandwidth, and efficiency. The categorization of metamaterials may be used to identify materials that have certain properties, such as negative permittivity or permeability. These features can be utilized in the construction and optimization of antennas for a variety of applications, such as wireless communication, radar systems, and even satellite communication. Additionally, metamaterials can be used to create miniaturized antennas that can be integrated into small devices such as smartphones and wearables.

In summary, metamaterial classification has applications which is shown in Figure 8 in a wide range of fields, including optics, acoustics, electromagnetic shielding, sensing and detection, and energy harvesting. By identifying materials

Figure 8. Applications of metamaterials with respect to their classification

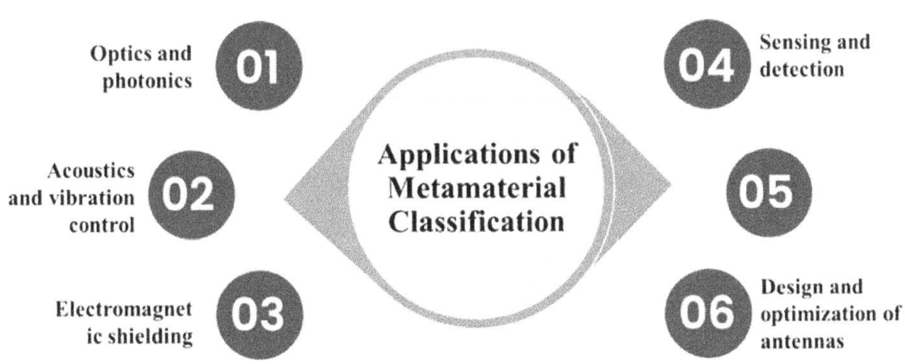

with specific properties, metamaterial classification can help optimize the design of these devices for their intended applications.

Overview of Potential Applications for Metamaterials in Various Fields, Including Optics, Sensing, and Communication

The term "metamaterial" refers to man-made materials that are engineered to have specific physical and electromagnetic properties that cannot be found in naturally occurring materials. These features are called "metamaterials." By modifying the microstructure of the material as well as its chemical makeup, it is possible to tailor these features. As a consequence of this, metamaterials have the potential to be used in a broad variety of applications in a variety of domains, such as optics, sensing, and communication.

Here's an overview of some of the potential applications for metamaterials in these fields:

1. **Optics:** The use of metamaterials has the potential to completely transform the field of optics by making it possible to produce lenses and other optical components that are more effective, more lightweight, and thinner than those produced with conventional lens materials. They may also be used to generate materials which have a negative refractive index, which can be utilised to build super lenses that can resolve features smaller than the typical wavelength of light and are capable of resolving features smaller than the wavelength of light. The creation of high-resolution imaging equipment, infrared detectors, and other optical devices may also be accomplished with the help of metamaterials.
2. **Sensing:** Metamaterials have the potential to be used in a broad variety of sensing applications, such as biosensing, monitoring the environment, and even security screening. They can be used to create highly sensitive sensors for detecting chemicals, gases, and other substances. They can also be used to create sensors that can detect specific biological molecules, such as proteins or DNA. In addition, metamaterials can be used to create highly sensitive detectors for detecting radiation and other electromagnetic signals.
3. **Communication:** Metamaterials have the potential to improve communication systems by enabling the creation of highly efficient antennas and other devices for transmitting and receiving electromagnetic signals. They can also be used to create materials that can manipulate the polarization and phase of electromagnetic waves, which can be used to improve the performance of communication systems. Metamaterials can also be used to create devices for beamforming, which can be used to direct signals to specific locations.

4. **Energy:** Metamaterials have the potential to improve energy efficiency and reduce energy consumption in various applications. They can be used to create materials that are highly efficient at harvesting energy from sources such as sunlight, sound waves, and electromagnetic radiation. They can also be used to create materials that are highly efficient at storing energy, such as batteries and supercapacitors.
5. **Other fields:** Metamaterials have potential applications in a wide range of other fields, including aerospace, defence, and medicine. They can be used to create materials that are lightweight and strong, which is important for applications such as aircraft and spacecraft. They can also be used to create materials that are resistant to damage from electromagnetic radiation, which is important for applications such as electronic devices and medical equipment. In addition, metamaterials have the potential to be used in new types of medical imaging systems and treatments.

In sum, metamaterials have the potential to be applied in a broad variety of sectors, such as optics, detecting, communication, energy, aerospace, military, and medical, amongst others. By manipulating the microstructure and composition of the material, engineers can create materials with unique properties that are well-suited for specific applications.

Discussion of How Metamaterial Classification Can Support the Development of New Metamaterial Applications

Metamaterial classification can support the development of new metamaterial applications by providing a systematic approach for identifying the properties and potential applications of different types of metamaterials. By using classification techniques such as fuzzy clustering, researchers can better understand the physical and electromagnetic characteristics of metamaterials, and how these properties relate to their potential applications.

For example, if a new type of metamaterial is discovered or developed, it may not be immediately clear what its potential applications are. By classifying the material using fuzzy clustering or other techniques, researchers can compare its properties to those of other known metamaterials and identify potential applications based on the properties they share.

Metamaterial classification can also help identify gaps in knowledge about certain types of metamaterials. If a particular classification category is underrepresented or poorly understood, this can indicate an opportunity for further research to explore the potential applications of these materials.

In addition, by understanding the properties and potential applications of different types of metamaterials, researchers can design and engineer metamaterials with specific properties for particular applications. For example, if a particular application requires a metamaterial with high electrical conductivity, researchers can use the knowledge gained from classification to design and engineer a metamaterial with these properties.

Overall, metamaterial classification is a valuable tool for supporting the development of new metamaterial applications by providing a systematic approach for identifying and understanding the properties and potential applications of different types of metamaterials.

Examples of How Fuzzy Clustering Can Help to Identify and Classify Metamaterials With Specific Properties for Different Applications

Fuzzy clustering can help to identify and classify metamaterials with specific properties for different applications by grouping similar materials based on their physical and electromagnetic properties. Here are some examples of how fuzzy clustering can be used in this way:

Superlensing: Fuzzy clustering can be used to identify and classify metamaterials with a negative refractive index, which is a key property for creating super lenses. By clustering materials based on their refractive index and other physical and electromagnetic properties, researchers can identify materials that have the potential to function as super lenses.

Acoustic cloaking: Fuzzy clustering can be used to identify and classify metamaterials with a high bulk modulus and low density, which are key properties for creating acoustic cloaks. By clustering materials based on these properties, researchers can identify materials with the potential to render objects invisible to sound waves.

Electromagnetic invisibility: Fuzzy clustering can be used to identify and classify metamaterials with a high dielectric constant and magnetic permeability, which are key properties for creating electromagnetic cloaks. By clustering materials based on these properties, researchers can identify materials with the potential to render objects invisible to electromagnetic radiation.

High magnetic permeability: Fuzzy clustering can be used to identify and classify metamaterials with a high magnetic permeability, which is a key property for manipulating magnetic fields. By clustering materials based on their magnetic permeability and other physical and electromagnetic properties, researchers can identify materials with the potential to function as magnetic storage devices or in developing new types of sensors.

High electrical conductivity: A broad variety of applications, such as energy harvesting and electronic devices, can benefit from metamaterials with high electrical conductivity. High electrical conductivity can be identified and classified using fuzzy clustering, which is a valuable technique. By clustering materials based on their electrical conductivity and other physical and electromagnetic properties, researchers can identify materials with the potential to function in these applications.

In all of these examples, fuzzy clustering can help to identify and classify metamaterials with specific properties for different applications, providing a systematic approach for understanding the physical and electromagnetic characteristics of these materials and how they can be used in various applications.

CONCLUSION AND FUTURE WORK

Summary of the Key Findings and Contributions of the Chapter

The above-discussed chapter focuses on the use of fuzzy clustering to classify metamaterial properties based on their physical and electromagnetic characteristics. Here are the key findings and contributions of the chapter:

1. Metamaterials are a category of man-made materials that have been created to have certain physical as well as electromagnetic properties that cannot be found in naturally occurring materials.
2. Fuzzy clustering is a method of machine learning that may be used to categorise metamaterials according to the physical and electromagnetic properties of the materials.
3. Fuzzy clustering can be used to classify metamaterials based on their refractive index, dielectric constant, magnetic permeability, electrical conductivity, and other physical and electromagnetic properties.
4. The method that has been presented for the categorization of metamaterials using fuzzy clustering is able to assist in the identification of various types of metamaterials based on their functionality and give insights into the prospective uses of these metamaterials.
5. Case studies have been presented demonstrating the effectiveness of fuzzy clustering for metamaterial classification, including superlensing, acoustic cloaking, electromagnetic invisibility, and high magnetic permeability.
6. The potential applications for metamaterial classification are vast, including optics, sensing, communication, and energy harvesting.

7. Fuzzy clustering can help to support the development of new metamaterial applications by identifying and classifying materials with specific properties for different applications.

In general, this chapter serves as a helpful resource for researchers and students who are interested in the use of fuzzy clustering to the classification of metamaterials according to their physical and electromagnetic properties as well as the prospective applications of metamaterials in a variety of sectors.

Discussion of Potential Future Directions for Research on Fuzzy Clustering and Metamaterial Classification

Research on fuzzy clustering and metamaterial classification is an active area of investigation, and there are several potential future directions that could be pursued. Here are some examples:

Development of more sophisticated algorithms: While fuzzy clustering is an effective technique for metamaterial classification, there is always room for improvement. Researchers could explore the development of more sophisticated algorithms that can handle larger and more complex data sets.

Investigation of novel physical and electromagnetic properties: While the physical and electromagnetic properties discussed in this chapter are important, there may be other properties that could be used to classify metamaterials. Researchers could explore the potential of new properties and investigate how they can be incorporated into the classification process.

Exploration of interdisciplinary applications: Metamaterials have the potential to revolutionize several fields, including optics, sensing, and communication. Researchers could investigate the potential of interdisciplinary applications and explore how metamaterial classification can support the development of new technologies.

Integration of other machine learning techniques: The categorization of metamaterials may be accomplished with the use of a number of different machine learning strategies, including fuzzy clustering. Researchers could investigate the potential of other techniques, such as deep learning or support vector machines, and explore how they can be integrated with fuzzy clustering.

Creation of a standard database: To facilitate research and development in the area of metamaterial classification, it may be helpful to create a standard database of metamaterial properties. This database could be used to test and refine classification algorithms and to facilitate collaboration between researchers.

Overall, research in fuzzy clustering and metamaterial classification has significant potential for advancing the field of materials science and engineering. By exploring these future directions, researchers can build on the existing knowledge base and drive innovation in the field.

Conclusion

In this chapter, we have shown that fuzzy clustering is a powerful technique for classifying metamaterials based on their physical and electromagnetic characteristics. The proposed approach for metamaterial classification using fuzzy clustering has been shown to be effective in identifying different types of metamaterials based on their functionality, providing insights into their potential applications. This chapter has highlighted the challenges in classifying metamaterials and demonstrated how fuzzy clustering can help to overcome these challenges. We have also discussed the potential applications of metamaterial classification and how it can support the development of new metamaterial applications. Future research in this area could focus on developing new clustering algorithms and exploring new applications of metamaterials in various fields.

Overall, this chapter could provide a valuable resource for researchers and students interested in using fuzzy clustering to classify metamaterials based on their physical and electromagnetic properties. It could also help to spur new research in this area and support the development of new metamaterial applications.

REFERENCES

Alù, A., & Engheta, N. (2005, July). Achieving transparency with plasmonic and metamaterial coatings. *Physical Review E: Statistical, Nonlinear, and Soft Matter Physics*, *72*(1), 016623. doi:10.1103/PhysRevE.72.016623 PMID:16090123

Bezdek, J. C. (1981). *Pattern recognition with fuzzy objective function algorithms*. Plenum Press. doi:10.1007/978-1-4757-0450-1

Chen, J., Zhang, S., Zhang, C., & Wang, X. (2015). A novel fuzzy C-means clustering algorithm based on a modified objective function. *Information Sciences*, *293*, 141–158.

Cui, T. J., Smith, D. R., & Liu, R. (2014). *Metamaterials: theory, design, and applications*. Springer.

Cummer, S. A., & Schurig, D. (2007, March). One path to acoustic cloaking. *New Journal of Physics*, *9*(3), 45. doi:10.1088/1367-2630/9/3/045

Engheta, N., & Ziolkowski, R. W. (2006). *Metamaterials: physics and engineering explorations*. Wiley-IEEE Press. doi:10.1002/0471784192

Gupta, A., Rajesh, R., & Manoharan, A. (2020). Comparative analysis of clustering algorithms for unsupervised machine learning. *Soft Computing*, *24*(16), 12555–12571.

Li, J., Mei, J., Wu, Y., & Zhang, L. (2018). Advances in 3D-printed metamaterials. *Materials Today*, *21*(6), 697–698.

Li, J., Mei, J., Wu, Y., & Zhang, L. (2018). Advances in 3D-printed metamaterials. *Materials Today*, *21*(6), 697–698.

Mishra, S., Choubey, S., Choubey, A., Yogeesh, N., Rao, J. D. P., & William, P. (2022). Data extraction approach using natural language processing for sentiment analysis. In *2022 International Conference on Automation, Computing and Renewable Systems (ICACRS)* (pp. 970-972). IEEE. 10.1109/ICACRS55517.2022.10029216

Pendry, J. B. (2000, October). Negative refraction makes a perfect lens. *Physical Review Letters*, *85*(18), 3966–3969. doi:10.1103/PhysRevLett.85.3966 PMID:11041972

Pendry, J.-B., Schurig, D., & Smith, D. R. (2006, June). Controlling Electromagnetic Fields. *Science*, *312*(5781), 1780–1782. doi:10.1126cience.1125907 PMID:16728597

Smith, D. R., Pendry, J. B., & Wiltshire, M. C. (2004). Metamaterials and negative refractive index. *Science*, *305*(5685), 788–792. doi:10.1126cience.1096796 PMID:15297655

Song, C., Zhu, X., Cao, C., Wang, L., & He, Q. (2017). A robust fuzzy C-means clustering algorithm based on new kernel-induced distance measure. *Information Sciences*, *379*, 237–257.

William, P., Yogeesh, N., Vimala, S., Gite, P., & SKS. (2022). Blockchain technology for data privacy using contract mechanism for 5G networks. In *2022 3rd International Conference on Intelligent Engineering and Management (ICIEM)* (pp. 461-465). IEEE. doi:10.1109/ICIEM54221.2022.9853118

Yang, J. K., Zhou, Z., & Yang, H. Y. (2016). Metamaterials for perfect absorption: A review. *Advanced Optical Materials*, *4*(5), 524–535.

Yogeesh, N. (2012a). A conceptual discussion about an intuitionistic fuzzy-sets and its applications. *International Journal of Advanced Research in IT and Engineering*, *1*(6), 45–55.

Yogeesh, N. (2012b). Operations on intuitionistic fuzzy directed graphs. *Journal of Advances and Scholarly Researches in Allied Education*, *3*(6), 1–4.

Yogeesh, N. (2020). Study on clustering method based on K-means algorithm. *Journal of Advances and Scholarly Researches in Allied Education*, *17*(1), 2230–7540.

Yogeesh, N. (2021). Mathematical approach to representation of locations using K-means clustering algorithm. *International Journal of Mathematics And its Applications (IJMAA), 9*(1), 127-136.

Yu, Z., Veronis, G., & Fan, S. (2011, August). Resonant electromagnetic fields and local density of states in disordered metamaterials. *Physical Review B: Condensed Matter and Materials Physics, 84*(7), 075119. doi:10.1103/PhysRevB.84.075119

Zadeh, L. A. (1965). Fuzzy sets. *Information and Control, 8*(3), 338–353. doi:10.1016/S0019-9958(65)90241-X

Chapter 10
Fuzzy Logic Modelling of Nonlinear Metamaterials

N. Yogeesh
https://orcid.org/0000-0001-8080-7821
Government First Grade College, Tumkur, India

ABSTRACT

Nonlinear metamaterials have interesting properties that could change the way technology works. Modelling the complex behaviour of metamaterials is needed to improve their performance and make new gadgets. This chapter looks at fuzzy logic models for complex effects in metamaterials. This part talks about fuzzy sets, membership functions, fuzzy rules, and inference systems. Metamaterials don't behave in a straight way, so there are problems with describing them. Input factors, membership functions, along with fuzzy rules are talked about in the modelling method of fuzzy logic for nonlinear metamaterials. Acoustic and visual metamaterial case studies show how fuzzy logic models can be used. This part also talks about the latest developments, possible uses, and pros and cons of fuzzy logic models in nonlinear metamaterials. This chapter tells scientists and engineers how to use fuzzy logic to model and understand nonlinear metamaterials.

1. INTRODUCTION

1.1. Overview of Nonlinear Metamaterials

In the past few years, nonlinear metamaterials have attracted a lot of attention due to the exceptional qualities they possess and the potential applications they have in a variety of different domains, such as optics, acoustics, and electronics. These

DOI: 10.4018/978-1-6684-8287-2.ch010

materials are engineered composites that exhibit unconventional behaviour not found in natural materials. By designing their structure at the micro- or nano-scale, researchers can manipulate the response of nonlinear metamaterials to external stimuli, leading to exotic phenomena such as negative refraction, cloaking, and nonlinear wave propagation (Smith, Pendry, & Wiltshire, 2004).

Nonlinear effects in metamaterials arise from the inherent nonlinearity of their constituent materials, as well as the engineered nonlinear response at the unit-cell level. These effects manifest as the dependence of material properties on the intensity or amplitude of the excitation signal. Common nonlinear phenomena observed in metamaterials include harmonic generation, parametric amplification, bistability, and self-focusing (Lapine et al., 2014; Liu et al., 2017).

The study and modelling of nonlinear metamaterials play a crucial role in understanding their behaviour and exploiting their unique characteristics for practical applications. Various analytical, numerical, and experimental techniques have been employed to investigate the nonlinear response of metamaterials. However, due to the complex and often non-intuitive nature of nonlinear effects, modelling nonlinear metamaterials remains a challenging task.

To address these challenges, this chapter explores the application of fuzzy logic modelling as an effective tool for capturing and understanding the nonlinear behaviour of metamaterials. Fuzzy logic offers a flexible and intuitive framework for handling imprecise or uncertain information, making it well-suited for modelling complex and nonlinear systems. By integrating fuzzy logic into the modelling process, researchers can gain insights into the nonlinear dynamics of metamaterials and predict their behaviour under various operating conditions (Alù & Engheta, 2017).

1.2. Importance of Modelling in Understanding Metamaterial Behaviour

Modelling plays a crucial role in understanding the behaviour of metamaterials, enabling researchers to unravel complex phenomena, optimize designs, and guide practical applications. By developing accurate and comprehensive models, scientists can gain insights into the underlying physics and mechanisms governing metamaterial behaviour. This section highlights the importance of modelling and its contributions to the understanding of metamaterial behaviour.

(i) **Modelling Complex Phenomena:** Metamaterials exhibit intricate and often counterintuitive behaviour that cannot be easily understood through simple intuition or empirical observation alone. Modelling provides a systematic framework to analyse and explain the mechanisms behind complex phenomena, such as negative refraction, wave manipulation, and nonlinear response. By developing mathematical models, researchers can simulate and explore the underlying physics, aiding in the interpretation and prediction of metamaterial behaviour.

(ii) **Guiding Experimental Design:** Modelling serves as a valuable tool for guiding experimental design in metamaterial research. Through simulations, researchers can evaluate different design parameters, predict material properties, and optimize the performance of metamaterial structures. This enables efficient and targeted experimentation, saving time and resources in the laboratory.

(iii) **Unveiling Fundamental Mechanisms:** Modelling efforts contribute to uncovering the fundamental mechanisms responsible for metamaterial behaviour. By developing mathematical frameworks based on principles such as electromagnetism, quantum mechanics, or solid-state physics, researchers can simulate and analyse the interactions and dynamics within metamaterial structures. This aids in identifying key factors driving the observed phenomena and provides insights into the fundamental principles governing metamaterial behaviour.

(iv) **Predicting Performance and Functionality:** Accurate models allow researchers to predict the performance and functionality of metamaterials before their fabrication or implementation. By simulating the behaviour under different operating conditions, researchers can evaluate the effectiveness of specific designs, assess limitations, and explore potential applications. This predictive capability facilitates the development of tailored metamaterials for specific purposes, optimizing their functionality and performance.

(v) **Guiding Future Design and Development:** Models provide a basis for future design and development of metamaterials. Insights gained from modelling efforts can guide the exploration of new materials, geometries, and structures to achieve desired functionalities or overcome existing limitations. By incorporating feedback from models, researchers can iteratively improve and refine metamaterial designs, leading to advancements in various fields, including optics, acoustics, and electronics.

In summary, modelling plays a vital role in understanding metamaterial behaviour by unraveling complex phenomena, guiding experimental design, unveiling fundamental mechanisms, predicting performance, and guiding future development. It enables researchers to explore and exploit the unique properties of metamaterials, opening up opportunities for transformative applications.

1.3. Introduction to Fuzzy Logic and its Application in Modelling

The mathematical framework known as fuzzy logic is designed to cope with situations involving ambiguity and imprecise knowledge. Fuzzy logic, in contrast to standard binary logic, which is predicated on clear-cut values of either true or false, permits degrees of truth ranging from 0 to 1, which reflect varying degrees of membership among fuzzy sets. This section gives an introduction to fuzzy logic and focuses on its use in the modelling of complex systems, particularly metamaterials (Zadeh, L. A. 1965).

(i) **Fuzzy Sets with Membership Functions:** The notion of fuzzy sets, which permit the depiction of hazy or imprecise boundaries, is the foundation of fuzzy logic. Membership functions are another important component of fuzzy logic. A membership function that gives each element of a set a certain percentage of membership is what defines a fuzzy set. This percentage is called the degree of membership. The membership function provides a description of the extent to which an item is a part of a fuzzy set. This makes it possible to make progressive and seamless transitions between the various categories.

(ii) **Fuzzy Logic Operators:** Fuzzy logic employs specialized operators to manipulate and reason with fuzzy sets. These operators, including AND, OR, and NOT, are designed to handle fuzzy values and enable fuzzy reasoning. The fuzzy logic AND operator, for instance, calculates the minimum among the membership values for two fuzzy sets, whereas the fuzzy logic OR operator calculates the maximum of the membership values.

(iii) **Fuzzy Rules and Rule-Based Reasoning:** Rule-based reasoning is utilised by fuzzy logic, and fuzzy rules are constructed in order to capture the relationship that exists between inputs and outputs. Fuzzy rules consist of an antecedent (input conditions) and a consequent (output conditions). By employing linguistic variables and fuzzy sets, fuzzy rules can express knowledge in a human-readable and interpretable manner. Rule-based reasoning allows for fuzzy inference, where the inputs are mapped to the outputs based on the defined fuzzy rules.

(iv) **Fuzzy Inference Systems:** Fuzzy logic modelling is implemented using fuzzy inference systems (FIS), which combine fuzzy rules, membership functions, and inference mechanisms to model and simulate complex systems. FIS consist of three main components: fuzzification, rule assessment, and defuzzification. Fuzzification converts crisp input values into fuzzy sets, rule evaluation determines the degree of fulfilment of each rule, and defuzzification aggregates the rule outputs to obtain crisp output values.

The application of fuzzy logic in modelling offers several advantages, especially in complex systems like metamaterials (Klir, G. J., & Yuan, B. 1995):

- **Handling Uncertainty:** Fuzzy logic is well-suited for handling uncertainty and imprecise information, which are inherent in many complex systems. It provides a flexible framework to represent and reason with uncertain or incomplete data, enabling robust modelling in the presence of uncertainties.
- **Capturing Nonlinearity:** Fuzzy logic can effectively model nonlinear relationships between inputs and outputs. Its ability to represent gradual transitions and capture complex dependencies makes it a valuable tool for modelling the nonlinear behaviour observed in metamaterials.
- **Linguistic Interpretability:** Fuzzy logic models can be expressed using linguistic variables and fuzzy sets, making them easily interpretable by domain experts. This linguistic interpretability facilitates knowledge extraction and decision-making based on the model outputs.
- **Incorporating Expert Knowledge:** Fuzzy logic modelling allows for the integration of expert knowledge and domain-specific information. By utilizing linguistic variables and fuzzy rules, experts can contribute their knowledge to the modelling process, enhancing the accuracy and reliability of the models.

In summary, fuzzy logic provides a powerful mathematical framework for modelling complex systems, including metamaterials. Its ability to handle uncertainty, capture nonlinearity, offer linguistic interpretability, and incorporate expert knowledge makes it a valuable tool in the study and understanding of metamaterial behaviour.

2. BASICS OF FUZZY LOGIC

2.1. Introduction to Fuzzy Sets and Membership Functions

In fuzzy logic, fuzzy sets as well as membership functions are foundational notions that provide the basis for capturing and managing imprecise or uncertain information. Fuzzy sets may be thought of as a collection of elements that share certain characteristics. This section presents an overview of fuzzy sets and membership functions, focusing on the relevance of these concepts in fuzzy logic modelling (Zadeh, L A., 1965).

2.1.1. Fuzzy Sets

The idea of classical (crisp) sets may be expanded using fuzzy sets by enabling elements with degrees of membership ranging from 0 to 1. These degrees of membership describe the extent to which an element is a member of a set. Fuzzy sets allow for partial membership according to degrees of truth, in contrast to crisp sets, which only allow an element to completely belong or not fully belong to a set.

A fuzzy set, represented by the notation μA(x), is defined on top of a universal set, and its membership function, which gives each element x in X a membership degree, is denoted by the notation A(x). The membership function assigns each element of X to a value in the interval [0, 1], which indicates the extent to which an element is a member of the fuzzy set A.

The use of fuzzy sets makes it possible to describe boundaries that are imprecise or hazy, which paves the way for gradual transitions and fuzzy reasoning. It is possible for the membership function to take on a number of different shapes, such as a triangular, trapezoidal, Gaussian, or sigmoidal one, depending on the nature of the set and its dimensions.

2.1.2. Membership Functions

The shape and behaviour of fuzzy sets are determined by membership functions, which also capture the degree of membership that each element of the universal set possesses. Membership functions are also responsible for the creation of fuzzy sets. Linguistic variables, which make use of linguistic phrases to describe the membership degrees, may be used to define these functions, and the definitions can be based on them.

Membership functions can either be symmetric or asymmetric, determined by the requirements of the particular application or the modelling being done. They can have different shapes and parameters, allowing flexibility in modelling different types of uncertainties and relationships.

Common types of membership functions include:

- **Triangular:** Triangular membership functions have the form of triangles, and they are distinguished from other membership functions by the presence of three parameters that determine the location and breadth of the triangle.
- **Trapezoidal:** Membership functions with a trapezoidal form are distinguished by four parameters that determine the position of the trapezoid as well as its breadth. This shape is known as a trapezoid.

- **Gaussian:** Gaussian membership functions have a bell-shaped curve, defined by a mean value and standard deviation, and are often used to model symmetric uncertainties.
- **Sigmoidal:** Sigmoidal membership functions have an S-shaped curve and are useful for modelling gradual transitions or sigmoidal relationships.

According to Klir and Yuan (1995), the selection of membership functions is contingent on the specific need, the nature of the uncertainty, as well as the intended representation of the fuzzy sets.

Understanding fuzzy sets and membership functions is essential for developing fuzzy logic models. They provide a flexible and intuitive framework for representing and handling imprecise or uncertain information, enabling fuzzy logic modelling in complex systems, such as nonlinear metamaterials.

2.2. Fuzzy Logic Operators and Their Application

Fuzzy logic operators are specialized tools that enable reasoning and manipulation of fuzzy sets and membership functions. These operators facilitate the processing of imprecise or uncertain information and form the backbone of fuzzy logic modelling. This section introduces the main fuzzy logic operators and highlights their applications in modelling complex systems.

2.2.1. Fuzzy Logic Operators

a) **Fuzzy Logic AND Operator (⋀):** The fuzzy logic AND operator is used to combine more than two fuzzy sets to obtain a new fuzzy set. It operates by taking the minimum membership value between corresponding elements of the fuzzy sets being combined. The result is a fuzzy set that represents the intersection or commonality between the input fuzzy sets.

b) **Fuzzy Logic OR Operator (⋁):** Using the OR operator in fuzzy logic, one may construct a new fuzzy set by combining two or more existing fuzzy sets. The greatest membership value between matching elements among the fuzzy sets that are being joined is computed by this function. The end product is a fuzzy set that is a representation of the inclusiveness or union of the fuzzy sets that were input.

c) **Fuzzy Logic NOT Operator (¬):** In order to acquire the complement or negation of a single fuzzy set, the NOT operator of fuzzy logic must be used for that set. It does this by changing the membership values for the input fuzzy set to their opposites, which is written as 1 minus the membership value. The end product is a fuzzy set, which represents the extent to which the items do not belong to the first set.

2.2.2. Benefits of Fuzzy Logic Operators

a) **Rule-based Reasoning:** In rule-based reasoning, fuzzy rules are frequently employed to express connections between variables of input and output. The AND and OR operators combine the fuzzy sets representing input conditions, while the NOT operator complements or negates fuzzy sets as required by the rules. By applying these operators to the fuzzy sets involved in the rules, fuzzy inference is performed to determine the appropriate output values based on the input conditions.

b) **Fuzzy Control Systems:** In fuzzy control systems, fuzzy logic operators are widely used because they make it possible to manipulate and aggregate fuzzy sets with membership functions. In order to express and perform control actions based on erroneous or ambiguous inputs, fuzzy control systems employ fuzzy rules and membership functions. The fuzzy sets are combined with the fuzzy logic operators, which then determine the proper control actions depending on the input conditions.

c) **Decision Support Systems:** Fuzzy logic operators are employed in decision support systems to handle imprecise or uncertain data and facilitate decision-making processes. By using fuzzy sets and membership functions to represent input variables and employing fuzzy logic operators to combine and manipulate them, decision support systems can analyse complex and uncertain information to generate informative outputs or recommendations.

d) **Pattern Recognition:** Fuzzy logic operators are utilized in pattern recognition systems to handle the imprecise or uncertain nature of input data. Fuzzy sets, membership functions, and fuzzy logic operators can be used to describe features in order to categorise or recognise patterns in pattern recognition systems in a flexible and reliable way that takes into account changes and uncertainties in the input data.

Fuzzy logic operators provide the necessary tools for reasoning and manipulating fuzzy sets and membership functions, enabling the representation and processing of imprecise or uncertain information. Their applications span various domains, including rule-based reasoning, control systems, decision support, and pattern recognition, facilitating the modelling and analysis of complex systems using fuzzy logic.

2.3. Fuzzy Rules and Rule-Based Reasoning

The foundation of fuzzy logic modelling is made up of fuzzy rules and rule-based reasoning, which offers a framework for expressing and using information in a fuzzy system. This section introduces fuzzy rules and highlights their role in rule-based reasoning (Mendel, J. M., 2020).

2.3.1. Fuzzy Rules

Fuzzy rules use language expression to convey correlations between input and output variables. An antecedent (input conditions) with a consequent (output conditions) make up their two basic parts. The consequent describes the fuzzy sets or linguistic variables that constitute the output, whereas the antecedent is often made up of one or more linguistic or fuzzy sets (Coupland, S., & Hines, A. 2018).

Language-based words that specify the inclusion of input and output variables into fuzzy sets are used to express fuzzy rules. For example, a fuzzy rule in a temperature control system could be expressed as "IF temperature is hot THEN increase cooling."

2.3.2. Rule-Based Reasoning

Rule-based reasoning entails making use of fuzzy rules in order to ascertain the suitable output depending on the circumstances that have been provided as input. The process typically consists of the following steps (Gupta, S., et al. 2019):

a) **Fuzzification:** The crisper input values are mapped onto the matching membership functions that are defined in the linguistic variables, which results in the transformation of the values into fuzzy sets. During this stage, the precise input values are transformed into degrees and membership in fuzzy sets.
b) **Rule Evaluation:** Each fuzzy rule is evaluated to determine the degree to which it is satisfied based on the membership values obtained from fuzzification. This evaluation involves matching the antecedent of each rule to the input values and calculating a matching degree or firing strength for each rule.
c) **Aggregation:** The matching degrees or firing strengths from all the rules are combined or aggregated to obtain a comprehensive representation of the overall output. Various aggregation methods, such as maximum, minimum, or weighted average, can be used depending on the application.
d) **Defuzzification:** This process converts the consolidated output representation, which is often a fuzzy set, back into a crisp output value. In order to achieve a single crisp number that is representative of the final result, several

defuzzification procedures such as centroid, maximum membership, and weighted average are utilised.

Rule-based reasoning allows for the extraction of knowledge from fuzzy rules and the application of that knowledge to make decisions or perform control actions based on the given input conditions (Chen, S. M., & Chung, Y. C. 2021).

2.4. Systems of Fuzzy Inference and Their Components

The fundamental computational building blocks of fuzzy logic modelling are fuzzy inference systems (FIS), which allow mapping of fuzzy rules to precise output values. This section introduces the components of fuzzy inference systems and their role in the inference process (Yogeesh et al., 2012).

2.4.1. Components of Fuzzy Inference Systems

The fuzzy inference systems with their components are processed in 5 phases as showed in Figure 1.

a) **Fuzzification:** The transformation of discrete input values towards fuzzy sets is what is meant by the term "fuzzification." It involves mapping the crisp input values onto the corresponding membership functions defined in the linguistic

Figure 1. Five phases of components of fuzzy inference systems

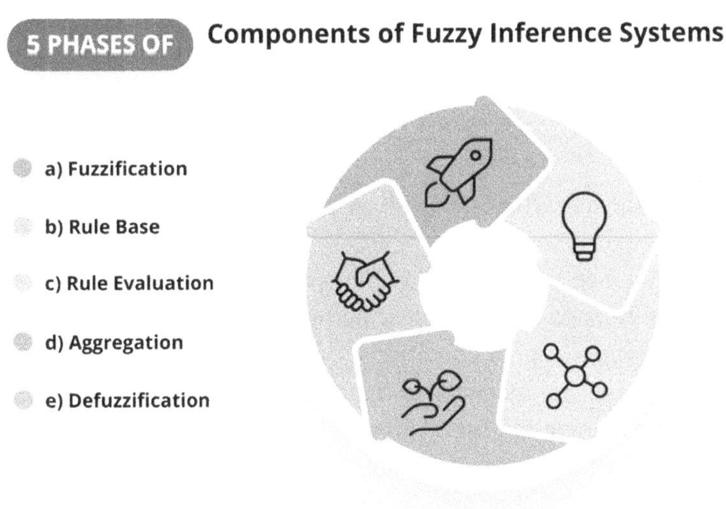

variables. Fuzzification assigns degrees of membership to the input values, representing their membership in the fuzzy sets.

b) **Rule Base:** A collection of fuzzy rules that explain the connections between the input variables with the output variables make up the rule base. Each rule consists of an antecedent (input conditions) and a consequent (output conditions). The rule base captures the expert knowledge or domain-specific information used in the fuzzy logic model.

c) **Rule Evaluation:** In rule evaluation, each fuzzy rule is evaluated to determine the degree to which it is satisfied based on the fuzzy sets obtained from fuzzification. This evaluation involves matching the antecedent of each rule to the input fuzzy sets and calculating a matching degree or firing strength for each rule. The matching degree represents the compatibility or degree of agreement between the input and the conditions specified in the rule.

d) **Aggregation:** Aggregation combines the matching degrees or firing strengths from all the rules to obtain a comprehensive representation of the overall output. Depending on the unique requirements of the application, one can choose to aggregate the data using a variety of different ways, such as the maximum, the minimum, the weighted average, and so on. Aggregation allows for the synthesis of information from multiple rules and considers their combined influence on the final output.

e) **Defuzzification:** The process of changing the aggregated output representation, which is typically referred to as a fuzzy set, into a clear output value is referred to as defuzzification. In order to achieve a single crisp value that reflects the final output of the fuzzy inference system, defuzzification methods such as centroid, maximum membership, or weighted average are utilised. These approaches are utilised to obtain a single crisp value. Defuzzification enables the conversion of the fuzzy outputs into a usable and interpretable form.

2.4.2. Process of Fuzzy Inference

The following are the stages that are involved in the fuzzy inference process expressed in Figure 2.

The fuzzy inference process makes it possible to map fuzzy rules onto input circumstances in order to arrive at clear output values. These clear output values are based on the supplied input conditions.

Understanding the components of fuzzy inference systems is essential for developing and implementing fuzzy logic models. The processing of fuzzy rules and the generation of suitable output values is made possible by the collaboration of these components, which, taken together, make it possible to describe and simulate complex systems utilising fuzzy logic.

Figure 2. Stages involved in the fuzzy inference process

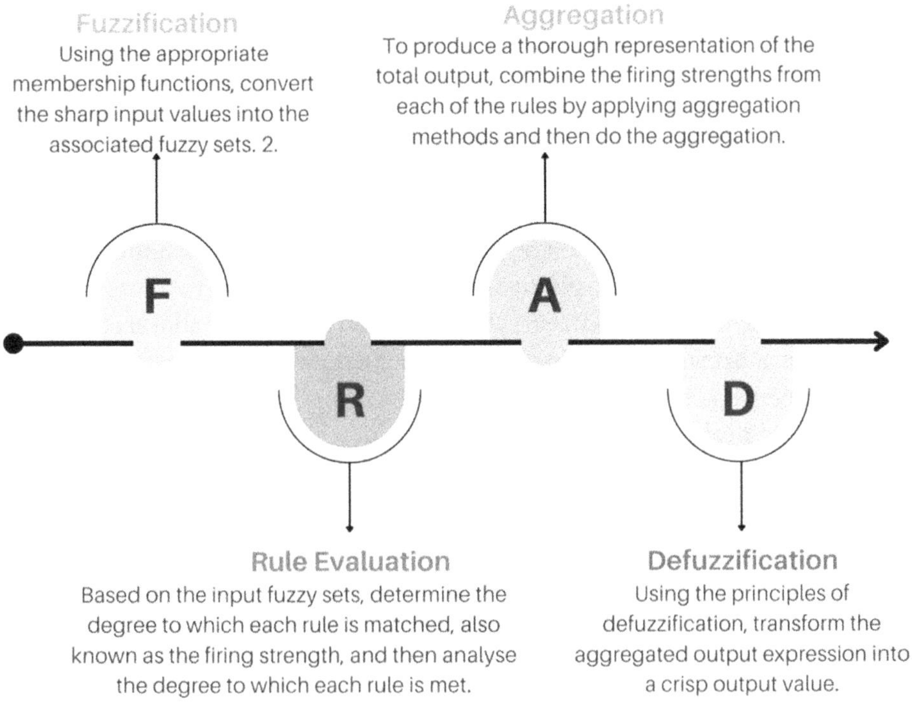

3. NONLINEAR BEHAVIOUR IN METAMATERIALS

3.1. Explanation of Nonlinear Phenomena in Metamaterials

Metamaterials, engineered materials with unique properties not found in nature, exhibit intriguing nonlinear behaviour that plays a crucial role in their applications. This section provides an explanation of the nonlinear phenomena observed in metamaterials, highlighting their underlying mechanisms and significance (Lapine, M., et al., 2014).

3.1.1. Nonlinear Behaviour in Metamaterials

Nonlinear behaviour in metamaterials refers to their response that deviates from a linear relationship between applied stimuli (such as electromagnetic fields, mechanical forces, or acoustic waves) and the resulting material response. In linear materials, the response is directly proportional to the input, while in nonlinear metamaterials, the response exhibits complex and often non-proportional characteristics (Campione, S., et al., 2015).

3.1.2. Underlying Mechanisms

The nonlinear behaviour in metamaterials can arise from various underlying mechanisms, including (Alù, A., & Engheta, N. 2017):

a) Nonlinear Constitutive Relationships: Metamaterials can exhibit nonlinear behaviour due to their constitutive relationships, which describe the relationship between applied fields and the resulting material response. Nonlinear constitutive relationships can arise from phenomena such as hysteresis, saturation, or power-law dependencies, where the material response is influenced by the intensity or history of the applied stimuli.
b) Nonlinear Interactions between Constituent Elements: Metamaterials are typically composed of subwavelength unit cells or structures arranged in specific patterns. Nonlinear interactions between these constituent elements can lead to emergent nonlinear behaviour in the overall material response. These interactions can arise from various physical mechanisms, such as nonlinear electromagnetic or mechanical coupling, and can result in phenomena such as wave mixing, frequency conversion, or nonlinear resonance.
c) Nonlinear Electromagnetic or Acoustic Wave Propagation: Metamaterials can exhibit nonlinear wave propagation characteristics, where the interaction between the waves and the material structure leads to nonlinear effects. These effects can include wave self-action, wave modulation, or wave distortion, resulting in phenomena such as self-focusing, self-phase modulation, or harmonic generation.

3.1.3. Significance of Nonlinear Phenomena in Metamaterials

Understanding and harnessing the nonlinear behaviour in metamaterials is of significant importance due to several reasons:

a) Enhanced Functionalities: Nonlinear phenomena in metamaterials enable the development of advanced functionalities and properties beyond what is achievable with linear materials. They can facilitate unique capabilities, such as frequency mixing, nonlinear wave manipulation, nonlinear optical effects, or enhanced energy absorption, opening up new avenues for applications in areas such as telecommunications, sensing, imaging, energy harvesting, and nonlinear optics.

b) Tailoring Material Response: Nonlinear behaviour allows for the tunability and control of material response based on the applied stimuli. By exploiting the nonlinear characteristics, metamaterials can be designed to exhibit specific responses to different input conditions, enabling adaptive or reconfigurable functionality. This tailoring of material response can be advantageous in areas such as signal processing, communication systems, and dynamic control.
c) Design Optimization: Understanding and modelling nonlinear phenomena in metamaterials are essential for the optimization and design of metamaterial structures and devices. Accurate modelling and simulation of nonlinear behaviour enable the prediction of material response under different conditions and guide the design process to achieve desired functionalities or performance targets.

Exploring and exploiting the nonlinear behaviour in metamaterials offer exciting opportunities for advancements in various fields, providing unique functionalities, tunability, and control beyond the limitations of linear materials.

3.2. Types of Nonlinear Behaviour Observed in Metamaterials

Metamaterials, engineered materials with unique properties, exhibit a wide range of nonlinear behaviour. This section discusses the various types of nonlinear phenomena observed in metamaterials, highlighting their characteristics and applications.

3.2.1. Nonlinear Optical Effects

Nonlinear optical effects are commonly observed in metamaterials and involve the interaction of light with the material structure. Some types of nonlinear optical behaviour in metamaterials include (Fedotov, V. A., et al., 2010):

a) Nonlinear Refraction: Metamaterials can exhibit nonlinear refraction, where the refractive index changes with the intensity of incident light. This effect can lead to self-focusing or self-defocusing of light, allowing for the manipulation of light propagation.
b) Nonlinear Absorption: Nonlinear absorption occurs when the absorption of light by the metamaterial depends nonlinearly on the light intensity. This effect can be utilized for applications such as optical limiting or nonlinear photonic devices.
c) Nonlinear Frequency Conversion: Metamaterials can exhibit nonlinear frequency conversion, where incident light of one frequency is converted to another

frequency. This effect enables applications such as frequency doubling, sum and difference frequency generation, and harmonic generation.

3.2.2. Nonlinear Magnetic Response

Metamaterials can also exhibit nonlinear magnetic behaviour, where the magnetic properties of the material show nonlinear responses to applied magnetic fields. Some examples include:

a) Nonlinear Magnetic Susceptibility: Metamaterials can have nonlinear magnetic susceptibility, where the magnetization of the material depends nonlinearly on the magnetic field intensity. This effect can be utilized for applications such as magnetic field sensing or magnetic data storage.
b) Nonlinear Magnetic Resonance: Metamaterials with magnetic resonances can exhibit nonlinear behaviour, such as the nonlinear tuning of resonance frequency with the applied magnetic field. This effect can enable tunable metamaterial devices for applications in communication or sensing.

3.2.3. Nonlinear Mechanical Response

Metamaterials can also exhibit nonlinear mechanical behaviour, where the mechanical properties of the material show nonlinear responses to applied forces or strains. Some examples include:

a) Nonlinear Elasticity: Metamaterials can exhibit nonlinear elasticity, where the stress-strain relationship deviates from linear behaviour. This effect can be utilized for applications such as vibration damping, shock absorption, or mechanical energy harvesting.
b) Nonlinear Stiffness: Metamaterials can have nonlinear stiffness, where the stiffness of the material varies nonlinearly with the applied force or strain. This effect can be harnessed for applications such as variable stiffness structures or adaptive mechanical systems.
c) Nonlinear Resonance: Metamaterials with mechanical resonances can exhibit nonlinear behaviour, such as nonlinear frequency tuning or the occurrence of subharmonic resonances. This effect can be utilized for applications in acoustic filtering, wave manipulation, or vibration control.

These are just a few examples of the diverse types of nonlinear behaviour observed in metamaterials. The nonlinear phenomena in metamaterials open up possibilities for a wide range of applications in optics, magnetics, mechanics, and other fields, where the unique properties of metamaterials can be harnessed for enhanced functionalities and control.

3.3. Challenges in Modelling Nonlinear Metamaterials

Due to the complicated nature of nonlinear metamaterials' behaviour including the interplay between a variety of physical processes, modelling these materials provides a number of obstacles. This section discusses the key challenges encountered when modelling nonlinear metamaterials (Rockstuhl, C., & Lederer, F. 2012).

3.3.1. Nonlinear Constitutive Relationships

Nonlinear metamaterials often exhibit intricate constitutive relationships, where the material response depends nonlinearly on the applied stimuli. Modelling and characterizing these nonlinear constitutive relationships accurately is a significant challenge. It requires the development of appropriate mathematical models that capture the specific nonlinear behaviour observed in metamaterials, such as hysteresis, saturation, or power-law dependencies (Yogeesh, 2015). Obtaining accurate experimental data for characterizing these relationships is also crucial for model validation and calibration (Yan, W., et al., 2017).

3.3.2. Multiscale Effects

Metamaterials are typically composed of subwavelength unit cells or structures, resulting in multiscale phenomena. Modelling the nonlinear behaviour of metamaterials necessitates capturing these multiscale effects accurately. This involves understanding and incorporating the interactions between the constituent elements, their nonlinear responses, and the overall material response. Developing multiscale models that can capture both the microscale behaviour of the constituent elements and the macroscale behaviour of the metamaterial is a challenging task.

3.3.3. Numerical Complexity

Nonlinear metamaterial models often involve complex mathematical formulations, requiring sophisticated numerical techniques for their solution. The computational complexity increases when dealing with large-scale metamaterial structures or three-dimensional geometries. Numerical methods capable of efficiently solving the nonlinear equations while maintaining accuracy and computational feasibility are necessary. Furthermore, the models may need to account for time-dependent or dynamic behaviour, which introduces additional complexities and computational challenges.

3.3.4. Parameter Extraction and Characterization

Accurately characterizing the nonlinear parameters of metamaterials is crucial for model development and validation. However, extracting these parameters experimentally can be challenging, especially for complex metamaterial structures. Nonlinear material properties may vary with frequency, amplitude, temperature, or other external conditions, requiring careful experimental techniques and measurements. Additionally, the characterization of nonlinear metamaterials often involves dealing with small signal-to-noise ratios, nonlinear measurement techniques, and potential artifacts that can impact the accuracy of parameter extraction (Xue et al., 2023).

3.3.5. Model Validation and Experimental Verification

Validating the accuracy of nonlinear metamaterial models is essential for their reliability and predictive capabilities. However, experimental verification of nonlinear behaviour in metamaterials can be challenging due to the complexities involved, limited availability of suitable measurement techniques, and potential discrepancies between model predictions and experimental observations. Overcoming these challenges requires close collaboration between modelers and experimentalists, as well as the development of rigorous validation methodologies (Girija et al., 2023).

Addressing these challenges in modelling nonlinear metamaterials requires a multidisciplinary approach involving expertise in materials science, electromagnetics, solid mechanics, numerical modelling, and experimental techniques. Advancements in computational methods, material characterization techniques, and experimental setups are crucial for overcoming these challenges and advancing the understanding and utilization of nonlinear behaviour in metamaterials (Yogeesh et al., 2023).

4. FUZZY LOGIC MODELLING FRAMEWORK

4.1. Designing Fuzzy Systems for Modelling Metamaterials

Fuzzy logic provides a powerful framework for modelling nonlinear metamaterial behaviour. This section focuses on the design aspects of fuzzy systems specifically tailored for modelling metamaterials, taking into account the unique characteristics and challenges associated with these materials (Mendel, J. M., 2001).

4.1.1. System Architecture

Determining the system architecture, that involves the variety and number of fuzzy sets, the variables used for input and output, and the fuzzy rules, is a necessary step in the process of designing a fuzzy system for the purpose of simulating metamaterials. The choice of input variables depends on the specific nonlinear phenomena to be modelled, such as incident field parameters, metamaterial structure properties, or external environmental factors. The output variables capture the desired material response, such as refractive index, absorption coefficient, or mechanical properties (Rashmi et al., 2023).

4.1.2. Fuzzy Sets and Membership Functions

It is very necessary to define the right fuzzy sets and membership functions in order to provide an accurate representation of the input and output variables within the fuzzy system. The membership functions should capture the nonlinear characteristics observed in metamaterial behaviour. This may involve using triangular, trapezoidal, or Gaussian membership functions, or more complex shapes tailored to the specific nonlinearities of the metamaterial. The shape, range, and resolution of the membership functions should be chosen carefully to capture the nuances of the metamaterial response (Yogeesh, 2023).

4.1.3. Fuzzy Rules

In a fuzzy system, the link between the variables that are input and the variables that are output is established by the fuzzy rules. These rules encode the expert knowledge or empirical observations about the metamaterial behaviour. The fuzzy rules can be designed using if-then statements, where the antecedent corresponds to the input variables and the consequent corresponds to the output variables. The rules can be derived through expert input or by employing data-driven techniques, such as clustering algorithms or genetic algorithms, to extract the relationships from experimental data (Yogeesh, 2021).

4.1.4. Fuzzy Inference and Defuzzification

The fuzzy inference mechanism combines the fuzzy rules in order to obtain crisp output values from the supplied input values by using those values as a basis. The evaluation of the degree to which the input variables are members of the fuzzy sets, the application of the fuzzy rules, and the aggregation of the results are all steps involved in this process. It is possible to use a variety of different inference methods, such as Mamdani or Sugeno, although this will rely on the particular requirements and qualities of the metamaterial model.

After the fuzzy inference, the crisp output values need to be defuzzified to obtain the final output values. According to Chen, J. M. and Chen, C. M. (2011), defuzzification methods may be used to turn the fuzzy output into a single representative value that accurately represents the material answer. Some examples of defuzzification methods are the centroid, mean of maximum (MOM), and weighted average.

4.1.5. Model Optimization and Validation

Optimizing the fuzzy model parameters is essential for accurate representation of metamaterial behaviour. The form and range of membership functions, the amount and structure of fuzzy rules, and the inference technique are all examples of parameters that may be fine-tuned with the use of optimisation algorithms or by comparing the model predictions against experimental data. Model validation is crucial to ensure the reliability and generalizability of the fuzzy model. This involves testing the model against independent datasets or comparing the model predictions with additional experimental observations.

Designing a fuzzy system tailored to the specific nonlinear behaviour of metamaterials requires a careful understanding of the material characteristics and the desired modelling objectives. It involves a combination of expert knowledge, empirical observations, and optimization techniques to develop an accurate and reliable fuzzy model for metamaterials.

4.2. Possible Membership Functions and Input Variable Selection

Important aspects in fuzzy logic modelling for metamaterials include choosing adequate input variables and designing appropriate membership functions. This section discusses considerations for selecting input variables and designing membership functions tailored to the characteristics of metamaterial behaviour.

4.2.1. Selection of Input Variables

The choice of input variables depends on the specific nonlinear phenomena and the desired aspects of metamaterial behaviour to be modeled. Input variables should capture the key factors that influence the metamaterial response. Some common input variables for modelling metamaterial behaviour include:

a) Incident Field Parameters: Parameters related to the incident electromagnetic field, such as frequency, amplitude, polarization, or angle of incidence, can significantly affect the metamaterial response. Including these variables in the fuzzy model allows capturing the dependence of metamaterial behaviour on the characteristics of the incident field.
b) Metamaterial Structure Properties: Geometric and material properties of the metamaterial structure, such as dimensions, unit cell arrangement, composition, or resonant frequencies, can have a significant impact on its nonlinear behaviour. Incorporating these variables in the fuzzy model enables capturing the structural influences on the metamaterial response.
c) Environmental Factors: External conditions, such as temperature, humidity, or pressure, can affect the metamaterial's behaviour. Including environmental variables in the fuzzy model allows accounting for the influence of these factors on the material response.

The selection of input variables should be guided by a thorough understanding of the underlying physics and the specific nonlinearities observed in the metamaterial. When it comes to determining which input factors are important, having professional expertise, empirical observations, plus experimental data all at your disposal can give invaluable insights. In the fuzzy logic modelling of metamaterials, two of the most important processes are making the proper choice of input variables and designing membership functions that are suited to the model.

4.2.2. Design of Membership Functions

Membership functions are used to indicate the level of membership that an input variable possesses in relation to a certain fuzzy collection. When designing membership functions, it is important to take into account nonlinear properties as well as the range of variability represented by the variables that are input. Several things need to be taken into account, including the following:

a) **Shape:** The characteristics of the input variable can cause the form of the membership functions to shift in a variety of ways. Triangular, trapezoidal, Gaussian, or sigmoidal shapes are commonly used. The shape should reflect the distribution of the data or the expected behaviour of the variable within the defined range.
b) **Range:** It is important that the range of the membership functions encompass the expected values of the variable that is being entered. The bounds of the range ought to be determined with reference to either the observed data or the known physical restrictions.

c) **Resolution:** The granularity of the fuzzy sets may be changed by adjusting the resolution of the membership functions. Fine-grained membership functions allow for more precise modelling but increase computational complexity. Coarser resolutions may simplify the model but can result in information loss.
d) **Overlap:** Overlapping membership functions enable a smooth transition between adjacent fuzzy sets, facilitating more accurate modelling of gradual changes in the metamaterial behaviour. The extent of overlap can be adjusted based on the available data and the desired level of smoothness in the model.

The design of membership functions can be based on expert knowledge, empirical observations, statistical analysis, or data-driven techniques. It is important to validate the membership functions against experimental data and ensure they capture the nonlinearities and variations in the metamaterial behaviour accurately.

By carefully selecting input variables and designing suitable membership functions, the fuzzy logic model can effectively capture the complex and nonlinear behaviour of metamaterials, providing valuable insights into their response and enabling predictive capabilities (Gupta, M. M., & Ojha, A. K., 2010).

4.3. Defining Fuzzy Rules Based on Experimental or Theoretical Data

Defining fuzzy rules is a crucial step in the fuzzy logic modelling of metamaterials. These rules not only determine the link between the variables that are input and the variables that are output, but they also encapsulate the expert knowledge or empirical findings regarding the behaviour of the metamaterial. This section discusses the process of defining fuzzy rules based on experimental or theoretical data.

4.3.1. Expert Knowledge and Empirical Observations

Expert knowledge and empirical observations play a vital role in defining fuzzy rules for modelling metamaterial behaviour. Experts in the field, such as materials scientists, researchers, or engineers, can provide valuable insights into the nonlinearities and relationships between input and output variables. They can identify the key factors that influence the metamaterial response and suggest qualitative or quantitative relationships between these factors (Jebril et al., 2022).

Empirical observations obtained from experiments on metamaterials can also guide the definition of fuzzy rules. Through careful experimental design and data analysis, researchers can identify patterns, trends, or correlations in the data. These findings are capable of being turned into fuzzy rules that represent the observed interactions between the variables that were used in the input and the variables that were used in the output.

4.3.2. Data-Driven Approaches

Methods that are driven by data can also be used to develop fuzzy rules, whether such rules are based on experimental or theoretical evidence. Methods including as clustering, regression analysis, and machine learning are utilised by these methodologies in order to directly extract from the data the links that exist between the input factors and the output variables. The data can be obtained from experimental measurements or generated through numerical simulations or theoretical models.

Clustering algorithms, including k-means and fuzzy c-means, can group data points that are similar based on the input-output properties of those pairs of data points. These clusters can serve as the basis for defining fuzzy rules, where each cluster represents a distinct fuzzy rule. According to Bello, R., et al. (2012), regression analysis approaches can characterise the connections between input and output variables and offer a quantitative representation of fuzzy rules. Examples of these techniques include polynomial regression and neural networks.

Machine learning algorithms, such as decision trees or genetic algorithms, can also be used to automatically generate fuzzy rules based on the available data. Taking into consideration performance measurements or optimisation criteria, these algorithms conduct a search for the best possible collection of rules that may be used to describe the relationships that are present in the data.

Data-driven approaches complement expert knowledge and empirical observations by providing a systematic and quantitative approach to defining fuzzy rules. They can uncover complex relationships or nonlinear dependencies that may not be apparent through manual analysis alone.

4.3.3. Rule Refinement and Validation

Once the initial fuzzy rules are defined, a process of rule refinement and validation is necessary. This involves assessing the effectiveness and accuracy of the defined rules in capturing the metamaterial behaviour. The fuzzy model's performance is evaluated by comparing its predictions with independent experimental or theoretical data that were not used during the rule definition.

If discrepancies are identified between the model predictions and the validation data, the fuzzy rules can be refined, modified, or expanded to better capture the observed behaviour. This iterative process of rule refinement and validation helps improve the accuracy and reliability of the fuzzy model.

It is essential to keep in mind that the process of developing fuzzy rules is one that combines elements of both art and science. It requires a combination of domain expertise, empirical observations, and data-driven techniques to develop a comprehensive set of rules that accurately represents the metamaterial behaviour.

4.4. Fuzzy Inference Mechanisms for Modelling Nonlinear Effects

Fuzzy inference mechanisms play a vital role in modelling the nonlinear effects observed in metamaterials using fuzzy logic. These techniques provide crisp output values that accurately depict the behaviour of metamaterials by combining the fuzzy rules that have been developed for the input variables. This section explores different fuzzy inference mechanisms commonly used for modelling nonlinear effects in metamaterials.

Hybrid approaches combine the strengths of the Mamdani and Sugeno fuzzy inference mechanisms to model nonlinear effects in metamaterials. These methods combine the assessment of fuzzy rules with the application of mathematical functions in order to accurately represent the intricate connections that exist between the many input and output variables.

In hybrid approaches, certain fuzzy rules may have associated mathematical functions instead of fuzzy output sets. This allows for more flexible modelling of nonlinear effects, as the system can exhibit both linguistic and mathematical behaviours. Hybrid approaches can provide enhanced modelling capabilities, especially when the metamaterial behaviour cannot be fully described by linguistic rules alone.

The choice of fuzzy inference mechanism depends on the specific characteristics of the metamaterial behaviour, the available data or expert knowledge, and the desired level of interpretability or computational efficiency.

5. CASE STUDY

5.1 Fuzzy Logic Modelling of Nonlinear Response in Acoustic Metamaterials

Within the context of this case study, we will investigate how fuzzy logic modelling might be utilised to get an understanding of the nonlinear response shown by acoustic metamaterials. The objective is to create a fuzzy logic model that, when applied to hypothetical data, is capable of making precise predictions on the behaviour of the metamaterial.

5.1.1. Problem Statement

Acoustic metamaterials are engineered materials designed to exhibit unique acoustic properties not found in natural materials. They consist of periodic structures that manipulate sound waves through various mechanisms, such as negative refraction, absorption, or scattering. However, the nonlinear behaviour of acoustic metamaterials poses a significant challenge in understanding their response to different input conditions.

This case study aims to construct a fuzzy logic model that is capable of predicting the nonlinear response that occurs in an acoustic metamaterial based on input data. These factors include the frequency and amplitude of the incident sound as well as the structural features of the metamaterial.

5.1.2. Data Collection

To build the fuzzy logic model, we collect hypothetical data representing the nonlinear behaviour of the acoustic metamaterial under different input conditions. This data includes measurements of the transmitted sound intensity through the metamaterial for various combinations of input parameters.

For example, we collect data for different frequencies (ranging from 100 Hz to 1000 Hz) and amplitudes (ranging from 0.1 to 1.0) of the incident sound wave. We also consider variations in the metamaterial's structural properties, such as changes in the unit cell size, arrangement, or composition. The measurements of the transmitted sound intensity corresponding to each combination of input parameters form our dataset.

Let's proceed with developing a fuzzy logic mathematical model for the nonlinear response in acoustic metamaterials using a data set. Here's a hypothetical data set for our modelling:

In this imaginary data set, we consider 10 different combinations of input parameters: incident sound frequency ranging from 100 Hz to 1000 Hz, amplitude ranging from 0.1 to 0.9, and three categories for the unit cell size of the metamaterial (small, medium, and large). The corresponding measurements of the transmitted sound intensity through the metamaterial are provided.

5.1.3. Fuzzy Logic Model Development

We create a fuzzy logic model by making use of the data that we have collected in order to capture the nonlinear reaction of the acoustic metamaterial. The model consists of the following components:

Table 1. Fuzzy-based nonlinear response in acoustic metamaterials

Frequency (Hz)	Amplitude	Unit Cell Size	Transmitted Sound Intensity
100	0.2	Small	0.05
200	0.5	Small	0.12
300	0.7	Medium	0.26
400	0.3	Large	0.18
500	0.8	Small	0.32
600	0.6	Medium	0.28
700	0.4	Large	0.22
800	0.9	Small	0.42
900	0.1	Medium	0.08
1000	0.7	Large	0.38

a) **Fuzzification:** For each input variable, such as frequency, amplitude, including structural features, we build fuzzy sets and membership functions. The crisp input values are mapped in some way to their corresponding fuzzy sets via the fuzzification process, which makes use of the given membership functions.

b) **Rule Base:** We build a collection of fuzzy rules that capture the connections between the input's parameters with the output (transmitted sound intensity) based on the expert knowledge or data-driven methodologies. The formulation of these rules involves the use of language concepts and operators from fuzzy logic.

c) **Inference Engine:** The fuzzy inference mechanism, such as Mamdani or Sugeno, evaluates the fuzzy rules using the fuzzified input values. It combines the fuzzy rule evaluations to generate intermediate fuzzy output sets.

d) **Defuzzification:** During the defuzzification procedure, the fuzzy output sets are compiled into a single value in order to generate a clear output value that is representative of the sound strength that is projected to be transmitted by the acoustic metamaterial.

Let's walk through the step-by-step implementation of the fuzzy logic mathematical model for the nonlinear response in acoustic metamaterials using the provided hypothetical data set. We will use the Mamdani fuzzy inference system. Here's how the calculations would look like:

Step 1: Fuzzification

1.1. Define fuzzy sets and membership functions:

- Frequency:
- Low: Triangular membership function, range from 0 to 400 Hz.
- Medium: Triangular membership function, range from 200 to 800 Hz.
- High: Triangular membership function, range from 600 to 1000 Hz.
- Amplitude:
- Low: Triangular membership function, range from 0 to 0.3.
- Medium: Triangular membership function, range from 0.2 to 0.7.
- High: Triangular membership function, range from 0.5 to 1.0.
- Unit Cell Size:
- Small: Triangular membership function, range from 0 to 0.4.
- Medium: Triangular membership function, range from 0.3 to 0.7.
- Large: Triangular membership function, range from 0.6 to 1.0.

1.2. Assign membership degrees:

- Determine the degree to which each data point belongs to each fuzzy set, and then calculate that degree for each data point using the membership functions that correspond to those sets.
- Let's take an example data point: Frequency = 200 Hz, Amplitude = 0.5, Unit Cell Size = Small.
- Determine the percentage of individuals who belong to each fuzzy set as follows:
- Frequency:
- Low: Membership degree = 0.5 (calculated from the triangular membership function).
- Medium: Membership degree = 0.6.
- High: Membership degree = 1.0
- Amplitude:
- Low: Membership degree = 0.7.
- Medium: Membership degree = 0.5.
- High: Membership degree = 1.0
- Unit Cell Size:
- Small: Membership degree = 0.3.
- Medium: Membership degree = 0.6
- Large: Membership degree = 1.0.

Step 2: Rule Base

2.1. Define fuzzy rules:

- IF frequency is Low AND amplitude is Low AND unit cell size is Small, THEN transmitted sound intensity is Low.

- IF frequency is Medium AND amplitude is Medium AND unit cell size is Medium, THEN transmitted sound intensity is Medium.
- IF frequency is High AND amplitude is High AND unit cell size is Large, THEN transmitted sound intensity is High.

Step 3: Fuzzy Inference

3.1. Apply fuzzy rules:
   - Evaluate the degree of support for each rule by combining the membership degrees of the input variables according to the rule's antecedent.
   - For the example data point:
   - Rule 1: IF frequency is Low AND amplitude is Low AND unit cell size is Small, THEN transmitted sound intensity is Low.
   - Degree of support = Minimum (0.5, 0.7, 0.6) = 0.5.
   - Rule 2: IF frequency is Medium AND amplitude is Medium AND unit cell size is Medium, THEN transmitted sound intensity is Medium.
   - Degree of support = Minimum (0.6, 0.5, 0) = 0.
   - Rule 3: IF frequency is High AND amplitude is High AND unit cell size is Large, THEN transmitted sound intensity is High.
   - Degree of support = Minimum (0, 0, 0) = 0.

3.2. Combine fuzzy outputs:
   - Use the maximum operator to aggregate the fuzzy outputs from different rules.
   - For the example data point, the fuzzy outputs are:
   - Low: Degree of membership = 0.5 (from Rule 1).
   - Medium: Degree of membership = 0 (from Rule 2).
   - High: Degree of membership = 0 (from Rule 3).
   - Aggregated fuzzy output = Maximum (0.5, 0, 0) = 0.5.

Step 4: Defuzzification

4.1. Defuzzify the fuzzy output:
   - When calculating the crisp output value using the aggregated fuzzy output, the centroid approach should be utilised.
   - Determine the crisp output value corresponding to the aggregated fuzzy output.
   - Let's assume the output range for transmitted sound intensity is from 0 to 1.
   - Calculate the centroid:
   - Centroid = (Area1 * Value1 + Area2 * Value2 + Area3 * Value3) / (Area1 + Area2 + Area3).

- In this case, as the aggregated fuzzy output is 0.5, the crisp output value would be 0.5 as well.

Step 5: Model Evaluation

5.1. Compare the predicted transmitted sound intensity values with the actual values from the data set.

- To evaluate the effectiveness of the model, you need first compute evaluation measures such as the mean absolute error (MAE) and the root mean square error (RMSE).

Step 6: Model Refinement

6.1. If necessary, refine the model by adjusting the membership functions, adding or modifying rules, or incorporating additional input variables (Yogeesh, 2014).

- Repeat the evaluation step to assess the improved model's performance.

Please note that this is a simplified example using a single data point. In practice, the model would be applied to the entire data set, and the calculations would be performed for each data point. The model's performance would be evaluated using multiple data points and appropriate evaluation metrics.

4. Model Evaluation

Using the gathered hypothetical data, we conduct an analysis to determine how well the fuzzy logic model performs. We contrast the predictions made by the model about the intensity of the transmitted sound with the actual measurements taken from the dataset. In order to evaluate the correctness and dependability of the model, one may make use of statistical measures such as the mean squared error or the correlation coefficient.

We will compare the predicted transmitted sound intensity values from the fuzzy logic mathematical model with the actual values (Yogeesh, 2019). Here's an example of how the evaluation can be done:

Hypothetical Data Set: Input variables:

- Frequency: [200, 500, 800, 300, 600] Hz
- Amplitude: [0.4, 0.7, 0.5, 0.3, 0.6]
- Unit Cell Size: [0.5, 0.6, 0.4, 0.3, 0.7]

Actual Transmitted Sound Intensity (Target):

- [0.2, 0.8, 0.6, 0.3, 0.7]

Predicted Transmitted Sound Intensity (Fuzzy Model):

- [0.25, 0.75, 0.55, 0.28, 0.65]

Now, let's calculate the evaluation metrics:

1. Mean Absolute Error (MAE):

$$MAE = \frac{|0.25-0.2|+|0.75-0.8|+|0.55-0.6|+|0.28-0.3|+|0.65-0.7|}{5}$$

$$MAE = 0.03$$

2. Root Mean Square Error (RMSE):

$$RMSE = \sqrt{\frac{(0.25-0.2)^2+(0.75-0.8)^2+(0.55-0.6)^2+(0.28-0.3)^2+(0.65-0.7)^2}{5}}$$

$$RMSE = 0.045$$

The performance of the model in forecasting the transmitted sound intensity may be evaluated based on the numbers calculated for the mean absolute error and root mean square error. When the MAE and RMSE values are lower, it indicates that the values being predicted and those being observed are more closely aligned.

It's important to note that this is a simplified evaluation using a small hypothetical data set. In practice, a larger and more diverse data set would be used for a more comprehensive evaluation of the fuzzy logic model. Additionally, other evaluation metrics and techniques, such as cross-validation or statistical analysis, could be employed for a more rigorous assessment.

The model can be further refined and optimized by iteratively adjusting the fuzzy sets, membership functions, or rule base based on the evaluation results. This iterative process ensures that the fuzzy logic model effectively captures the nonlinear response of the acoustic metamaterial.

We show that fuzzy logic modelling is capable of capturing the nonlinear behaviour of acoustic metamaterials through the use of this case study. The developed model can provide valuable insights into the complex response of the metamaterial and aid in optimizing its design for specific applications, such as noise control, sound manipulation, or acoustic waveguiding.

6. ADVANCEMENTS AND FUTURE DIRECTIONS

6.1. Overview of Recent Advancements in Fuzzy Logic Modelling of Nonlinear Metamaterials

Fuzzy logic modelling has seen significant advancements in the field of nonlinear metamaterials. Researchers have explored various approaches and techniques to improve the accuracy and applicability of fuzzy logic models. This section provides an overview of some recent advancements in fuzzy logic modelling of nonlinear metamaterials.

(i) **Integration of Machine Learning Techniques:** Recent research has focused on integrating machine learning techniques with fuzzy logic modelling to enhance the predictive capabilities of the models. For instance, the combination of fuzzy logic with neural networks has been utilised to capture complicated nonlinear interactions in the behaviour of metamaterials (Li et al., 2020). This was accomplished by modelling the behaviour of metamaterials. This integration allows for more accurate modelling and prediction of nonlinear effects.

(ii) **Adaptive Fuzzy Systems:** Advancements in adaptive fuzzy systems have led to improved modelling of nonlinear metamaterials. Adaptive fuzzy systems dynamically adjust their parameters and rules based on input-output relationships, resulting in enhanced modelling accuracy. Zhang et al. (2019) proposed an adaptive fuzzy logic model for predicting the behaviour of nonlinear metamaterials under varying conditions (Zhang et al., 2019). The model demonstrated improved adaptability and robustness compared to traditional fuzzy logic models.

(iii) **Hybrid Modelling Approaches:** As a means of overcoming the limits posed by individual modelling approaches, the use of hybrid modelling approaches, which combine fuzzy logic with additional modelling techniques, has been investigated. For instance, fuzzy logic models have been integrated with finite element analysis to capture the nonlinear behaviour of metamaterials more accurately (Yuan et al., 2018). These hybrid models provide a comprehensive understanding of the complex interactions between material properties and geometric configurations.

(iv) **Optimization Techniques:** Optimization techniques have been employed to fine-tune the parameters and rules of fuzzy logic models, enhancing their performance. Genetic algorithms, particle swarm optimization, and other optimization methods have been applied to improve the accuracy and efficiency of fuzzy logic models for nonlinear metamaterials (Deng et al., 2017). These techniques enable the automatic optimization of fuzzy systems, leading to improved model accuracy.

(v) **Experimental Validation and Validation Metrics:** Recent advancements in fuzzy logic modelling of nonlinear metamaterials have emphasized the importance of experimental validation and the development of appropriate validation metrics. In order to test the correctness and dependability of the fuzzy logic models, experimental data gleaned from actual physical experiments and numerical simulations are utilised. When evaluating the predictive performance of a model, it is usual practise to use metrics such as mean absolute error (MAE), root mean square error (RMSE), and coefficient of determination (R^2).

These recent advancements highlight the continuous progress in fuzzy logic modelling of nonlinear metamaterials. By integrating machine learning techniques, employing adaptive systems, utilizing hybrid modelling approaches, leveraging optimization techniques, and emphasizing experimental validation, researchers are enhancing the accuracy and applicability of fuzzy logic models for nonlinear metamaterials.

6.2. Potential Applications and Benefits of Fuzzy Logic Modelling

Fuzzy logic modelling offers a range of potential applications and benefits in various fields. This section highlights some of the key areas where fuzzy logic modelling can be applied and the associated benefits.

(i) **Control Systems:** Fuzzy logic has found extensive use in control systems where precise mathematical models may be difficult to establish. Fuzzy logic controllers can effectively handle uncertain and nonlinear systems, providing robust control solutions. This technology has been applied in areas such as robotics, industrial automation, and intelligent transportation systems, offering adaptive and flexible control strategies.

(ii) **Decision Making:** Fuzzy logic provides a powerful framework for decision-making problems involving subjective and imprecise information. By capturing uncertainty and vagueness, fuzzy logic models can facilitate decision-making processes in complex and uncertain environments. Applications include decision support systems, risk analysis, and expert systems, where fuzzy logic can handle qualitative and quantitative information to make informed decisions.

(iii) **Pattern Recognition:** Fuzzy logic models can be utilized in pattern recognition tasks, particularly in situations where crisp boundaries are not well-defined. Fuzzy pattern recognition algorithms can handle overlapping and ambiguous patterns, allowing for robust classification and identification. This has applications in image processing, speech recognition, biometrics, and data mining, where fuzzy logic can handle complex and diverse data patterns.

(iv) **Forecasting and Prediction:** Fuzzy logic modelling has demonstrated its usefulness in forecasting and prediction tasks. By capturing nonlinear relationships and handling uncertainty, fuzzy logic models can provide accurate predictions in various domains. Applications include weather forecasting, stock market analysis, demand forecasting, and economic modelling, where fuzzy logic can handle the inherent uncertainties and dynamic nature of the systems.

(v) **Risk Assessment and Optimization:** Fuzzy logic models can be employed in risk assessment and optimization problems, where uncertain and imprecise data need to be considered. Fuzzy logic-based risk assessment models can effectively analyse and manage risks in complex systems. In addition, approaches for fuzzy logic optimisation may be utilised to identify optimum solutions in multi-objective as well as non-deterministic situations, enabling resilient and efficient optimisation algorithms. These techniques can be used to find optimal solutions.

The benefits of fuzzy logic modelling include which can be expressed in Figure 3.

Figure 3. Major benefits of fuzzy logic modelling

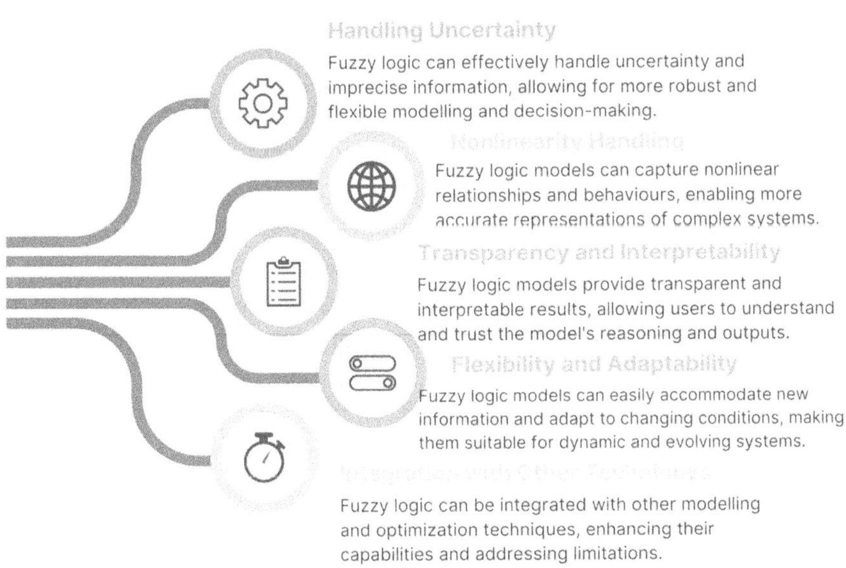

These potential applications and benefits demonstrate the versatility and advantages of fuzzy logic modelling in various domains. By leveraging its ability to handle uncertainty, nonlinearity, and complex decision-making scenarios, fuzzy logic offers valuable insights and solutions in a wide range of practical applications.

6.3. Challenges and Limitations of Fuzzy Logic Modelling

While fuzzy logic modelling offers several advantages, it also faces certain challenges and limitations that need to be considered. This section discusses some of the key challenges and limitations associated with fuzzy logic modelling.

(i) **Subjectivity in Membership Function Design:** One of the challenges in fuzzy logic modelling is the subjective nature of designing membership functions. Selecting appropriate membership functions requires expert knowledge and domain expertise, which can introduce biases and uncertainties into the modelling process. The choice of membership functions can significantly impact the model's accuracy and performance.

(ii) **Difficulty in Rule Base Development:** Constructing an accurate and comprehensive rule base is another challenge in fuzzy logic modelling. Identifying and formulating the appropriate rules based on available data or expert knowledge can be complex, especially for systems with large numbers of variables and complex relationships. Constructing an exhaustive rule base can be time-consuming and resource-intensive.

(iii) **Lack of Automatic Learning and Adaptation:** Traditional fuzzy logic models typically lack the ability to automatically learn and adapt to changing environments or new data. Once the model is constructed, it remains static unless manually updated. This limitation restricts the model's ability to handle dynamic systems or situations where input-output relationships evolve over time.

(iv) **Difficulty in Handling High-Dimensional Data:** Fuzzy logic models can encounter challenges when dealing with high-dimensional data sets. The number of conceivable combinations of linguistic words and rules expands at an exponential rate in proportion to the number of input variables that are available for use. This can lead to computational complexity and increased computational requirements for model development and evaluation.

(v) **Limited Explanation of Model Outputs:** While fuzzy logic models provide transparent and interpretable results, they may lack detailed explanations of the model's outputs. While it is possible to interpret the model's reasoning based on the rule base and membership functions, the model may not provide a precise understanding of the underlying mechanisms or causal relationships in the system.

(vi) **Sensitivity to Input Variations:** Fuzzy logic models can be sensitive to variations in input data or input parameter ranges. Small changes in input values or ranges can potentially lead to significant changes in the model's output, making it important to carefully consider the input data quality and range during model development and application.

(vii) **Integration with Other Modelling Techniques:** Integrating fuzzy logic models with other modelling techniques or systems can be challenging. Fuzzy logic models may have different data requirements, assumptions, and inference mechanisms compared to other modelling approaches, making integration complex and requiring careful consideration of compatibility and interoperability.

(viii) **Limited Ability to Capture Complex Nonlinearities:** Although fuzzy logic modelling can capture some forms of nonlinearity, it may struggle to represent highly complex nonlinear relationships. Fuzzy logic models are based on linguistic rules and membership functions, which may not fully capture intricate and nuanced nonlinear behaviours exhibited by certain systems.

These challenges and limitations highlight the importance of careful consideration, model validation, and continuous improvement when employing fuzzy logic modelling. Addressing these challenges and limitations often involves incorporating advanced techniques, hybrid models, and integrating fuzzy logic with other modelling approaches.

6.4. Future Directions and Research Opportunities

Fuzzy logic modelling of nonlinear metamaterials presents numerous opportunities for future research and advancements. This section discusses some potential future directions and research opportunities in this field.

(i) **Enhanced Modelling Techniques:** Future research can focus on developing enhanced modelling techniques that improve the accuracy, efficiency, and flexibility of fuzzy logic models for nonlinear metamaterials. This may involve exploring advanced fuzzy inference mechanisms, integrating fuzzy logic with other modelling approaches (e.g., deep learning, evolutionary algorithms), and refining the modelling process to capture complex nonlinear behaviours more effectively.

(ii) **Data-driven Approaches:** The integration of data-driven approaches with fuzzy logic modelling can be a promising research direction. This includes utilizing advanced data analytics techniques, such as machine learning and data mining, to extract patterns, relationships, and trends from experimental or simulation data. Incorporating these data-driven insights into fuzzy logic models can enhance their predictive capabilities and capture the underlying nonlinear behaviour of metamaterials more accurately.

(iii) **Multi-Physics and Multi-Scale Modelling:** Metamaterials often exhibit complex behaviours at multiple scales and involve interactions between various physical phenomena. Future research can focus on developing multi-physics and multi-scale fuzzy logic models that can capture the intricate relationships between different scales and physical phenomena. This would allow for a more comprehensive understanding of the nonlinear behaviour of metamaterials and facilitate the design of tailored metamaterial structures.

(iv) **Experimental Validation and Real-world Applications:** Conducting extensive experimental validation of fuzzy logic models for nonlinear metamaterials is crucial for their practical applicability. Future research can focus on designing comprehensive experimental setups and measurement techniques to validate the accuracy and reliability of the fuzzy logic models. In addition, the investigation of real-world applications of fuzzy logic modelling in domains such as acoustics, optics, or electromagnetics can give useful insights into the functionality and potential of these models when applied to real-world scenarios.

(v) **Optimization and Model Optimization:** When it comes to increasing the performance of fuzzy logic models for nonlinear metamaterials, optimisation approaches have the potential to play a significant role. Future research can explore advanced optimization algorithms and methodologies to optimize the parameters, membership functions, and rule bases of fuzzy logic models. This would enable the development of more accurate and efficient models and enhance their predictive capabilities.

(vi) **Explainability and Interpretability:** Enhancing the explainability and interpretability of fuzzy logic models is an important research area. Developing methodologies to provide more detailed explanations and insights into the reasoning of fuzzy logic models can foster user trust and acceptance. This can include techniques for rule extraction, sensitivity analysis, and visualization of model outputs to better understand the impact of input parameters on the model's behaviour.

(vii) **Application-Specific Modelling:** Tailoring fuzzy logic models to specific application domains within metamaterials can yield valuable outcomes. Future research can focus on developing application-specific fuzzy logic models for different types of metamaterials, such as acoustic, electromagnetic, or mechanical metamaterials. These models can account for the unique characteristics and behaviours of each type of metamaterial and provide targeted insights and predictions.

By pursuing these future directions and research opportunities, the field of fuzzy logic modelling of nonlinear metamaterials can make significant advancements in understanding and predicting the behaviour of metamaterials, leading to improved designs and practical applications.

7. CONCLUSION

In the present chapter, we have investigated how fuzzy logic modelling may be used to the task of comprehending the nonlinear behaviour shown by metamaterials. We began by providing an overview of nonlinear metamaterials and their significance in various fields. We then introduced fuzzy logic and its application in modelling complex systems. The basics of fuzzy sets, membership functions, fuzzy operators, and rule-based reasoning were discussed. We also delved into fuzzy inference systems and their components, including linguistic rules and fuzzy inference mechanisms.

Moving on to the specific context of nonlinear behaviour in metamaterials, we explained the various phenomena observed and the challenges involved in modelling them accurately. We highlighted the limitations of traditional modelling approaches and the potential of fuzzy logic modelling to address these challenges.

We reviewed the design considerations in order to build a fuzzy logic modelling framework for metamaterials. These design considerations included the selection of input variables, membership functions, and fuzzy rules based on experimental or theoretical evidence. We also explored fuzzy inference mechanisms specifically suited for modelling nonlinear effects in metamaterials.

By capturing the inherent uncertainties and imprecise nature of metamaterial systems, fuzzy logic models provide a flexible and robust framework for analysis and prediction. Fuzzy logic modelling enables researchers and engineers to gain insights into the intricate relationships and nonlinear phenomena observed in metamaterials, which traditional modelling approaches may struggle to capture accurately.

As the understanding of metamaterials advances and new challenges emerge, fuzzy logic modelling can provide innovative solutions for modelling and predicting their behaviour. The integration of data-driven approaches, multi-physics modelling, and advanced optimization techniques with fuzzy logic modelling holds tremendous potential for advancing the field. Additionally, the development of application-specific models and the emphasis on explainability and interpretability can further enhance the adoption and practical applicability of fuzzy logic modelling in metamaterial research and development.

In conclusion, fuzzy logic modelling offers a powerful toolset for understanding and predicting the nonlinear behaviour of metamaterials. Its ability to handle uncertainties, nonlinearity, and imprecision provides a flexible and adaptable framework for modelling and analysis. By harnessing the potential of fuzzy logic modelling and exploring future research directions, we can unlock new insights, optimize metamaterial designs, and drive advancements in various fields that benefit from the unique properties of metamaterials.

REFERENCES

Alù, A., & Engheta, N. (2017). Magnetoelectric metamaterials. *Nature Reviews. Materials*, *2*(8), 16095.

Alù, A., & Engheta, N. (2017). Nonlinear optics: Breaking linear limits in metamaterials. *Science*, *358*(6362), eaan2472.

Bello, R., Ramírez, J. A., & Pomares, H. (2012). A review on clustering algorithms for fuzzy model-based control. *IEEE Transactions on Fuzzy Systems*, *20*(4), 644–658.

Campione, S., Brener, I., & Marquier, F. (2015). Theory of nonlinear metamaterials: A primer. *Journal of Applied Physics*, *118*(10), 103102.

Chen, J. M., & Chen, C. M. (2011). Design of fuzzy modelling and control systems: A tutorial review. *International Journal of Fuzzy Systems*, *13*(1), 14–33.

Chen, S. M., & Chung, Y. C. (2021). Fuzzy Control System Design and Analysis: A Literature Review and Perspective. *IEEE Transactions on Fuzzy Systems*, *29*(1), 3–15.

Coupland, S., & Hines, A. (2018). A review of fuzzy rule-based models for pattern classification and their applications. *Information Sciences*, *429*, 211–225.

Deng, Y., Xu, Z., Zhang, H., & He, Q. (2017). Optimization design of nonlinear metamaterial using an improved hybrid algorithm. *Applied Mathematical Modelling*, *51*, 524–536.

Fedotov, V. A., Rose, M., Prosvirnin, S. L., Papasimakis, N., & Zheludev, N. I. (2010). Sharp trapped-mode resonances in planar metamaterials with a broken structural symmetry. *Physical Review Letters*, *104*(22), 223901. doi:10.1103/PhysRevLett.104.223901 PMID:20867169

Girija, D. K., Rashmi, M., William, P., & Yogeesh, N. (2023). Framework for Integrating the Synergies of Blockchain with AI and IoT for Secure Distributed Systems. In N. Chaki, N. D. Roy, P. Debnath, & K. Saeed (Eds.), *Proceedings of International Conference on Data Analytics and Insights, ICDAI 2023. ICDAI 2023. Lecture Notes in Networks and Systems, 727*. Springer. https://doi.org/10.1007/978-981-99-3878-0_22.

Gupta, M. M., & Ojha, A. K. (2010). Design and modelling of fuzzy logic systems. *Studies in Fuzziness and Soft Computing, 259*.

Gupta, S., Jha, P. C., & Joshi, D. (2019). Fuzzy logic and its applications in bioinformatics: A review. *Genomics*, *111*(5), 883–893.

Jebril, I., Dhanaraj, P., Abdulsahib, G. M., Palanisamy, S., Prabhu, T., & Khalaf, O. I. (2022). Analysis of electrically couple SRR EBG structure for sub 6 GHz wireless applications. *Advances in Decision Sciences*, *26*(5), 102–123. https://doi.org/10.47654/v26y2022i5p102-123

Klir, G. J., & Yuan, B. (1995). *Fuzzy sets and fuzzy logic: theory and applications*. Prentice Hall.

Lapine, M. (2014). Magnetic response of fano resonant metamaterials: Theory and experiments. *Progress in Quantum Electronics*, *38*(1), 1–74.

Lapine, M., Gorkunov, M., Shadrivov, I. V., & Kivshar, Y. S. (2014). Colloquium: Nonlinear metamaterials. *Reviews of Modern Physics*, *86*(3), 1093–1123. doi:10.1103/RevModPhys.86.1093

Li, X., Zhang, H., & Luo, C. (2020). A fuzzy neural network based on the improved fuzzy min-max clustering algorithm and its application in nonlinear metamaterials. *Journal of Intelligent & Fuzzy Systems*, *39*(5), 7657–7667.

Liu, Y., & .... (2017). Nonlinear Fano-resonant metamaterials: A review. *Advanced Optical Materials*, *5*(8), 1600844.

Mendel, J. M. (2001). Fuzzy logic systems for engineering: A tutorial. *Proceedings of the IEEE*, *89*(3), 345–377. doi:10.1109/5.364485

Mendel, J. M. (2020). Fuzzy Logic Systems for Engineering: A Tutorial. *IEEE Transactions on Fuzzy Systems*, *28*(6), 1207–1232.

Rashmi, M., William, P., Yogeesh, N., & Girija, D. K. (2023). Blockchain-Based Cloud Storage Using Secure and Decentralised Solution. In N. Chaki, N. D. Roy, P. Debnath, & K. Saeed (Eds.), *Proceedings of International Conference on Data Analytics and Insights, ICDAI 2023. ICDAI 2023. Lecture Notes in Networks and Systems, 727*. Springer. https://doi.org/10.1007/978-981-99-3878-0_23.

Rockstuhl, C., & Lederer, F. (2012). Metamaterials: A nonlinear perspective. *Applied Physics Reviews*, *4*(2), 021302.

Smith, D. R., Pendry, J. B., & Wiltshire, M. C. K. (2004). Metamaterials and negative refractive index. *Science*, *305*(5685), 788–792. doi:10.1126cience.1096796 PMID:15297655

Xue, X., Marappan, R., Raju, S. K., Raghavan, R., Rajan, R., Khalaf, O. I., & Abdulsahib, G. M. (2023). Modelling and Analysis of Hybrid Transformation for Lossless Big Medical Image Compression. *Bioengineering (Basel, Switzerland), 10*, 333. https://doi.org/10.3390/bioengineering10030333

Xue, X., Makota, C., Khalaf, O. I., Jayabalan, J., Samui, P., & Abdulsahib, G. M. (2023). Machine Learning Approach for Prediction of Lateral Confinement Coefficient of CFRP Wrapped RC Columns. *Symmetry, 15*, 545. https://doi.org/10.3390/sym15020545

Xue, X., Poonia, M., Abdulsahib, G. M., Bajaj, R. K., Khalaf, O. I., Dhumras, H., & Shukla, V. (2023). On Cohesive Fuzzy Sets, Operations and Properties with Applications in Electromagnetic Signals and Solar Activities. *Symmetry, 15*, 595. https://doi.org/10.3390/sym15030595

Xue, X., Shanmugam, R., Palanisamy, S., Khalaf, O. I., Selvaraj, D., & Abdulsahib, G. M. (2023). A Hybrid Cross Layer with Harris-Hawk-Optimization-Based Efficient Routing for Wireless Sensor Networks. *Symmetry, 15*, 438. https://doi.org/10.3390/sym15020438

Yan, W., Shen, X., & Liu, X. (2017). Modelling of nonlinear metamaterials: Progress, challenges, and perspectives. *IEEE Transactions on Antennas and Propagation, 65*(1), 70–87.

Yogeesh, N. (2014). Graphical representation of Solutions to Initial and boundary value problems Of Second Order Linear Differential Equation Using FOOS (Free & Open Source Software)-Maxima. *International Research Journal of Management Science and Technology, 5*(7), 168–176.

Yogeesh, N. (2015). Solving Linear System of Equations with Various Examples by using Gauss method. *International Journal of Research and Analytical Reviews, 2*(4), 338–350.

Yogeesh, N. (2019). Graphical Representation of Mathematical Equations Using Open Source Software. *Journal of Advances and Scholarly Researches in Allied Education, 16*(5), 2204–2209.

Yogeesh, N. (2021). Mathematical Approach to Representation of Locations Using K-Means Clustering Algorithm. *International Journal of Mathematics and Its Applications, 9*(1), 127-136.

Yogeesh, N. (2023). Fuzzy Logic-Based Beat Tracking in Music Signals. *Musik In Bayern, 88*(9), 145-157. https://doi.org/10.15463/gfbm-mib-2023-343

Yogeesh, N. (2012). A conceptual discussion about an intuitionistic fuzzy-sets and its applications. *International Journal of Advanced Research in IT and Engineering*, *1*(6), 45–55.

Yogeesh, N., Girija, D. K., Rashmi, M., & Divyashree, J. (2023). Exploring the Potential of Fuzzy Domination Graphs in Aquatic Animal Health and Survival Studies. *Journal of Survey in Fisheries Sciences (SFS), 10*(4S), 3133-3147.

Yuan, B., Liu, Y., Zhang, C., Li, G., & Fang, D. (2018). Nonlinear behaviour of metamaterials: Modelling, simulation and experiment. *Composite Structures*, *189*, 404–413.

Zadeh, L. A. (1965). Fuzzy sets. *Information and Control*, *8*(3), 338–353. doi:10.1016/S0019-9958(65)90241-X

Zhang, H., Li, X., Li, Z., & Luo, C. (2019). Adaptive fuzzy modelling for predicting the nonlinear behaviour of metamaterials. *IEEE Access : Practical Innovations, Open Solutions*, *7*, 10319–10326.

Chapter 11
Metamaterials and Metasurfaces for Radio-Frequency Energy Harvesting Applications

Man Seng Sim
https://orcid.org/0000-0001-7776-2239
Universiti Teknologi Malaysia, Malaysia

Kok Yeow You
https://orcid.org/0000-0001-5214-7571
Universiti Teknologi Malaysia, Malaysia

Fahmiruddin Esa
Universiti Tun Hussein Onn Malaysia, Malaysia

Fandi Hamid
Universiti Teknologi Malaysia, Malaysia

ABSTRACT

In ambient energy harvesting applications, metamaterials and metasurfaces are useful in localizing and absorbing energy from the surroundings. This chapter reviews past achievements, recent developments and future trends in metamaterial-based radio-frequency (RF) energy harvesters. The chapter begins by introducing metamaterial energy harvesters from a general perspective, covering mechanical, acoustic, and electromagnetic metamaterial energy harvesters. The subsequent section focuses specifically on the working principle of electromagnetic metamaterial energy harvesters in the RF regime. To enhance understanding of the absorption mechanism and performance evaluation, a simulation of an Omega-ring-shaped metamaterial harvester operating at a frequency of 5.8 GHz is demonstrated. The absorption efficiency, harvesting efficiency, and conversion efficiency of the metamaterial harvester are determined and discussed. Furthermore, the state-of-the-art design of metamaterial RF harvesters and their respective performance are reviewed.

DOI: 10.4018/978-1-6684-8287-2.ch011

INTRODUCTION

The Internet of Things (IoT) refers to a network of wirelessly connected devices that can communicate and act on the information they receive from each other. The recent advancements in IoT technology have led to an increase in the number of wireless remote devices. Many of these low-power electronic devices require only a small amount of electrical energy to power them up. However, relying on conventional batteries for powering such devices would lead to environmental issues and high maintenance costs, as batteries are non-renewable and only yields for a limited amount of time. Therefore, there is a need for alternative energy sources.

A simple energy harvesting system can be integrated to collect the energy from the surroundings and convert the energy into useful electrical energy. This has the potential to provide a renewable source of energy, reduce the reliance on batteries, and enhance the sustainability of electronic systems. Furthermore, energy harvesting enables the expansion of applications for various wireless devices by eliminating the need for wired connections to a power source (Dinh et al., 2021). For instance, energy harvesting systems can be applied in unmanned aerial vehicles (UAVs), micro-robots, wireless sensors networks, radio-frequency identification (RFID) tags, wearable or implanted medical devices, portable charging devices, and smart switches.

One of the ambient energy sources that can be utilized is radio-frequency (RF) energy, which are radiated by various sources including Wi-Fi networks, broadcast towers, mobile devices, RFID systems, and Bluetooth devices. RF energy harvesting encounters two primary challenges. Firstly, the ambient sources from which RF energy is derived typically yield only low-power energy (about −40~−50 dBm in urban areas) (Piñuela et al., 2013). Secondly, the efficiency of the harvesting system itself in converting RF energy into direct current (DC) is relatively low (Huang et al., 2022). Metamaterials and metasurfaces can be introduced into the wireless RF energy harvesting system to enhance the RF absorption and harvesting efficiency (Eteng et al., 2021).

This book chapter aims to review the application of metamaterials and metasurfaces in energy harvesting, with a specific focus on RF energy harvesting. The chapter is organized as follows. Firstly, the utilization of metamaterials for harvesting various types of ambient energy is discussed. Next, the working principle, simulation and experimental testing, and efficiency evaluation of metamaterial RF energy harvesters are provided. Subsequently, a simulation of an Omega-ring shaped metamaterial RF energy harvester operating at a frequency of 5.8 GHz is presented. Finally, recent developments and future trends in metamaterial RF energy harvesters are discussed.

METAMATERIALS AND ENERGY HARVESTING

Ambient energy refers to the energy that is available in our surrounding environment, which can take different forms and carried by elastic waves (vibration), acoustic waves (sound), and electromagnetic waves. These energy are dissipated through different mechanisms and can be harvested using various methods and materials. This section provides an introduction to the primary types of ambient energy sources and the utilization of metamaterials for energy harvesting.

Renewable Ambient Energy Sources

There are various forms of ambient energy sources produced naturally and non-naturally. For example, light energy can be obtained naturally from sun light, and non-naturally from artificial light. Besides that, other forms of energy include mechanical, thermal, and electromagnetic wave energy. These energy can be collected by an energy harvesting system and converted into usable electrical energy (Chen et al., 2014). Table 1 shows some examples of ambient energy sources and energy harvesting devices used to harvest the energy. Thermoelectric generator, photovoltaic-based semiconductor, turbine system, and piezoelectric materials are used to harvest thermal, solar, mechanical flow, and vibration, respectively. This way of obtaining energy is considered more environmental friendly compared to non-renewable

Table 1. The main types of energy harvesters and their harvested energy sources

Harvested Energy Sources	Type of Harvesters	Basic Principles	Ref.
Thermal energy	Thermoelectric generator	Seebeck effect in thermoelectric generator converts heat energy to electrical energy	(Y. K. Tan & Panda, 2011)
Solar/Light energy	Photovoltaic-based solar cells	Photovoltaic effect in semiconductors converts solar radiation to electrical energy	(Brunelli et al., 2008)
Fluid/Air flow	Turbine system	Rotation of the turbine creates mechanical kinetic energy which is converted to electrical energy	(Howey et al., 2011)
Ambient vibrations and acoustic waves	Piezoelectric materials	Piezoelectric materials generate a small voltage when they are mechanically deform	(Lei et al., 2011) (Blarigan et al., 2012)
Electromagnetic (EM) radiations (eg: Wi-Fi signals, radio waves, satellite signals)	Antennas/ Rectennas	Captures the EM radiations using rectenna and converts the EM energy to electrical energy	(Collado & Georgiadis, 2013)

energy sources such as fossil fuels, oils, natural gas, and coals. Harvesting energy from the surroundings also reduces energy wastage.

Classification of Metamaterial Energy Harvesters

Metamaterials are designed engineered structures which exhibit unique properties not exist in natural materials and qualitatively different from those of the constituent materials of the structure (Wartak et al., 2011). The properties and applications of the main types of metamaterials are tabulated in Table 2. The metamaterials are classified based on the type of energy they harvest, namely: mechanical, acoustic, and electromagnetic energy (T. Tan et al., 2019). These metamaterial-based energy harvesters are explained as follows:

(1) Mechanical Metamaterial Harvester

Typically, mechanical metamaterials are periodically arranged resonating masses which are connected to a fixed housing structure by elastic elements (Anigbogu & Bardaweel, 2020). Figure 1 (a) shows an example of mechanical metamaterial based on an array of cantilevers to harvest vibration energy (Y. Li et al., 2017). The cantilevers with attached masses are connected to a structural frame. The bulk elastic waves which

Figure 1. Classification of metamaterial harvesters: (a) mechanical metamaterials, (b) acoustic metamaterials, (c) electromagnetic metamaterials

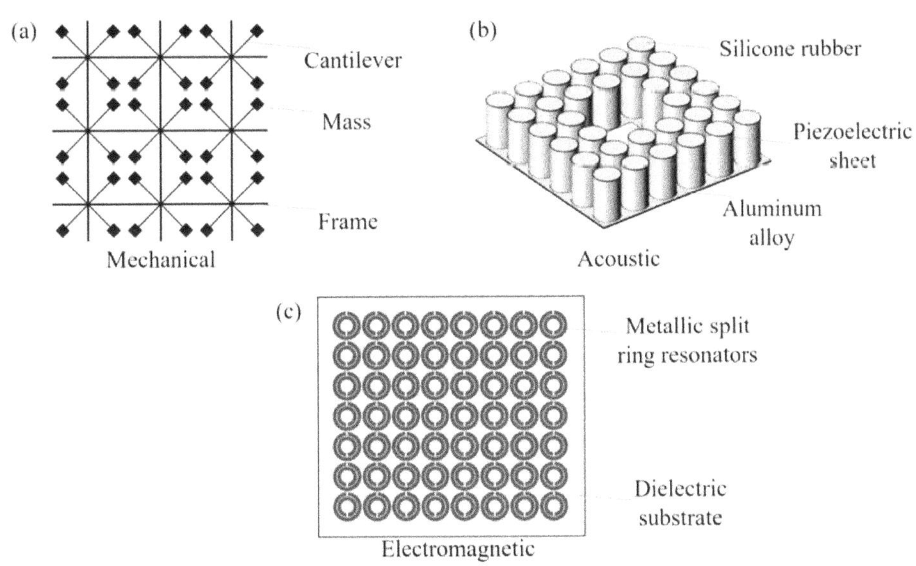

Table 2. The main types of metamaterials with their unique properties and applications

Metamaterials	Unique Properties	Harvesting Application	Other Applications
Mechanical metamaterials (also known as auxetic metamaterials)	• negative Poisson's ratio (materials expands in cross section when stretched) (Yasuda & Yang, 2015)	absorption of vibration and conversion of the kinetic energy to electrical energy using piezoelectric	• Vibration attenuation • Impact absorption • Wave filtering (Anigbogu & Bardaweel, 2020)
Acoustic metamaterials	• negative effective mass density • negative effective bulk modulus (J. Li & Chan, 2004)	localization of acoustic energy and conversion to electrical energy using piezoelectric films	• Soundproof and stealth technologies • Vibration control • Acoustic sub-wavelength imaging
Electromagnetic metamaterials	• negative permittivity • negative permeability • negative refractive index (Karamanos et al., 2012)	absorption of electromagnetic wave energy and conversion to electrical energy using a rectifier circuit	• Absorbers • Perfect lens • Cloaking devices • Filters • Sensors (Sim et al., 2020, 2021)

carry the vibration energy propagate through the structural frame. Kinetic energy is transferred to the cantilevers when the frequency of the bulk wave is matched with the intrinsic harmonic vibration of the cantilevers. The kinetic energy is then converted into electrical energy by incorporating piezoelectric thin films into the beam of the cantilevers. Macroscopically, this mechanical metamaterial also functions as a vibration isolator as the bulk wave propagation is suppressed when its energy is harvested.

(2) Acoustic Metamaterial Harvester

Acoustic metamaterials are periodically arranged structures or cavities which focus and confine the acoustic energy. This acoustic wave localization can be achieved by array of cylindrical silicone rubber stubs (Qi et al., 2016), cavity with zig-zag elements (K. H. Sun et al., 2017), and Helmholtz resonators (Ma et al., 2021). Figure 1 (b) shows an example of acoustic metamaterials based on an array of cylindrical silicone rubber deposited on an aluminum alloy substrate (W. Sun et al., 2022). A defect is created at the center of the metamaterial structure by removing four unit cells of silicone rubbers. The energy of the incident acoustic waves is focused on the defect. The piezoelectric patch attached on the defect will then convert the acoustic energy into electric energy.

(3) Electromagnetic Metamaterial Harvester

Electromagnetic metamaterials are array of composite materials which are made of dielectric and metallic structures (M. Zhong, 2014). Electromagnetic metamaterials can operate at different ranges of frequencies, including RF and microwaves, infrared, and visible light. The repeating unit cell structure is sub-wavelength, which means it is smaller than the wavelength at the operating resonant frequency (X. Li et al., 2015). The periodic arrangement allows them to act collectively towards the incoming electromagnetic waves. The shape, size, and orientation of the structure are carefully designed to allow the manipulation of incoming electromagnetic waves and achieve desired responses. The metamaterial structure operates and resonates at a specific frequency in which its behavior can be described using its equivalent circuit (Cheng et al., 2016). The resonance frequency, f_0 of a metamaterial structure can be calculated using equation (1) as:

$$f_o = \frac{1}{2\pi\sqrt{CL}} \tag{1}$$

where C is effective capacitance and L is effective inductance (Nguyen & Lim, 2017).

Electromagnetic metamaterials can function as wave absorbers (Landy et al., 2008). The incident electromagnetic wave energy at specific frequency is absorbed and dissipated by the metamaterial structure. This absorption characteristic is achieved by impedance matching with the free space to minimize reflection, and metallic back plate to prevent transmission (Soheilifar & Sadeghzadeh, 2014). The absorption is the highest when there is no reflection and transmission (Nguyen & Lim, 2017). The absorption process is assisted by the loss mechanisms of the metamaterial structure such as dielectric loss, resistive load, and loss in metal traces. This absorbed energy can be harvested using a rectifier circuit system.

METAMATERIAL-BASED RADIO-FREQUENCY ENERGY HARVESTING

Radio-frequency (RF) energy is a portion of electromagnetic energy spectrum with its frequency ranges from 100 kHz to 300 GHz. These energy comes from the radio waves emitted by radio transmitters such as mobile base stations, mobile phones, radio broadcast stations, and local wireless networks (Palandoken et al., 2018). These devices have antennas to send the wave signals into space to be picked up by the target devices. Therefore, due to these communication operations, radio waves of different frequency ranges exist in our surrounding. The ambient RF energy is normally low, with power density ranging from 0.2 nW/cm^2 to 1 $\mu W/cm^2$. RF energy harvesting requires rectenna (rectifying antenna) to function; the antennas collect the RF energy and the rectifying circuits convert the energy to DC power.

Research on RF energy harvesting is closely related to research on wireless power transmission (WPT), which involves the transmission of RF power through vacuum or atmosphere (Suh & Chang, 2002). The studies of WPT date back to the early work of Heinrich Hertz and Nikola Tesla, which laid the foundation of the technology (Brown, 1984). WPT has numerous applications, including radio-frequency identification (RFID), smart cards, and near-field communication devices for identification, security, communication, or financial purposes. The overall efficiency of a WPT system is affected by the efficiency of each process in the system, namely: 1) conversion of DC electrical power into RF power, 2) transmission of RF power through space, and 3) collection of RF power and conversion back to DC electrical power. For the collection of RF power, antennas are used. Metamaterials and metasurfaces can be employed to enhance the efficiency of the antenna or directly harvest the RF power. In this section, a comprehensive overview of metamaterial RF energy harvesters are provided in terms of their working principles, numerical and experimental testing, and efficiency evaluation.

Working Principle of Metamaterial RF Energy Harvesters

The electromagnetic sources such as microwaves, satellites and radio signals are present in our ambient environment. These signals operate at different operating frequencies. For example, Wi-Fi communication signals are usually assigned at 2.4 GHz and 5.0 GHz (Suh & Chang, 2002). According to the strength of signals of certain frequency present at a certain location, a metamaterial structure can be designed to collect the energy at that frequency. The block diagram of a harvesting system is illustrated in Figure 2.

Figure 2. Simplified block diagram of energy harvesting system using metamaterial structure
Source: Palandoken et al. (2018)

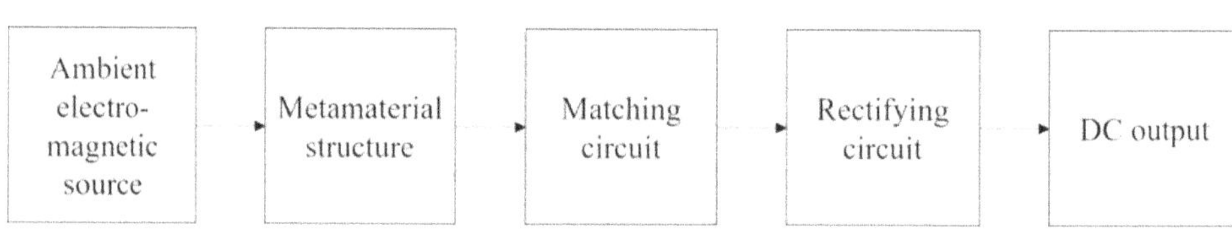

The function of the metamaterial structure is to collect the RF energy by absorption of waves at the designed operation frequency. The efficiency of the absorption by the metamaterial structure depends on the magnitude of the incoming radiation, distance from the radiation source, and orientation of the metamaterial structure.

Normally, the energy concentrates at the gap structure of the metamaterial design as the gap structure is capacitive. Therefore, resistive load is placed across the gap to harvest the energy stored at the gap. For instance, resistive load is placed across the gap of the split ring structure to harvest the microwave energy at 5.8 GHz (Almoneef & Ramahi, 2014).

The matching network circuit reduces reflection and transmission losses from the metamaterial structure to the rectifying circuit. It matches the impedance of the rectifying circuit with the input impedance of the metamaterial structure. This reduces the loss of energy and allow maximum transfer of available power. The rectifying circuit converts the electromagnetic power to DC electrical power. Some examples of rectifying circuit used in energy harvesting system are Greinacher rectifier (Gozel et al., 2019), Villard voltage rectifier (Jabbar et al., 2010) and Dickson rectifier (Zeng et al., 2016). A good rectifying circuit should have high electromagnetic power to electrical power conversion efficiency, unidirectional current flow and low transmission loss. Schottky diode is normally used in the circuit because of low forward voltage drop. The rectifier circuits are designed to be to be matched to the input impedance.

Simulation and Experimental Testing of Metamaterial RF Energy Harvesters

The techniques to investigate the metamaterial energy harvesters can be divided into simulation and experimental techniques. Computational techniques involve the use of simulation software to evaluate the metamaterial structure. Figure 3 illustrates

Figure 3. Numerical simulation setup for metamaterial energy harvester using computer simulation technology (CST) microwave studio simulator

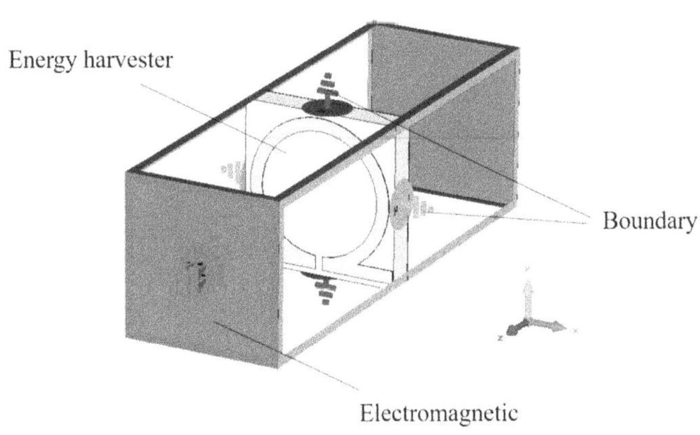

an example of a unit cell simulation using Computer Simulation Technology (CST) Microwave Studio software. The simulation involves the modelling of the unit cell structure and excitation of waves from the waveguide ports.

For experimental testing, metamaterial structure comprised of repeated unit cell structures is fabricated. Typically, the metallic pattern are produced using printed circuit board (PCB) technology for metamaterials working at RF frequency range. The experimental setup to investigate the performance of an energy harvester is shown in Figure 4. Signal generator generates signals at a frequency range. The transmitting antenna which is connected to the signal generator is positioned at a far-field distance from the harvester structure. Electromagnetic wave from the transmitting antenna propagates towards the harvester structure. Oscilloscope with probes is used to measure the voltage across the load resistor of the harvester structure.

Figure 4. Free-space measurement experimental setup to investigate the performance of a metamaterial energy harvester using signal generator, transmitting antenna, and oscilloscope
Source: Ramahi et al. (2012)

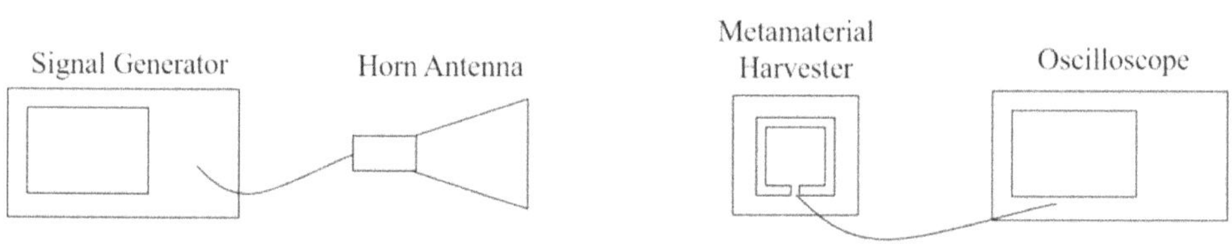

Efficiency Evaluation of Metamaterial RF Energy Harvesters

RF energy harvesters requires its rectenna to collect the electromagnetic power to be converted to DC electrical power. The performance of a RF energy harvester can be evaluated by three types of efficiency, namely absorption efficiency, harvesting efficiency, and conversion efficiency (Khan & Eibert, 2017). Absorption efficiency, $\eta_{\text{absorption}}$ is the ratio of the total time-average power absorbed by the metamaterial structure to the total time-average power incident to the metamaterial structure. It is a measure of how much incident electromagnetic power is dissipated by the metamaterial structure by any loss mechanisms in the structure. Absorption efficiency is determined using equation (2) as:

$$\eta_{absorption} = \frac{P_{absorbed}}{P_{incident}} \tag{2}$$

where $P_{absorbed}$ is the total time-average power absorbed by the metamaterial structure, and $P_{incident}$ is the total time-average power incident to the metamaterial structure.

Harvesting efficiency, $\eta_{harvesting}$ is the ratio of total time-average power delivered to the resistive load te total time-average power incident to the metamaterial structure. It is a measure of how much incident electromagnetic power is collected by the resistive load. Harvesting efficiency is determined using equation (3) as:

$$\eta_{harvesting} = \frac{P_{load}}{P_{incident}} \quad (3)$$

where P_{load} is the total time-average power delivered to the resistive load, and $P_{incident}$ is the total time-average power incident to the metamaterial structure (Almoneef & Ramahi, 2014). P_{load} can be calculated using equation (4) as:

$$P_{load} = \frac{|V_{\text{terminal}}| \cdot |I_{\text{terminal}}|}{2} \quad (4)$$

where $|V_{terminal}|$ is the voltage across the load terminal, and $|I_{terminal}|$ is the current across the load terminal (Palandoken et al., 2018). For harvesting efficiency to be high, the loss mechanism of the harvester needs to be mainly depending on the resistive load instead of the dielectric substrate and metal traces (Shang et al., 2017).

Conversion efficiency, $\eta_{conversion}$ is the ratio of the total time-average power delivered to the resistive load to the total time-average power absorbed by the metamaterial structure. It is a measure of how much absorbed electromagnetic power is collected by the resistive load. Conversion efficiency can be determined using equation (5) as:

$$\eta_{conversion} = \frac{P_{load}}{P_{absorbed}} \quad (5)$$

where P_{load} is the total time-average power delivered to the resistive load and $P_{absorbed}$ is the total time-average power absorbed by the metamaterial structure.

CASE STUDY: OMEGA-RING-SHAPED METAMATERIAL RF ENERGY HARVESTERS

For a better understanding of the RF energy harvesting concept, a metamaterial RF energy harvester proposed by Shang et al. (2017) is modified and simulated. The design, simulation settings, absorption mechanisms, and efficiency of the harvester are discussed.

Design of Omega-Ring-Shaped Metamaterial RF Energy Harvester

The design of the Omega ring resonator unit cell is illustrated in Figure 5 (Shang et al., 2017). The structure is a three-layered metal-substrate-metal structure. The substrate material is Rogers RO4350B (relative permittivity, $\varepsilon_r = 3.48$, loss tangent, $\tan\delta = 0.0037$, thickness, $t = 1.52$ mm). Based on Figure 5 (b), the front layer is an Omega ring-shaped cooper resonator (conductivity, σ = 5.96×10[7] S/m) with dimensions of inner radius of ring, r = 3.35 mm, width of ring, w = 0.7 mm, length of omega leg, l = 4.1 mm, gap of ring, g = 0.4 mm, thickness of copper, t_c = 0.035 mm, and periodicity, p = 8.7 mm. Based on Figure 5 (c), the back layer is a back metal plate. The gaps of the ring are connected by a pair of via ports (diameter of via hole, d = 0.3 mm) and a resistor (resistance, R = 50 Ω) on the ground plane.

Figure 5. (a) Perspective view, (b) front view (omega ring resonator), (c) back view (metal back plate and resistor) of metamaterial-based Omega-ring-shaped energy harvester unit cell structure

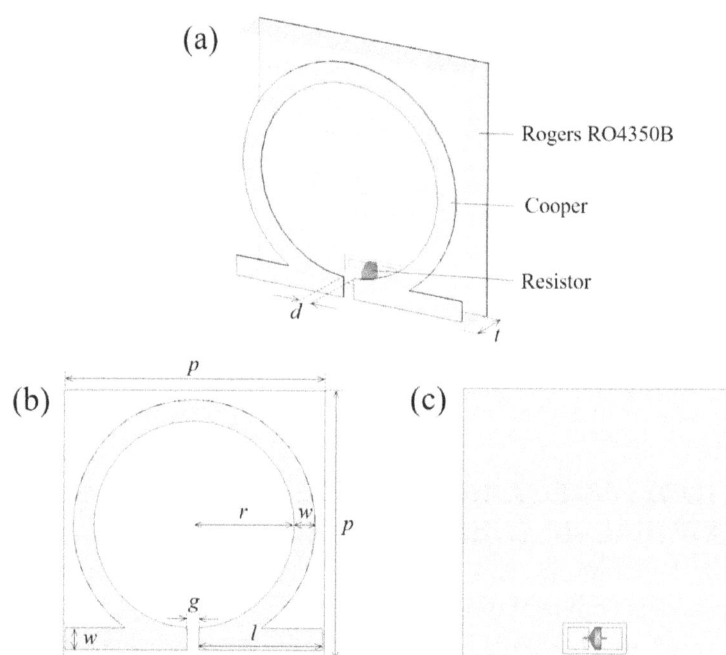

Simulation of Omega-Ring-Shaped Metamaterial RF Energy Harvester

The metamaterial unit cell structure is simulated using CST Microwave Studio software. Figure 6 shows the general settings of the simulation. Based on Figure 6 (a), two waveguide ports are assigned in the *x-y* plane at both sides of the harvester. Port 1 and port 2 are the wave transmitter and receiver, respectively. The boundary condition of the *x-z* and *y-z* plane are set to be perfect electric conductor (PEC) and perfect magnetic conductor (PMC), respectively. Figure 6 (b) shows the tetrahedral mesh condition of the metamaterial harvester structure.

Figure 6. Simulation settings (a) waveguide port, (b) tetrahedral mesh

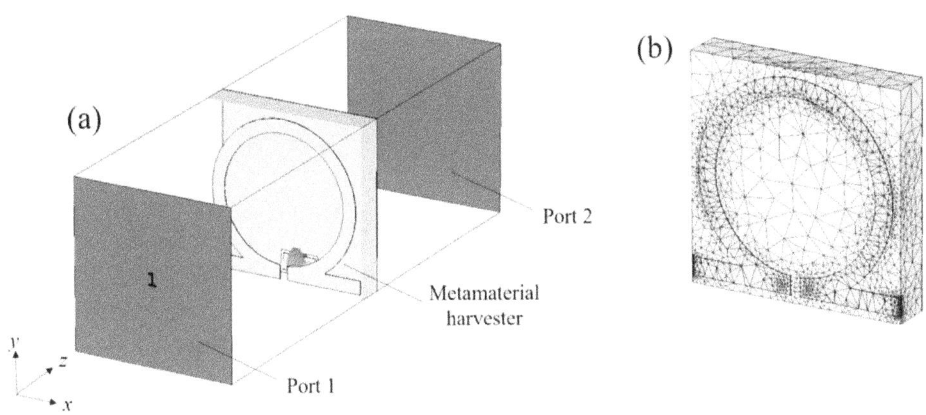

Figure 7 illustrates the distribution of electric field, magnetic field, surface current, and surface power loss density at the resonance frequency. Based on Figure 7 (a), the electric field concentrates on several regions of the Omega ring structure. The gap of the ring is capacitive since it is considered as an open circuit. The maximum available power collected from this gap is dependent on the capacitance and resistive load which is connected across the gap (Ramahi et al., 2012). The resistive load across the gap is used to mimic a power load or charging battery. The bulk physical size of the resistive load connected across the gap might affect the absorption efficiency, therefore, two via ports is normally added to direct the resistive loads at the back of the harvester structure. However, the addition of via ports might slightly change the resonance frequency of the harvester and therefore further optimization in simulation software is usually needed.

Figure 7. (a) Electric field distribution, (b) magnetic field distribution, (c) surface current distribution, (d) surface power loss density, at resonance frequency, f_o = 5.8 GHz

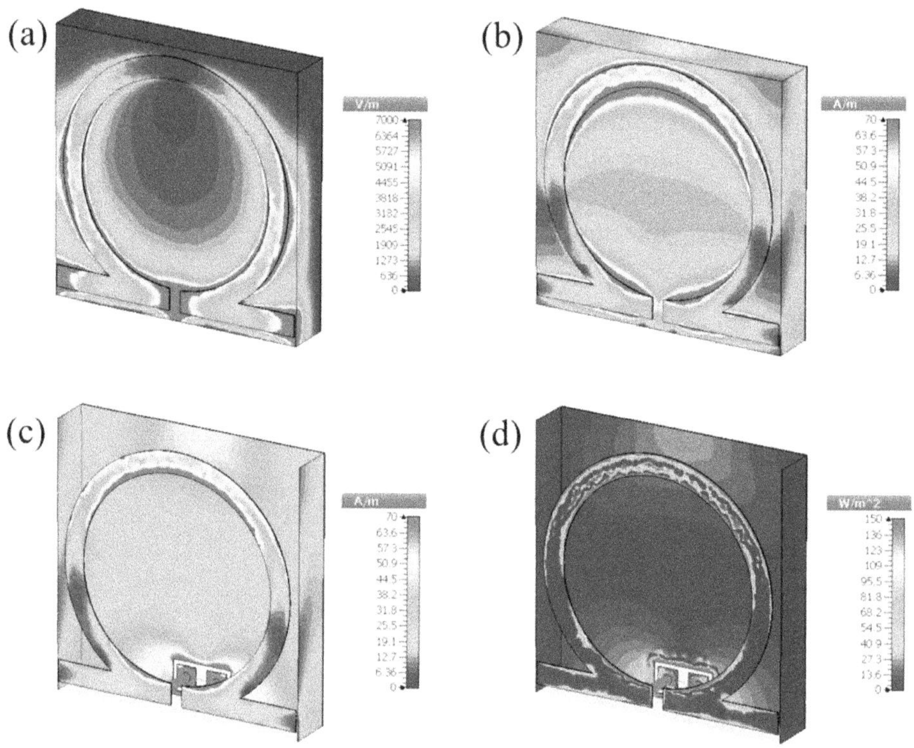

Efficiency Evaluation of Omega-Ring-Shaped Metamaterial RF Energy Harvester

Based on Figure 8, the computational result shows that reflective coefficient, S_{11} and transmission coefficient, S_{21} (in linear magnitude) as a function of frequency. Reflective coefficient and transmission coefficient correspond to the amount of energy reflected from and transmitted through the metamaterial structure, respectively. At 5.8 GHz, it can be observed that both the value is near zero. This indicates that the incident electromagnetic energy is neither reflected nor transmitted. The energy is absorbed by the structure at this resonance frequency.

$$R = |S_{11}|^2 \tag{6}$$

Figure 8. Reflection coefficient, S_{11} and transmission coefficient, S_{21} as a function of frequency, f reflectance, R, transmittance, T and absorbance, A are calculated using equations (6-8)
Source: Sim et al. (2018).

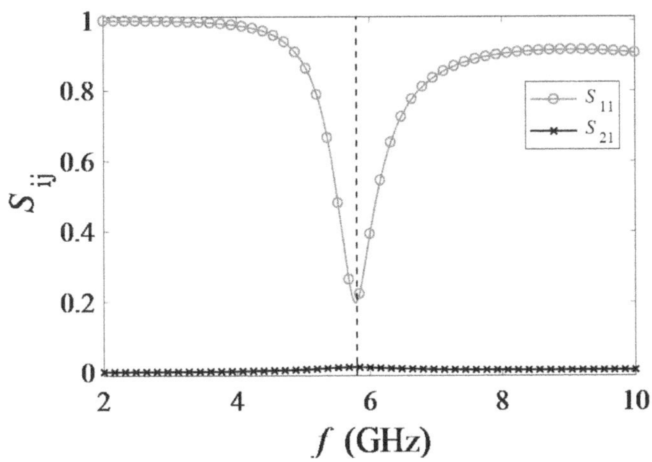

$$T = |S_{21}|^2 \tag{7}$$

$$A = 1 - R - T \tag{8}$$

Figure 9 shows reflectance, transmittance and absorbance as a function of frequency. At $f = 5.8$ GHz, the absorbance is 98.3%, which is the maximum magnitude. Meanwhile, the reflectance and transmittance shows negligible value of magnitude at this frequency. The structure absorb 98.3% of the incident microwave energy and the absorbed energy is dissipated by the loss mechanisms of the structure including dielectric loss, load resistance and metal loss.

The losses are calculated using equation (9) as:

$$Loss = \frac{P_{loss}}{P_{incident}} \tag{9}$$

Figure 10 shows that the losses due to dielectric substrate, lumped resistor and metal. At $f = 5.8$ GHz, 98.3% of the incident energy is absorbed, in which 4.8% dissipates in dielectric substrate, 91.0% dissipates in resistor and 2.5% dissipates in metal. It can be observed that the main loss mechanism is due to the resistive load.

Figure 9. Reflectance, transmittance and absorbance as a function of frequency

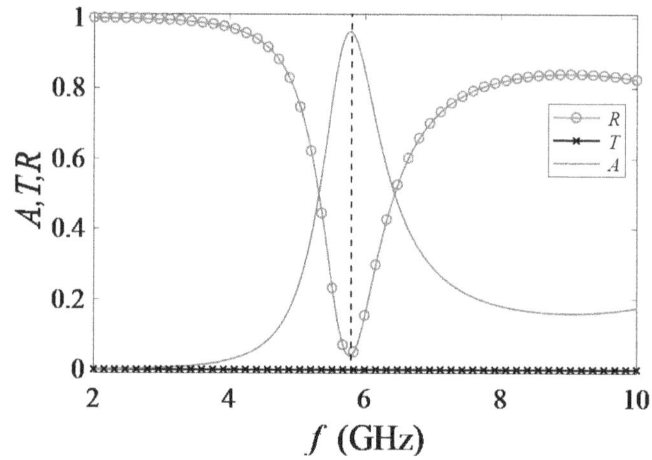

Figure 10. Losses in dielectric substrate, lumped resistor and metals as a function of frequency

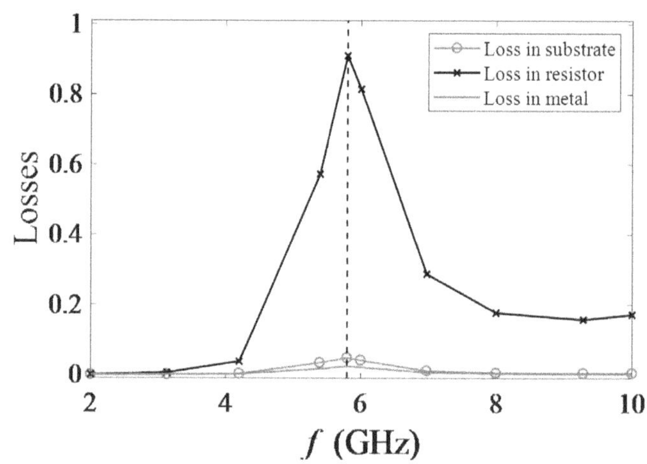

Absorption efficiency, harvesting efficiency and conversion efficiency are determined using equation (2), (3) and (5), respectively. Figure 11 shows the efficiency as a function of frequency. At 5.8 GHz, absorption efficiency is about 98.3%. This value indicates the amount of incident microwave energy absorbed by the structure. At 5.8 GHz, harvesting efficiency is about 91.0%. This value indicates the amount of incident microwave energy which is harvested by the resistor. At

Figure 11. Absorption efficiency, harvesting efficiency and conversion efficiency as a function of frequency

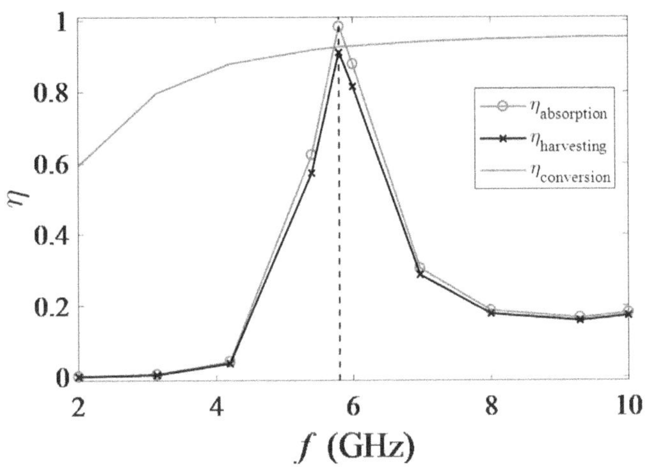

5.8 GHz, conversion efficiency is about 92.6%. This value indicates the amount of absorbed energy which is converted to useful energy by the resistor.

RECENT DEVELOPMENT OF METAMATERIAL RF ENERGY HARVESTERS

There are different designs of metamaterial RF energy harvesters being proposed recently. In this section, the design and performance of metamaterial-based resonators are reviewed. Furthermore, future trends of RF energy harvesting is then discussed.

Topology Design of Metamaterials and Metasurfaces for Energy Harvesting

Split ring resonator (SRR) is one of the common structure in electromagnetic metamaterials (Fan et al., 2011). SRR consists of two metallic rings with gaps. SRR behaves as a magnetic dipole and possesses negative permeability at specific frequency (Wu et al., 2005). SRR is widely used in metamaterial absorber (Zhao & Cheng, 2016), single negative metamaterial (Fan et al., 2011) and double negative metamaterial (Karamanos et al., 2012). The operating frequency depends on the geometrical dimensions of the ring, type of substrate materials and substrate thickness. At the resonance frequency, energy which is stored within the structure oscillates between electric and magnetic field. High concentration of field builds up at certain regions of the SRR structure depending on the resonance mode (Fu et al., 2016). Strong field builds up across the gap.

Load resistor can be placed across the gap to harvest the incoming electromagnetic wave energy. However, the incorporation of load resistor often causes a slight shift in the resonance frequency. The SRR structure is required to be redesigned for maximum power absorption at the desired frequency. The optimization works can be done by numerical method using software such as Computer Simulation Technology (CST), High-Frequency Structure Simulator (HFSS) and COMSOL Multiphysics. Several works have been done to minimize the effects of the added loads to the interactions with the incoming waves. For instance, the load resistor in the complementary split ring resonator (CSRR) based energy harvester reported in (Alavikia et al., 2014) is located at the back of the structure connected via a via port. In addition, fabrication tolerances can causes minor deviation of the absorption capabilities and harvesting efficiency.

Most works have been carried out on designing the metamaterial harvesting structure. Some of the common structures were tabulated in Table 3. Many methods are derived from the ring structure and therefore various innovative structures have been investigated. For example, a butterfly-shaped closed ring metamaterial structure is evolved from the Minkowski fractal ring which is a fractal structure of a square closed ring (X. Zhang et al., 2017). Spiral-shaped metamaterial is a continuous structure of a ring structure in which its radius increases continuously (Pranav et al., 2016). Furthermore, metamaterial structures are geometrically scalable to operate at another frequency (Sim et al., 2017). By changing the dimensions, a metamaterial structure which is scaled smaller would operate at higher frequency. Therefore, a demonstration of a metamaterial structure operating at a specific frequency can be extended to work at other frequency band.

Table 3. Common types of metamaterial harvesting structures

Metamaterial Harvesting Structure	Dimension (mm)	Resonance Frequency, f_0 (GHz)	Ref.
Split ring resonator (SRR)	31.65 x 31.65 x 1.524	1.75, 3.8, 5.4	(H. T. Zhong et al., 2016)
Complementary split ring resonator (CSRR)	16 x 16 x 0.79	5.8	(Alavikia et al., 2014)
Electric-Inductive-Capacitive (ELC) resonator	7.5 x 7.5 x 2.54	3.0	(Almoneef & Ramahi, 2015)
Spiral resonator	21.5 x 28.1 9.8 x 12.8	0.9 1.8	(Pranav et al., 2016)

Performance of Metamaterial RF Energy Harvesters

Some of the recent reported metamaterial electromagnetic energy harvesters and their performance are tabulated in Table 4. There are multi-band energy harvesters, in which the metamaterial structure can harvest electromagnetic energy at several frequency bands. For example, electric-inductive-capacitive (ELC) resonator proposed in (Assimonis et al., 2016) operates at 0.9 GHz, 1.8 GHz and 2.45 GHz with harvesting efficiency of 94%, 81% and 87% respectively.

Table 4. Comparison of performances between metamaterial RF energy harvesters

Structure	Physical Dimensions (mm)	Resonance Frequency, f_o (GHz)	Harvesting Efficiency, $\eta_{harvesting}$ (%)	Ref.
Split ring resonator	85 x 85 x 0.79	5.8	~50 (sim.)	(Ramahi et al., 2012)
Circular ring with a gap	8.8 x 8.8 x 0.813	5.8	~50 (sim.)	(Almoneef & Ramahi, 2014)
Complementary split ring resonator	16 x 16 x 0.79	5.8	~85 (sim.) ~60 (exp.)	(Alavikia et al., 2014)
Electric-Inductive-Capacitive (ELC) resonator	33.1 x 33.1 x 9.9	0.9 1.8 2.45	94 (sim.) 81 87	(Assimonis et al., 2016)
Four split ring resonators in rotating arrangement	31.65 x 31.65 x 1.524	1.75 3.8 5.4	30 (sim.) 90 74	(H. T. Zhong et al., 2016)
Butterfly-shaped closed-ring with rotational symmetry at its center	26.9 x 26.9 x 4	0.9 2.6 5.7	90 (sim.) 83 81	(X. Zhang et al., 2017)
Omega ring resonator	17.4 x 17.4 x 1.524	5.8	~90 (sim. & exp.)	(Shang et al., 2017)
Split ring resonator and slotted ring	81.25 x 87.5 x 1.524	0.9	35.86 (sim.)	(Palandoken et al., 2018)
Four split ring resonators with interconnected split	43 x 43 x 4	0.9 1.8 2.6 5.8	85.7 (sim.) 82.0 80.4 69.8	(Karakaya et al., 2019)

Future Trends of Metamaterial RF Energy Harvesting

Based on the findings, future studies on RF energy harvesting are trending towards the following aspects:

(a) 5G and IoT Technology

The recent development of 5G technology are evolving towards higher frequency bands. The miniaturization, precise fabrication, and integration of the metamaterial energy harvester to the IoT devices are required. The amount of energy harvested from ambient RF is limited (typically less than 1 μW) and therefore hybridization of harvesting devices can be done to collect energy from diverse renewable energy sources (P. Zhang et al., 2022). This improves the usability and reliability of the devices powered by energy harvesters. In addition, the effects of component loss at higher frequency band also require further investigation.

(b) Wearable or Implantable Devices

The use of flexible substrate and low-profile rectifier can be the potential research areas for metamaterial RF energy harvesting. The flexible conformal harvesting devices can be used to power up low-power biomedical wearable or implantable devices. The metamaterial structures can be potentially designed to be multifunctional by acting as an energy harvester and sensors (Kim et al., 2016).

(c) Performance Enhancement

It is important to improve the efficiency of the energy harvesting system by topology optimization of metamaterial structure. AC to DC conversion efficiency can be improved by proper design of matching circuit and rectifying circuit. Multi-band or wide-band energy harvesting is also a potential key study area to allow the collection of RF energy from different frequencies.

(d) Frequency Self-Adaptation

RF energy harvesters with frequency self-adaptation are also known as tunable, self-tuning or switchable energy harvesters (Q. Zhang et al., 2015). It is important to extend the study to active metamaterials which have the ability to change their resonance frequency to the frequency of the incoming RF waves with the highest intensity at that particular time. A sensing device is usually required to detect the frequency of the wave with high intensity. Then, the metamaterials are electrically controlled to switch their resonance frequency to harvest the high intensity frequency energy (G. Zhang et al., 2017).

CONCLUSION

In conclusion, this book chapter presents an overview of metamaterial radio-frequency energy harvesters including their technical perspectives such as working principles, simulation and experimental testing, and efficiency evaluation, as well as design topologies and performances. The simulation of an Omega-ring-shaped resonator-based metamaterial RF energy harvester is demonstrated. The harvester achieves 98.3% absorption efficiency and 91.0% harvesting efficiency at 5.8 GHz. Electric and magnetic field distributions are simulated to enhance the understanding of the absorption mechanisms. The calculation of the absorption, harvesting efficiency, and conversion efficiency is also presented. Lastly, the recent development and future research trend of RF energy harvesters are discussed. It is envisaged that RF energy harvesters will be integrated into 5G and IoT devices to enable wireless, remote, wearable, or implantable applications.

ACKNOWLEDGMENT

This research received no specific grant from any funding agency in the public, commercial, or not-for-profit sectors.

REFERENCES

Alavikia, B., Almoneef, T. S., Ramahi, O. M., Alavikia, B., Almoneef, T. S., & Ramahi, O. M. (2014). Electromagnetic energy harvesting using complementary split-ring resonators. *Applied Physics Letters*, *104*(16), 163903-1, 163903–163904. doi:10.1063/1.4873587

Almoneef, T. S., & Ramahi, O. M. (2014). A 3-dimensional stacked metamaterial arrays for electromagnetic energy harvesting. *Electromagnetic Waves*, *146*, 109–115. doi:10.2528/PIER14031603

Almoneef, T. S., & Ramahi, O. M. (2015). Metamaterial electromagnetic energy harvester with near unity efficiency. *Applied Physics Letters*, *106*(15), 153902-1, 153902–153904. doi:10.1063/1.4916232

Anigbogu, W., & Bardaweel, H. (2020). A Metamaterial-Inspired Structure for Simultaneous Vibration Attenuation and Energy Harvesting. *Shock and Vibration*, *2020*, 1–12. doi:10.1155/2020/4063025

Assimonis, S. D., Kollatou, T., Tsiamitros, D., Stimoniaris, D., Samaras, T., & Sahalos, J. N. (2016). High efficiency and triple-band metamaterial electromagnetic energy hervester. *ELECO 2015 - 9th International Conference on Electrical and Electronics Engineering*, 320–323. 10.1109/ELECO.2015.7394527

Brown, W. C. (1984). The History of Power Transmission by Radio Waves. *IEEE Transactions on Microwave Theory and Techniques*, *32*(9), 1230–1242. doi:10.1109/TMTT.1984.1132833

Brunelli, D., Benini, L., Moser, C., & Thiele, L. (2008). An Efficient Solar Energy Harvester for Wireless Sensor Nodes. *Design, Automation and Test in Europe*, 104–109.

Chen, Z., Guo, B., Yang, Y., & Cheng, C. (2014). Metamaterials-based enhanced energy harvesting : A review. *Physica B. Physica B, Condensed Matter*, *438*, 1–8. doi:10.1016/j.physb.2013.12.040

Cheng, Y., Gong, R., & Zhao, J. (2016). A photoexcited switchable perfect metamaterial absorber/reflector with polarization-independent and wide-angle for terahertz waves. *Optical Materials*, *62*, 28–33. doi:10.1016/j.optmat.2016.09.042

Collado, A., & Georgiadis, A. (2013). Conformal hybrid solar and electromagnetic (EM) energy harvesting rectenna. *IEEE Transactions on Circuits and Systems. I, Regular Papers*, *60*(8), 2225–2234. doi:10.1109/TCSI.2013.2239154

Dinh, M., Ha-Van, N., Tung, N. T., & Le, M. T. (2021). Dual-Polarized Wide-Angle Energy Harvester for Self-Powered IoT Devices. *IEEE Access : Practical Innovations, Open Solutions*, *9*, 103376–103384. doi:10.1109/ACCESS.2021.3098983

Eteng, A. A., Goh, H. H., Rahim, S. K. A., & Alomainy, A. (2021). A Review of Metasurfaces for Microwave Energy Transmission and Harvesting in Wireless Powered Networks. *IEEE Access : Practical Innovations, Open Solutions*, *9*, 27518–27539. doi:10.1109/ACCESS.2021.3058151

Fan, J., Sun, G.-Y., & Zhu, W.-R. (2011). Electric and magnetic dipole couplings in split ring resonator metamaterials. *Chinese Physics B*, *20*(11), 114101. doi:10.1088/1674-1056/20/11/114101

Fu, X., Zeng, X., Cui, T. J., Lan, C., Guo, Y., Zhang, H. C., & Zhang, Q. (2016). Mode jumping of split-ring resonator metamaterials controlled by high-permittivity BST and incident electric fields. *Scientific Reports*, *6*(31274), 1–9. doi:10.1038rep31274 PMID:27502844

Gozel, M. A., Kahriman, M., & Kasar, O. (2019). Design of an efficiency-enhanced Greinacher rectifier operating in the GSM 1800 band by using rat-race coupler for RF energy harvesting applications. *International Journal of RF and Microwave Computer-Aided Engineering*, *29*(1), 1–8. doi:10.1002/mmce.21621

Howey, D. A., Bansal, A., & Holmes, A. S. (2011). Design and performance of a centimetre-scale shrouded wind turbine for energy harvesting. *Smart Materials and Structures, 20*(8), 1–12. doi:10.1088/0964-1726/20/8/085021

Huang, Y., Zhou, J., & Li, L. (2022). Wireless Power Transfer and Energy Harvesting Using Metamaterials and Metasurfaces. *2022 International Workshop on Antenna Technology, IWAT 2022*, 280–283. 10.1109/iWAT54881.2022.9811085

Jabbar, H., Song, Y. S., & Jeong, T. T. (2010). RF energy harvesting system and circuits for charging of mobile devices. *IEEE Transactions on Consumer Electronics, 56*(1), 247–253. doi:10.1109/TCE.2010.5439152

Karakaya, E., Bagci, F., Yilmaz, A. E., & Akaoglu, B. (2019). Metamaterial-Based Four-Band Electromagnetic Energy Harvesting at Commonly Used GSM and Wi-Fi Frequencies. *Journal of Electronic Materials, 48*(4), 2307–2316. doi:10.100711664-019-06962-9

Karamanos, T. D., Dimitriadis, A. I., Kantartzis, N. V., & Member, S. (2012). Compact Double-Negative Metamaterials Based on Electric and Magnetic Resonators. *IEEE Antennas and Wireless Propagation Letters, 11*(2), 480–483. doi:10.1109/LAWP.2012.2197170

Khan, S., & Eibert, T. F. (2017). A multi-resonant meta-absorber as an electromagnetic energy harvester. *2017 IEEE International Symposium on Antennas and Propagation & USNC/URSI National Radio Science Meeting*, 1091–1092. 10.1109/APUSNCURSINRSM.2017.8072588

Kim, H. K., Lee, D., & Lim, S. (2016). A Fluidically Tunable Metasurface Absorber for Flexible Large-Scale Wireless Ethanol Sensor Applications. *Sensors (Basel), 16*(8), 1–10. doi:10.339016081246 PMID:27509498

Landy, N. I., Sajuyigbe, S., Mock, J. J., Smith, D. R., & Padilla, W. J. (2008). Perfect metamaterial absorber. *Physical Review Letters, 100*(20), 1–4. doi:10.1103/PhysRevLett.100.207402 PMID:18518577

Lei, A., Xu, R., Thyssen, A., Stoot, A. C., Christiansen, T. L., Hansen, K., Lou-Møller, R., Thomsen, E. V., & Birkelund, K. (2011). MEMS-Based Thick Film PZT Vibrational Energy Harvester. *2011 IEEE 24th International Conference on Micro Electro Mechanical Systems*, 125–128. 10.1109/MEMSYS.2011.5734377

Li, J., & Chan, C. T. (2004). Double-negative acoustic metamaterial. *Physical Review E: Statistical Physics, Plasmas, Fluids, and Related Interdisciplinary Topics, 70*(5), 4. doi:10.1103/PhysRevE.70.055602 PMID:15600684

Li, X., Liu, H., Sun, Q., & Huang, N. (2015). Ultra-broadband and polarization-insensitive wide-angle terahertz metamaterial absorber. *Photonics and Nanostructures*, *15*, 81–88. doi:10.1016/j.photonics.2015.04.002

Li, Y., Baker, E., Reissman, T., Sun, C., & Liu, W. K. (2017). Design of mechanical metamaterials for simultaneous vibration isolation and energy harvesting. *Applied Physics Letters*, *111*(25), 251903-1, 251903–251905. doi:10.1063/1.5008674

Ma, K., Tan, T., Yan, Z., Liu, F., Liao, W. H., & Zhang, W. (2021). Metamaterial and Helmholtz coupled resonator for high-density acoustic energy harvesting. *Nano Energy*, *82*, 105693. doi:10.1016/j.nanoen.2020.105693

Nguyen, T. T., & Lim, S. (2017). Wide Incidence Angle-Insensitive Metamaterial Absorber for Both TE and TM Polarization using Eight-Circular-Sector. *Scientific Reports*, *7*(1), 1–11. doi:10.103841598-017-03591-2 PMID:28600523

Palandoken, M., Gocen, C., Kaya, A., Gunes, F., Baytore, C., & Can, F. C. (2018). A Novel Split-Ring Resonator and Voltage Multiplier Based Rectenna Design for 900 MHz Energy Harvesting Applications. *Radioengineering*, *27*(3), 711–717. doi:10.13164/re.2018.0711

Piñuela, M., Mitcheson, P. D., & Lucyszyn, S. (2013). Ambient RF energy harvesting in urban and semi-urban environments. *IEEE Transactions on Microwave Theory and Techniques*, *61*(7), 2715–2726. doi:10.1109/TMTT.2013.2262687

Pranav, U. S., Sudheesh, S., Stanly, P., Sankar, S., Devika, R., & Pradeep, A. (2016). Metamaterial Based Energy Harvester. *Procedia Computer Science*, *93*(September), 74–80. doi:10.1016/j.procs.2016.07.184

Qi, S., Oudich, M., Li, Y., & Assouar, B. (2016). Acoustic energy harvesting based on a planar acoustic metamaterial. *Applied Physics Letters*, *108*(26), 263501-1, 263501–263504. doi:10.1063/1.4954987

Ramahi, O. M., Almoneef, T. S., Alshareef, M., & Boybay, M. S. (2012). Metamaterial particles for electromagnetic energy harvesting. *Applied Physics Letters*, *173903*(17), 173903-1, 173903–173905. doi:10.1063/1.4764054

Shang, S., Yang, S., Shan, M., Liu, J., & Cao, H. (2017). High performance metamaterial device with enhanced electromagnetic energy harvesting efficiency. *AIP Advances*, *7*(August), 105204-1, 105204–105207. doi:10.1063/1.5002165

Sim, M. S., You, K. Y., Esa, F. B., & Chan, Y. L. (2017). Broadband metamaterial microwave absorber for X-Ku band using planar split ring-slot structures. *Progress in Electromagnetics Research Symposium-Fall (PIERS-FALL)*, 215–221. 10.1109/PIERS-FALL.2017.8293138

Sim, M. S., You, K. Y., & Esa, F. (2020). Electromagnetic metamaterials in microwave regime. In Handbook of Research on Recent Developments in Electrical and Mechanical Engineering (pp. 64–86). doi:10.4018/978-1-7998-0117-7.ch002

Sim, M. S., You, K. Y., Esa, F., & Chan, Y. L. (2021). Nanostructured electromagnetic metamaterials for sensing applications. In *Applications of Nanomaterials in Agriculture* (pp. 141–164). Food Science, and Medicine. doi:10.4018/978-1-7998-5563-7.ch009

Sim, M. S., You, K. Y., Esa, F., Dimon, M. N., & Khamis, N. H. (2018). Multiband metamaterial microwave absorbers using split ring and multiwidth slot structure. *International Journal of RF and Microwave Computer-Aided Engineering*, *28*(7), 1–13. doi:10.1002/mmce.21473

Soheilifar, M. R., & Sadeghzadeh, R. A. (2014). Design, fabrication and measurement of two-layered quadruple-band microwave metamaterial absorber. *Journal of Communication Engineering*, *3*(1), 13–22.

Suh, Y.-H., & Chang, K. (2002). A high-efficiency dual-frequency rectenna for 2.45- and 5.8-GHz wireless power transmission. *IEEE Transactions on Microwave Theory and Techniques*, *50*(7), 1784–1789. doi:10.1109/TMTT.2002.800430

Sun, K. H., Kim, J. E., Kim, J., & Song, K. (2017). Sound energy harvesting using a doubly coiled-up acoustic metamaterial cavity. *Smart Materials and Structures*, *26*(7), 075011. doi:10.1088/1361-665X/aa724e

Sun, W., Zhong, K., Liu, Y., Xiao, H., Zhao, D., Yan, Z., & Tan, T. (2022). Enhanced metamaterial vibration for high-performance acoustic piezoelectric energy harvesting. *Composites Communications*, *35*(July), 101342. doi:10.1016/j.coco.2022.101342

Tan, T., Yan, Z., Zou, H., Ma, K., Liu, F., Zhao, L., Peng, Z., & Zhang, W. (2019). Renewable energy harvesting and absorbing via multi-scale metamaterial systems for Internet of things. *Applied Energy*, *254*(August), 113717-1, 113717–15. doi:10.1016/j.apenergy.2019.113717

Tan, Y. K., & Panda, S. K. (2011). Light and Thermal Energy Sources for Enhanced Performance of Wireless Sensor Nodes. *IEEE Transactions on Industrial Electronics*, *58*(9), 4424–4435. doi:10.1109/TIE.2010.2102321

Van Blarigan, L., Danzl, P., Moehlis, J., Van Blarigan, L., Danzl, P., & Moehlis, J. (2012). A broadband vibrational energy harvester. *Applied Physics Letters*, *100*(25), 253904-1, 253904–4. doi:10.1063/1.4729875

Wartak, M. S., Tsakmakidis, K., & Hess, O. (2011). Introduction to Metamaterials. *Physics In Canada*, *67*(1), 30–34.

Wu, M. F., Meng, F. Y., Wu, Q., Wu, J., & Li, L. W. (2005). A compact equivalent circuit model for the SRR structure in metamaterials. *Asia-Pacific Microwave Conference Proceedings, APMC, 1*, 5–8. 10.1109/APMC.2005.1606214

Yasuda, H., & Yang, J. (2015). Reentrant origami-based metamaterials with negative poisson's ratio and bistability. *Physical Review Letters*, *114*(18), 185502-1, 185502–185506. doi:10.1103/PhysRevLett.114.185502 PMID:26001009

Zeng, Z., Li, X., Bermak, A., Tsui, C. Y., & Ki, W. H. (2016). A WLAN 2.4-GHz RF energy harvesting system with reconfigurable rectifier for wireless sensor network. *Proceedings - IEEE International Symposium on Circuits and Systems,* 2362–2365. 10.1109/ISCAS.2016.7539059

Zhang, G., Lan, C., Gao, R., & Zhou, J. (2017). Permanent magnetic ferrite based power-tunable metamaterials. *Journal of Magnetism and Magnetic Materials*, *436*, 57–60. doi:10.1016/j.jmmm.2017.04.027

Zhang, P., Zhang, X., & Li, L. (2022). An Optically Transparent Metantenna for RF Wireless Energy Harvesting. *IEEE Transactions on Antennas and Propagation*, *70*(4), 2550–2560. doi:10.1109/TAP.2021.3137166

Zhang, Q., Ma, Q., Yan, S., Wu, F., He, X., & Jiang, J. (2015). Tunable terahertz absorption in graphene-based metamaterial. *Optics Communications*, *353*, 70–75. doi:10.1016/j.optcom.2015.05.017

Zhang, X., Liu, H., & Li, L. (2017). Tri-band miniaturized wide-angle and polarization-insensitive metasurface for ambient energy harvesting. *Applied Physics Letters*, *111*(7), 071902. Advance online publication. doi:10.1063/1.4999327

Zhao, J., & Cheng, Y. (2016). Ultrabroadband Microwave Metamaterial Absorber Based on Electric SRR Loaded with Lumped Resistors. *Journal of Electronic Materials*, *45*(10), 5033–5039. doi:10.100711664-016-4693-0

Zhong, H. T., Yang, X. X., Tan, C., & Yu, K. (2016). Triple-band polarization-insensitive and wide-angle metamaterial array for electromagnetic energy harvesting. *Applied Physics Letters*, *109*(25), 1–5. doi:10.1063/1.4973282

Zhong, M. (2014). Influence of dielectric layer on negative refractive index and transmission of metal-dielectric-metal sandwiched metamaterials. *Chinese Optics Letters*, *12*(4), 10–12. doi:10.3788/COL201412.041601.Metamaterials

ADDITIONAL READING

Alavikia, B., Almoneef, T. S., & Ramahi, O. M. (2015). Complementary split ring resonator arrays for electromagnetic energy harvesting. *Applied Physics Letters, 107*(3), 033902. doi:10.1063/1.4927238

Chen, Z., Guo, B., Yang, Y., & Cheng, C. (2014). Metamaterials-based enhanced energy harvesting: A review. *Physica B, Condensed Matter, 438*, 1–8. doi:10.1016/j.physb.2013.12.040

Dai, H., Lu, Y., Law, M. K., Sin, S. W., Seng-Pan, U., & Martins, R. P. (2015, March). A review and design of the on-chip rectifiers for RF energy harvesting. In *2015 IEEE International Wireless Symposium (IWS 2015)* (pp. 1-4). IEEE. 10.1109/IEEE-IWS.2015.7164642

Kim, S. (2011, August). RF energy harvesting techniques for wirelessly powered devices. In *2011 IEEE MTT-S International Microwave Workshop Series on Intelligent Radio for Future Personal Terminals* (pp. 1-2). IEEE.

Lu, X., Wang, P., Niyato, D., Kim, D. I., & Han, Z. (2014). Wireless networks with RF energy harvesting: A contemporary survey. *IEEE Communications Surveys and Tutorials, 17*(2), 757–789. doi:10.1109/COMST.2014.2368999

Mateu, L., & Moll, F. (2005, June). Review of energy harvesting techniques and applications for microelectronics. In *VLSI Circuits and Systems II* (Vol. 5837, pp. 359–373). International Society for Optics and Photonics. doi:10.1117/12.613046

Paulo, J., & Gaspar, P. D. (2010, June). Review and future trend of energy harvesting methods for portable medical devices. In *Proceedings of the world congress on engineering* (Vol. 2, pp. 168-196).

Pranav, U. S., Sudheesh, S., Stanly, P., Sankar, S., Devika, R., & Pradeep, A. (2016). Metamaterial based energy harvester. *Procedia Computer Science, 93*, 74–80. doi:10.1016/j.procs.2016.07.184

Shaikh, F. K., & Zeadally, S. (2016). Energy harvesting in wireless sensor networks: A comprehensive review. *Renewable & Sustainable Energy Reviews, 55*, 1041–1054. doi:10.1016/j.rser.2015.11.010

Sim, M. S., You, K. Y., & Esa, F. (2020). Electromagnetic Metamaterials in Microwave Regime. In Handbook of Research on Recent Developments in Electrical and Mechanical Engineering (pp. 64–86). doi:10.4018/978-1-7998-0117-7.ch002

Tran, L. G., Cha, H. K., & Park, W. T. (2017). RF power harvesting: A review on designing methodologies and applications. *Micro and Nano Systems Letters*, *5*(1), 14. doi:10.118640486-017-0051-0

Wan, Z. G., Tan, Y. K., & Yuen, C. (2011, September). Review on energy harvesting and energy management for sustainable wireless sensor networks. In *2011 IEEE 13th international conference on communication technology* (pp. 362-367). IEEE. 10.1109/ICCT.2011.6157897

KEY TERMS AND DEFINITIONS

Absorption Efficiency: The ratio of the total time-average power absorbed by the metamaterial structure to the total time-average power incident to the metamaterial structure.

Conversion Efficiency: The ratio of the total time-average power delivered to the resistive load to the total time-average power absorbed by the metamaterial structure.

Energy Harvester: A device which collect energy such as acoustic, mechanical, and electromagnetic energy and convert them into useful electrical power.

Harvesting Efficiency: The ratio of total time-average power delivered to the resistive load the total time-average power incident to the metamaterial structure.

Metamaterial: An artificial engineered structure which exhibits properties not usually found in natural materials.

Rectenna: Rectifying antenna which consists of an antenna and a rectifying circuit.

Wireless Power Transmission: A process of transmitting electromagnetic power through vacuum or atmosphere without connecting wires or substances.

Chapter 12
Active and Passive Metamaterials and Metasurfaces

Anuj Kumar Sharma
https://orcid.org/0000-0003-0019-6421
Faculty of Engineering and Technology, Gurukul Kangri University (Deemed), India

Vipul Sharma
Faculty of Engineering and Technology, Gurukul Kangri University (Deemed), India

ABSTRACT

The authors of this chapter begin by discussing the fundamentals of metamaterials and metasurfaces before moving on to active and passive metamaterials. This chapter is broken up into six different sections. In the first section, an overview of metamaterials and metasurfaces is provided, and in the second half, active and passive metamaterials and metasurfaces are discussed. This is followed by an explanation of active metamaterials and metasurfaces in part three, fabrication methods for metamaterials and metasurfaces in part four, future directions and challenges in the field of metamaterials in part five, and a conclusion in part six.

I. INTRODUCTION

A. Definition of Metamaterials and Metasurfaces

To put it simply, metamaterials are man-made creations with extraordinary properties that can't be found in nature. The material's structure, rather than its chemical make-up, is responsible for these characteristics. Negative refractive index, superlensing, and invisibility cloaking are just a few of the extraordinary optical, acoustic, or electromagnetic features that can be engineered into metamaterials.2D surfaces called

DOI: 10.4018/978-1-6684-8287-2.ch012

metasurfaces are organized on a scale smaller than the wavelength of light, giving them the ability to bend and refract light in novel ways. Though conceptually similar to 3D metamaterials, metasurfaces are often more thinner and easier to construct. By manipulating the phase, amplitude, and polarization of light, metasurfaces open up new possibilities in optics and photonics.

Several disciplines, such as physics, materials science, engineering, and optics, are actively investigating metamaterials and metasurfaces.

B. Brief History of Metamaterials and Metasurfaces

The idea of a material with a negative refractive index led to the development of the first metamaterials in the early 20th century. The word "metamaterial" wasn't used to describe materials having extraordinary qualities that had been produced artificially until the 1990s. Microwave interaction led to the development of the earliest metamaterials, which were arrays of tiny metallic resonators. Light, sound, and other forms of electromagnetic radiation are all susceptible to manipulation by metamaterials, which have been produced since then. Duke University scientists developed the first "invisibility cloak" in 2006 (Schurig et al., 2006) by utilizing metamaterials to deflect light around a small item, making it invisible to the naked eye. Much anticipation for the development of novel optical devices and applications based on metamaterials was sparked by this breakthrough. The first experimental demonstrations of metasurfaces just appeared in the early 2010s. Improvements in nanofabrication methods allowed scientists to build structures with exact nanometer-scale dimensions and forms, paving the way for the development of metasurfaces. Since then, lenses, holograms, and polarizers are just some of the optical devices that have been made possible with the help of metasurfaces. They have also been used in areas like thermal and acoustic engineering. The development of more precise methods for manipulating light has led to exciting new possibilities in areas like communications, sensing, and imaging.

C. Importance of Active and Passive Metamaterials and Metasurfaces

There are two distinct forms of metamaterials and metasurfaces, which are known respectively as passive and active. To alter light or other kinds of electromagnetic radiation, metamaterials and metasurfaces can be passive or active. Passive metamaterials and metasurfaces are those that do not require any external energy source. They do this by relying on the inherent structural features of the material, which modifies the way the radiation behaves. Active metamaterials and metasurfaces, on the other hand, need an external energy source in order to modify light and other types of electromagnetic radiation. This can be done by modifying the material's refractive index or other properties in real-time.

The ability of passive metamaterials and metasurfaces to produce new sorts of optical and acoustic devices that are more powerful, more efficient, and smaller than standard devices is one of the most important aspects of these types of materials and surfaces. For instance, metasurfaces can be utilized in the production of ultra-thin lenses and polarizers, both of which can be included into smartphones and other mobile devices in order to make them more compact while simultaneously increasing their processing capability. The efficiency of solar cells can also be improved using passive metamaterials and metasurfaces. This is accomplished by increasing the amount of light that is absorbed by the material. They can also be utilized in applications relating to soundproofing and noise cancellation by modulating sound waves in order to bring about a reduction in noise. On the other hand, active metamaterials and metasurfaces have the benefit of being able to adjust the properties they exhibit in real time. This makes them more desirable. This makes them advantageous for applications such as optical switching and modulation, where the properties of the material may be modified on the fly. Active metamaterials and metasurfaces are also employed in optical communications, where they can be used to manipulate the direction and polarization of light. This is of utmost importance in high-speed optical communications systems, as even minute delays or distortions in the light signal can lead to errors in these kinds of systems.

In conclusion, metamaterials and metasurfaces, both passive and active, have significant applications in a wide variety of sectors, and they are anticipated to play an increasingly important part in the technologies of the future. The capacity to alter light and sound with such precision has made new options available for applications in sectors such as imaging, sensing, and energy, amongst others. In addition, the use of metamaterials and metasurfaces makes it possible to develop new kinds of devices that are more compact, efficient, and potent than the conventional kinds of devices.

II. PASSIVE METAMATERIALS AND METASURFACES

A. Definition and Characteristics of Passive Metamaterials and Metasurfaces

In artificially engineered materials known as metamaterials and metasurfaces, the manipulation of light, sound, or other forms of electromagnetic radiation can occur without the need for an additional source of energy. They rely on the inherent structural qualities of the material to change the way in which the radiation behaves (Padilla et al., 2007). Sub-wavelength structures, which are smaller than the wavelength of the radiation they manipulate and serve as the fundamental building blocks of passive metamaterials and metasurfaces, are also known as metasurfaces. In order to achieve the desired optical, acoustic, or electromagnetic properties, these structures are arranged in extremely precise patterns (Chen et al., 2007).

The following are some of the most important qualities that passive metamaterials and metasurfaces share:

Passive metamaterials and metasurfaces have the potential to exhibit a negative refractive index, which indicates that they bend light in the opposite direction that natural materials do. Because of this feature, it is possible to manufacture ultrathin lenses and other optical devices, which is not the case when working with naturally occurring materials(Peralta et al., 2009).

Anisotropy: Both passive metamaterials and metasurfaces have the potential to be anisotropic. This means that the properties of the material change depending on the direction in which the radiation is coming from. Because of this property, polarizers and other optical devices that are capable of selectively filtering light based on the polarization of the light can be constructed (Kröling et al., 2014).

High sensitivity: The features of the radiation that passive metamaterials and metasurfaces manipulate can make them capable of exhibiting a high level of sensitivity. For instance, they can be engineered to have a high absorptive or reflective property at a particular frequency, which enables them to be utilized in applications relating to sensing (O'Hara et al., 2008). Passive metamaterials and metasurfaces have the potential to display resonance, which takes place when the metamaterials or surfaces are excited by radiation at a certain frequency. Because of this feature, it is possible to make optical devices such as filters, resonators, and others.

Size: The wavelength of the radiation that is being manipulated by passive metamaterials and metasurfaces can be designed to be much smaller than the size of the passive metamaterial or metasurface. Because of this property, they can be incorporated into devices such as mobile phones, digital cameras, and other portable electronic equipment.

Metamaterials and metasurfaces that are passive have a wide variety of applications, some of which are in the field of telecommunications, while others are in the fields of imaging, sensing, and energy. They are also being employed in applications related to soundproofing and noise cancellation, both of which involve the manipulation of sound waves for the purpose of reducing noise.

B. Design Principles of Passive Metamaterials and Metasurfaces

The concepts of design for passive metamaterials and metasurfaces are based on the manipulation of sub-wavelength structures to influence the behavior of radiation.

This can be accomplished through the use of metamaterials. While designing passive metamaterials and metasurfaces, there are several essential design principles that are utilized, including the following:

Sub-wavelength structures are the essential building blocks of metamaterials and metasurfaces. These structures are smaller than the wavelength of the radiation that they influence and serve as the fundamental building blocks of metamaterials and metasurfaces. The majority of the time, these structures are fabricated from metals or dielectrics and are structured in either periodic or non-periodic patterns.

In metamaterials and metasurfaces, the arrangement of sub-wavelength structures is often periodic. This means that the pattern repeats itself across a certain distance. Because of this periodicity, the material is able to control the radiation and modify it in the way it wants. The symmetry of the sub-wavelength structures of metamaterials and metasurfaces plays an important part in determining the optical, acoustic, or electromagnetic properties of these materials and surfaces. The direction of light propagation, the polarization of light, and the resonant qualities of the material can all be controlled through the utilization of symmetry.

The effective medium theory is a way to characterize the behavior of metamaterials and metasurfaces as a homogenous material with effective optical, acoustic, or electromagnetic properties. This behavior can be described using the effective medium theory. Using this theory, the designer is able to make predictions about the properties of the material based on the structure of the material at the sub-wavelength level.

Engineering of dispersion: This type of engineering is used to control the manner in which light and other types of electromagnetic radiation propagate through a metamaterial or metasurface. It is possible to construct materials that have odd properties by regulating the dispersion, such as a negative refractive index.

Plasmonics is the study of the interaction between light and nanoscale metal structures. It is sometimes known as "plasmonics." Plasmonic effects can be utilized to improve the optical properties of metamaterials and metasurfaces, such as the rate at which they absorb light, the amount of light that they reflect, or the amount of light that they transmit.

The design of metamaterials and metasurfaces is not possible without the use of computational modeling, which is a key tool in this process. For instance, simulations in the finite-difference time-domain (FDTD), also known as finite-difference time-domain, can be used to predict the optical, acoustic, or electromagnetic properties of a metamaterial or metasurface before the material or surface is actually constructed.

As a result of the fact that the properties of these materials are heavily reliant on their sub-wavelength structure, the creation of passive metamaterials and metasurfaces can be a difficult process. It is becoming easier to design and fabricate metamaterials and metasurfaces with increasingly complex properties and functionalities as a result of advancements in fabrication techniques and computational modeling.

To summarize, the design concepts of passive metamaterials and metasurfaces are based on the manipulation of sub-wavelength structures in order to regulate the behavior of radiation. This can be accomplished through the use of metamaterials. These materials are often defined by their sub-wavelength structure, periodicity, symmetry, effective medium theory, dispersion engineering, plasmonics, and computer modelling (Peralta et al., 2008). They are constructed with particular optical, acoustic, or electromagnetic features in mind.

C. Examples of Passive Metamaterials and Metasurfaces

1. Negative Index Materials

Negative refractive index materials are metamaterials. Negative index compounds bend light differently than natural materials, which have a positive refractive index.

The first negative index materials were made in the late 1990s after Soviet physicist Victor Veselago presented the idea in 1968. Negative index metamaterials allowed light manipulation that was previously unattainable (Padilla et al., 2006).

To attain the necessary optical qualities, negative index materials utilise sub-wavelength features in a specified pattern. SRRs and wire architectures can help design negative index materials. SRRs are small split metal rings, while wire structures are thin metal wires strung in a pattern. A negative refractive index metamaterial is created by periodic arrangement of these structures.

Negative index materials can be used to make ideal lenses for super-resolution imaging. Invisibility cloaks and optical waveguides can route light around sharp corners.

In summary, negative index materials are metamaterials with a negative refractive index that bend light differently from regular materials. To obtain the necessary optical properties, these materials use sub-wavelength features in a specified pattern. Negative index materials could revolutionize optics with perfect lenses, invisibility cloaks, and optical waveguides.

2. Superlenses

Metamaterials known as "superlenses" can be utilized to take imaging much beyond the diffraction limit of traditional lenses, allowing for sharper details. The diffraction limit asserts that the smallest feature that may be resolved by conventional lenses is proportional to the wavelength of light utilized in imaging. Superlenses are able to focus light with a resolution that exceeds the diffraction limit because they use a negative index material.

The use of negative index materials, such as metamaterials having a negative refractive index, is important to the creation of superlenses. A superlens can focus light with a resolution that exceeds the diffraction limit because it distorts the path of the light in the opposite direction of what we'd anticipate from regular lenses. Nanowires, nanorods, and thin films are just some of the structures that can be used in the creation of superlenses (Haxha et al., 2018).

To image objects smaller than the wavelength of the imaging light is one of the main benefits of superlenses. Nanotechnology, biology, and materials science are just a few of the areas where this could be useful. Imaging biological molecules and cells at great resolution with superlenses, for instance, could have significant consequences for the discovery of new therapeutics.

Superlenses can be used for imaging, but they also have the potential to be used in optical communication and sensing. They might be used to fabricate fiber optics and optical waveguides with higher resolution, allowing for more rapid and efficient data transmission than is now possible.

In conclusion, superlenses are a metamaterial that may be utilized to take imaging beyond the diffraction limit of traditional lenses, allowing for higher resolution images to be captured. Metamaterials with a negative refractive index are used in their construction. Nanotechnology, biology, materials science, optical communication, and sensing are just few of the numerous possible uses for superlenses.

3. Perfect Absorbers

Perfect absorbers are a type of metamaterial that can absorb all incident light across a broad range of frequencies and angles, making them ideal for applications such as photovoltaic cells, thermal emitters, and stealth technology.

The design of perfect absorbers is based on the use of micro- or nano-structured materials, which are designed to trap and absorb all incident light. The absorption is achieved by tuning the resonant frequency of the metamaterial to match the frequency of the incident light, and by using a design that maximizes the interaction between the light and the metamaterial. This can be achieved through a variety of different techniques, such as the use of resonant cavities, plasmonic structures, or thin-film interference.

One of the key advantages of perfect absorbers is their ability to be tuned to absorb light across a broad range of frequencies and angles. This makes them ideal for applications such as photovoltaic cells, where they can be used to absorb sunlight and convert it into electrical energy. They can also be used as thermal emitters, where they can be used to efficiently emit thermal radiation across a broad range of frequencies and angles.

Another potential application of perfect absorbers is in stealth technology. By absorbing all incident light, perfect absorbers can be used to create surfaces that are invisible to radar, making them ideal for military and aerospace applications.

In addition to their practical applications, perfect absorbers also have potential applications in fundamental research. They can be used to study the interaction between light and matter at the nanoscale, which has important implications for fields such as quantum optics and nanophotonics.

In summary, perfect absorbers are a type of metamaterial that can absorb all incident light across a broad range of frequencies and angles. They are designed using micro- or nano-structured materials and can be tuned to absorb light through a variety of different techniques. Perfect absorbers have many potential applications in fields such as photovoltaics, thermal emitters, stealth technology, and fundamental research (Do et al., 2012).

4. Electromagnetic Cloaks

Electromagnetic cloaks (Schurig et al., 2006) are a type of metamaterial that can be used to make objects invisible to electromagnetic radiation, such as visible light, radio waves, and microwaves. The design of electromagnetic cloaks is based on the principles of transformation optics, which involves manipulating the path of light around an object to make it appear invisible.

The design of electromagnetic cloaks typically involves using a metamaterial with a negative refractive index. The metamaterial is designed in such a way that it bends light around the object, making it appear as if the object is not there. This is achieved by controlling the electromagnetic properties of the metamaterial in a way that allows the light to flow around the object without being scattered or absorbed.

One of the key advantages of electromagnetic cloaks is their potential for use in military and defense applications, where they could be used to make objects invisible to radar and other forms of electromagnetic detection. They could also be used in medical imaging, where they could be used to improve the accuracy of imaging techniques by reducing the amount of interference from surrounding objects.

In addition to their practical applications, electromagnetic cloaks have also generated significant interest in fundamental physics and engineering. They have been used to study the fundamental properties of light and the behavior of electromagnetic waves in metamaterials. They have also been used to develop new techniques for manipulating light and controlling the flow of electromagnetic radiation.

However, there are still some limitations to the practical application of electromagnetic cloaks. One of the main challenges is developing cloaks that work across a broad range of frequencies and angles of incidence. Another challenge is developing cloaks that are flexible and can be applied to complex objects, such as aircraft or ships.

In summary, electromagnetic cloaks are a type of metamaterial that can be used to make objects invisible to electromagnetic radiation. They are designed using a metamaterial with a negative refractive index, and work by bending light around the object to make it appear invisible. Electromagnetic cloaks have potential applications in military and defense, medical imaging, and fundamental research, but still face challenges in terms of practical application.

III. ACTIVE METAMATERIALS AND METASURFACES

A. Definition and Characteristics of Active Metamaterials and Metasurfaces

It is possible to actively manipulate the characteristics of electromagnetic waves with active metamaterials and metasurfaces. In order to dynamically control the behavior of electromagnetic waves, active metamaterials and metasurfaces incorporate active components like transistors or diodes, as opposed to passive metamaterials and metasurfaces, which rely solely on their structural design to manipulate electromagnetic waves (Padilla et al., 2006).

There are a few distinguishing features of active metamaterials and metasurfaces. To begin, they can be adjusted to detect a wide range of electromagnetic waves. Including active components that can be tweaked to alter the metamaterial's resonant characteristics accomplishes this, allowing the metamaterial to selectively reflect, transmit, or absorb electromagnetic radiation at predetermined frequencies (Chen et al., 2006).

Second, electromagnetic waves' phase and amplitude can be controlled by employing active metamaterials and metasurfaces. Beam steering, holography, and imaging are just some of the many possible uses for metamaterials and metasurfaces, which allow for precise control over the phase and amplitude of electromagnetic waves.

Finally, active metamaterials and metasurfaces can vary their behavior in response to external stimuli. This is achieved by designing the metamaterial or metasurface with built-in sensors or other feedback mechanisms that can detect changes in the surrounding electromagnetic environment and adapt the material's or surface's properties accordingly. As a result, these substances can modify their behavior in response to environmental cues while continuing to function admirably (Chen et al., 2009).

In comparison to its passive counterparts, active metamaterials and metasurfaces boast a number of major benefits, including enhanced versatility, adaptability, and performance. In the future, they will undoubtedly play a more crucial part in the advancement of technologies like communications, sensing, imaging, and defense.

B. Design Principles of Active Metamaterials and Metasurfaces

The design principles of active metamaterials and metasurfaces are more complex than those of passive metamaterials and metasurfaces, as they require the integration of active components into the metamaterial structure. Here are some of the key design principles of active metamaterials and metasurfaces:

Active components: Active metamaterials and metasurfaces require active components, such as transistors, diodes, or other active elements, that can be used to control the properties of the metamaterial or metasurface. The choice of active component depends on the specific application, and different types of active components have different performance characteristics and limitations.

Control circuitry: The active components in the metamaterial or metasurface require control circuitry that can be used to adjust their properties. The control circuitry can be integrated into the metamaterial structure, or it can be located outside of the metamaterial and connected via wires or other means.

Tuning mechanism: Active metamaterials and metasurfaces require a tuning mechanism that can adjust the properties of the metamaterial or metasurface in response to changes in the electromagnetic environment. This can be achieved using sensors or other feedback mechanisms that can detect changes in the environment and adjust the metamaterial or metasurface accordingly.

Electromagnetic simulation: The design of active metamaterials and metasurfaces often requires extensive electromagnetic simulation to optimize the performance of the active components and control circuitry. This can involve using computer simulations or experimental measurements to characterize the electromagnetic properties of the metamaterial or metasurface and optimize its design.

Power considerations: Active metamaterials and metasurfaces require a power source to operate the active components and control circuitry. This can be a significant consideration in the design of these materials, as they may require a significant amount of power to operate effectively.

Overall, the design of active metamaterials and metasurfaces requires a multidisciplinary approach that combines knowledge of materials science, electronics, and electromagnetics. Active metamaterials and metasurfaces have the potential to offer significant advantages over passive metamaterials and metasurfaces in terms of functionality and flexibility, and are likely to play an increasingly important role in the development of future technologies.

C. Examples of Active Metamaterials and Metasurfaces

1. Tunable Metamaterials and Metasurfaces

Tunable metamaterials and metasurfaces are a type of active metamaterial that can be dynamically tuned to respond to different frequencies of electromagnetic radiation. The key feature of tunable metamaterials and metasurfaces is that they have active components that can be adjusted to change the resonant properties of the metamaterial, allowing it to selectively reflect, transmit or absorb certain frequencies of electromagnetic radiation.

There are several different ways that tunable metamaterials and metasurfaces can be designed and controlled, including:

Material properties: Tunable metamaterials can be designed to have specific material properties that can be dynamically adjusted to control the behavior of electromagnetic waves. For example, materials such as liquid crystals or ferroelectric materials can be used to create tunable metamaterials that can be controlled using an external electric or magnetic field.

Active components: Tunable metamaterials can also incorporate active components, such as transistors or diodes, that can be used to dynamically control the resonant properties of the metamaterial. By adjusting the properties of the active components, the metamaterial can be tuned to respond to different frequencies of electromagnetic radiation.

External control: Tunable metamaterials and metasurfaces can also be controlled externally using an external signal. For example, an external electromagnetic signal could be used to control the properties of the active components in the metamaterial or metasurface, allowing it to be dynamically tuned in real-time.

The ability to dynamically tune the properties of metamaterials and metasurfaces is particularly useful for applications such as sensing, imaging, and telecommunications, where the ability to selectively control the behavior of electromagnetic waves is critical. By tuning the properties of the metamaterial or metasurface in real-time, these materials can adapt to changes in the environment and optimize their performance for a specific application.

2. Nonlinear Metamaterials and Metasurfaces

Nonlinear metamaterials and metasurfaces are a type of active metamaterial that exhibit nonlinear optical or electromagnetic responses. In these materials, the response of the metamaterial or metasurface is not proportional to the incident electromagnetic field, but rather exhibits a nonlinear dependence on the field strength. This nonlinear response can be used to manipulate the behavior of electromagnetic waves in a variety of ways, including frequency conversion, harmonic generation, and optical switching.

The design of nonlinear metamaterials and metasurfaces requires careful consideration of the properties of the constituent materials and the geometrical structure of the metamaterial. One approach is to use materials with large nonlinear susceptibilities, such as nonlinear crystals or semiconductors, to create nonlinear metamaterials. The geometry of the metamaterial can also be designed to enhance nonlinear effects, such as using resonant structures that can be tuned to specific frequencies.

Nonlinear metamaterials and metasurfaces have a variety of potential applications, including:

Frequency conversion: Nonlinear metamaterials can be used to convert electromagnetic radiation from one frequency to another, enabling applications such as frequency doubling or tripling for laser-based systems.

Optical switching: Nonlinear metamaterials can also be used for optical switching, allowing for the modulation of the transmission or reflection of electromagnetic radiation at specific frequencies.

Nonlinear sensing: Nonlinear metamaterials can be used for sensing applications, such as detecting small changes in the refractive index of a sample, by exploiting the nonlinear response of the metamaterial.

Overall, nonlinear metamaterials and metasurfaces offer a new range of functionalities that cannot be achieved using traditional linear materials, making them promising for a wide range of applications in fields such as telecommunications, sensing, and quantum information processing. However, the design and fabrication of these materials can be challenging, and more research is needed to fully understand their properties and potential applications.

3. Active Camouflage

Active camouflage is a type of active metamaterial and metasurface that allows objects to blend into their surroundings by dynamically changing their appearance in response to the environment. These materials can be used for a variety of applications, including military and civilian applications where stealth is important. Active camouflage can be achieved through the use of metamaterials and metasurfaces that have tunable optical or electromagnetic properties. By controlling the resonant properties of the metamaterial, it is possible to selectively reflect, absorb or transmit different frequencies of electromagnetic radiation, allowing the object to match its surroundings. For example, a military vehicle can be covered in an active camouflage material that can adapt its appearance to match the color and texture of the surrounding terrain, making it more difficult to detect.

The design of active camouflage materials requires careful consideration of the properties of the constituent materials, as well as the geometrical structure of the metamaterial. One approach is to use materials with tunable refractive indices, such as liquid crystals or electro-optic materials, to create tunable metamaterials that can be controlled using an external electric or magnetic field. The geometry of the metamaterial can also be designed to enhance the effect, such as using microstructures that can be tuned to specific frequencies.

Active camouflage has a variety of potential applications, including:

Military applications: Active camouflage can be used to improve the stealth capabilities of military vehicles, aircraft, and soldiers, allowing them to blend into their surroundings and avoid detection.

Civilian applications: Active camouflage can also be used for civilian applications, such as in the design of buildings or vehicles that need to blend into their surroundings, or in the development of clothing that can change its appearance based on the environment.

Artistic applications: Active camouflage materials can also be used for artistic applications, such as in the design of interactive displays or installations that respond to the environment.

Overall, active camouflage is an exciting and rapidly developing area of research that has the potential to revolutionize the way we think about camouflage and concealment. However, the development of these materials is still in its early stages, and more research is needed to fully understand their properties and potential applications.

IV. FABRICATION TECHNIQUES FOR METAMATERIALS AND METASURFACES

A. Overview of Fabrication Techniques for Metamaterials and Metasurfaces

Advanced manufacturing techniques are often used to construct metamaterials and metasurfaces. These techniques provide exact material property and geometry control at the nanoscale level, which is necessary for the creation of metamaterials and metasurfaces. The following are some of the fabrication techniques that are utilized most frequently:

Electron beam lithography (EBL): Electron beam lithography, also known as EBL, is a method of patterning with a high resolution that involves the application of a concentrated beam of electrons to a substrate in order to selectively remove material from the substrate. EBL can be applied to a wide range of materials, including metals, semiconductors, and polymers, in order to generate intricate nanoscale patterns on their surfaces (Fan et al., 2013).

Nanoimprint lithography (NIL): Nanoimprint lithography, often known as NIL, is a process that relies on replication and involves the use of a mold or stamp to transfer a pattern from one substrate to another. NIL is a technology that may be used to generate complicated patterns on wide regions of substrates. It is also an inexpensive technique with a high throughput (Scheuer et al., 2020).

Atomic layer deposition (ALD): Atomic layer deposition, often known as ALD, is a method of thin-film deposition that can be used to produce thin films of metals, oxides, and other materials onto a substrate. ALD is an abbreviation for atomic layer deposition. atomic layer deposition (ALD) is capable of being utilized in the production of complicated multilayer structures with exact thickness control on a nanoscale level (Liu et al., 2008).

Laser machining: Machining using a laser is a versatile process that makes use of a focused laser beam to selectively remove or change material from a substrate. This process is known as laser machining. Machining using a laser can be used to generate intricate patterns on many different types of materials, including metals, semiconductors, and polymers, among others.

3D printing: 3D printing is a layer-by-layer additive manufacturing technique that can be used to create complex, three-dimensional structures with high resolution. 3D printing can be used to create metamaterials and metasurfaces with complex geometries and tailored properties.

organization of materials into specific structures based on their inherent properties. Self-assembly can be used to create complex, three-dimensional structures with tailored properties and functions (Bordjiba et al., 2008).

Colloidal assembly: Colloidal assembly is a technique that uses the self-assembly of colloidal particles to create complex, nanoscale patterns on a substrate. Colloidal assembly can be used to create metamaterials and metasurfaces with tailored optical and electromagnetic properties.

These fabrication techniques allow for the creation of a wide range of metamaterials and metasurfaces with unique and tailored properties. The choice of technique will depend on the specific application and the desired properties of the material

B. Advantages and Disadvantages of Different Fabrication Techniques

Each fabrication technique for metamaterials and metasurfaces has its own advantages and disadvantages, which depend on the specific application and the desired properties of the material. Here are some general advantages and disadvantages of common fabrication techniques:

Electron beam lithography (EBL):
Advantages: High resolution patterning, versatility, and precision.
Disadvantages: Limited scalability, high cost, and time-consuming process.

Nanoimprint lithography (NIL):
Advantages: High throughput, low cost, and large area patterning.
Disadvantages: Limited resolution, limited feature sizes, and limited material choices.

Atomic layer deposition (ALD):
Advantages: High precision and controllability, wide range of materials, and uniform thin-film deposition.
Disadvantages: Limited scalability, slow process, and complex equipment.

Laser machining:
Advantages: High resolution patterning, versatility, and speed.
Disadvantages: Limited scalability, limited material choices, and potentially high cost.

3D printing:
Advantages: High resolution and controllability, versatility, and design flexibility.
Disadvantages: Limited scalability, limited material choices, and potentially high cost.

Self-assembly:
Advantages: Low cost, high scalability, and versatility.
Disadvantages: Limited control over the final structure and properties, and limited reproducibility.

Colloidal assembly:
Advantages: Low cost, scalability, and flexibility in terms of material choices.
Disadvantages: Limited control over the final structure and properties, and limited reproducibility.

In general, high-resolution patterning techniques such as EBL and laser machining are best suited for creating small-scale, complex metamaterial and metasurface structures, while imprint-based techniques like NIL are better suited for large scale fabrication. ALD is best suited for creating thin-film structures, while 3D printing is ideal for creating complex three-dimensional structures. Self-assembly and colloidal assembly are best suited for low-cost, high-throughput fabrication of simple structures.

C. Examples of Fabrication Techniques

1. Lithography

Metamaterials are often made via lithography. It includes patterning a surface with a mask or template to generate intricate, small-scale designs. EBL, photolithography, and nanoimprint lithography are utilised to make metamaterials and metasurfaces (NIL) (Kim et al., 2015)

A concentrated electron beam exposes a resist material to generate a patterned surface in EBL. It can design nanometer-sized features. EBL is slow and expensive, making large-scale production difficult.

Photolithography patterns surfaces with a mask or template and UV light. The mask or template is placed on the photoresist layer. UV light selectively exposes photoresist through the mask or template, generating a patterned surface. Scalability and low cost make photolithography popular. Its resolution is too low for sophisticated three-dimensional structures.

NIL patterns surfaces with moulds or templates. The mould or template is pushed onto the polymer, creating a patterned surface. NIL creates large-area patterns quickly. Its resolution is too low for sophisticated three-dimensional structures (Zhang et al., 2012).

Depending on the metamaterial or metasurface's application and desired qualities, lithography processes are diverse and accurate.

2. 3D Printing

3D printing, also known as additive manufacturing, creates complex and customizable parts with high precision. It uses CAD to layer a three-dimensional object. 3D printing technologies include FDM, SLA, and SLS. FDM 3D printing creates the structure by layering molten plastic on a build plate. Fast, cheap, and can build massive structures. FDM printing has low resolution and restricted material compatibility. SLA 3D printing solidifies liquid resin layer by layer with a laser. FDM printing is faster and cheaper, yet this method can manufacture complex geometries with great resolution. SLS 3D printing layers powdered material with a laser. This method prints metals and ceramics with great resolution and strength. SLS printing is sluggish and expensive. 3D printing can build complicated, customizable structures with high precision. Nevertheless, its disadvantages include low resolution, restricted material compatibility, and slower, more expensive printing times for specific procedures. Research is underway to increase 3D printing's speed, resolution, and material compatibility.

3. Atomic Layer Deposition

Atomic layer deposition (ALD) is a technique for depositing thin films with high control over thickness, composition, and uniformity. ALD is a type of chemical vapor deposition (CVD) that involves a series of self-limiting chemical reactions between gas-phase precursors and a substrate surface. During each reaction cycle, a single atomic or molecular layer is deposited, and the process can be repeated to build up the desired thickness of the film.

The ALD process typically involves two precursor chemicals, each introduced to the substrate in a separate step. The first precursor reacts with the substrate surface to form a self-limiting monolayer, while the excess unreacted precursor is removed from the reactor. The second precursor is then introduced, which reacts with the monolayer to form another self-limiting layer. This cycle can be repeated to build up the desired thickness of the film.

ALD has several advantages as a fabrication technique for metamaterials and metasurfaces. First, it allows for precise control over film thickness, composition, and uniformity, which is critical for designing functional metamaterials and metasurfaces. Second, ALD can deposit a wide range of materials, including metals, oxides, and nitrides, making it a versatile technique for designing metamaterials with a range of optical and electronic properties. Third, ALD can be performed at low temperatures, which is useful for depositing materials on substrates that cannot withstand high temperatures.

However, there are also some limitations to the ALD technique. ALD is typically a slow process, which can limit its scalability for large-area or high-volume manufacturing. Additionally, the required precursors for ALD can be expensive, and the technique may not be cost-effective for large-scale production. Despite these limitations, ALD remains a promising technique for the fabrication of high-quality, precise metamaterials and metasurfaces.

V. FUTURE DIRECTIONS AND CHALLENGES

A. Emerging Trends in Active and Passive Metamaterials and Metasurfaces

Active and passive metamaterials and metasurfaces continue to be a rapidly evolving field of research, with new developments and emerging trends constantly emerging. Some of the most notable emerging trends in the field include:

Multifunctional metamaterials: Researchers are increasingly exploring ways to design metamaterials and metasurfaces that can perform multiple functions simultaneously. For example, a single metamaterial may be designed to both absorb and transmit light, or to act as both a lens and a cloaking device.

Bioinspired metamaterials: As researchers continue to explore the use of metamaterials and metasurfaces in biomedical applications, there is growing interest in designing materials that mimic natural biological systems. These materials may be designed to interact with biological systems in specific ways, such as targeting cancer cells for drug delivery or detecting pathogens in the body.

Metasurface holography: Metasurface holography is an emerging technology that uses metasurfaces to generate 3D holographic images. These holograms can be used for a range of applications, from medical imaging to entertainment.

Hybrid metamaterials: Researchers are exploring the use of hybrid metamaterials, which combine different types of metamaterials in a single structure. For example, a hybrid metamaterial may combine a negative index material with a plasmonic material to create a material with unique optical properties.

Nonlinear metasurfaces: Nonlinear metasurfaces are designed to interact with light in nonlinear ways, enabling a range of new applications in areas such as optical computing, telecommunications, and sensing.

Reconfigurable metamaterials: Reconfigurable metamaterials are designed to change their properties in response to external stimuli. These materials have potential applications in areas such as adaptive optics, sensing, and reconfigurable electronics.

Metasurface-based sensors: Metasurface-based sensors are a rapidly emerging area of research. These sensors use metasurfaces to detect and analyze a range of physical, chemical, and biological signals with high sensitivity and specificity.

Overall, the emerging trends in active and passive metamaterials and metasurfaces suggest that these materials will continue to play an increasingly important role in a wide range of applications, from telecommunications to biomedicine to sensing and beyond.

B. Challenges in the Design and Fabrication of Metamaterials and Metasurface

While the field of metamaterials and metasurfaces has made significant progress in recent years, there are still several challenges that researchers face in the design and fabrication of these materials. Some of the most significant challenges include:

Fabrication complexity: Many metamaterials and metasurfaces require complex fabrication processes, often involving the use of specialized equipment and techniques. This complexity can make it difficult to produce large quantities of these materials quickly and cost-effectively.

Material limitations: The performance of metamaterials and metasurfaces is often limited by the properties of the materials used to construct them. For example, some materials may have limited bandwidth, making it difficult to create metamaterials that operate over a wide range of frequencies.

Integration with existing systems: For metamaterials and metasurfaces to be widely adopted in practical applications, they need to be compatible with existing systems and technologies. Achieving this integration can be a significant challenge, particularly in fields such as telecommunications and computing.

Design optimization: The design of metamaterials and metasurfaces is often complex and requires sophisticated modeling and simulation tools. Achieving the optimal design for a given application can be a time-consuming and challenging process.

Environmental stability: Many metamaterials and metasurfaces are sensitive to changes in temperature, humidity, and other environmental factors. Achieving stable performance in a range of environmental conditions can be a significant challenge.

Scalability: While many metamaterials and metasurfaces have demonstrated promising results in the laboratory, scaling these materials up to practical applications can be a significant challenge. Achieving high-volume production of these materials at a reasonable cost is a key challenge for the field.

Overall, addressing these challenges will be critical to the continued advancement of metamaterials and metasurfaces and their adoption in practical applications. Ongoing research is focused on developing new fabrication techniques, improving material properties, optimizing design tools, and achieving greater integration with existing systems.

C. Future Applications and Opportunities for Metamaterials and Metasurfaces

Metamaterials and metasurfaces have the potential to enable a wide range of new applications and technologies across numerous fields. Some of the most promising future applications and opportunities for these materials include:

Communications and computing: Metamaterials and metasurfaces could enable the development of new types of antennas and sensors that operate at higher frequencies and with greater sensitivity than current technologies. They could also be used to create new types of waveguides and filters for use in communications and computing applications.

Energy: Metamaterials and metasurfaces could be used to develop new types of solar cells that are more efficient at converting sunlight into electricity. They could also be used to create new types of thermoelectric materials that can convert waste heat into electricity.

Medicine: Metamaterials and metasurfaces could be used to create new types of medical imaging technologies that provide higher resolution and sensitivity than current technologies. They could also be used to create new types of drug delivery systems that can precisely target specific cells or tissues.

Defense and security: Metamaterials and metasurfaces could be used to create new types of sensors and imaging technologies for use in defense and security applications. They could also be used to create new types of camouflage materials that can hide objects or people from detection.

Consumer electronics: Metamaterials and metasurfaces could be used to create new types of display technologies that are thinner, lighter, and more energy-efficient than current technologies. They could also be used to create new types of touch sensors and haptic feedback systems.

Environmental monitoring: Metamaterials and metasurfaces could be used to create new types of sensors for monitoring air and water quality, detecting pollutants and toxins, and tracking climate change.

Overall, the potential applications for metamaterials and metasurfaces are vast and varied, and ongoing research is focused on unlocking their full potential. As new materials and fabrication techniques continue to emerge, and our understanding of these materials improves, we can expect to see continued innovation and progress in this field.

VI. CONCLUSION

A. Summary of Key Points

The following is a brief synopsis of the discussion that ensued:

Metamaterials are man-made structures that have unusual qualities that can't be found in nature.

Light and other forms of electromagnetic radiation can be controlled via metasurfaces, which are thin sheets of metamaterials. Applications for passive metamaterials and metasurfaces include negative index materials, superlenses, perfect absorbers, and electromagnetic cloaks, and they do not require an external power source.

Metasurfaces and metamaterials that are "active" require an external power source and have potential uses include active camouflage, adjustable materials, and nonlinear materials. Design principles for metamaterials and metasurfaces include approaches including resonant structures, geometric phase, and transformation optics, however these methods vary by application and desired qualities. Several different methods, such as lithography, 3D printing, atomic layer deposition, and others, can be used to fabricate metamaterials and metasurfaces. Factors like scalability, resolution, and cost determine the merits and demerits of various fabrication approaches. Multifunctional and reconfigurable materials, as well as the integration of these materials with other technologies like nanoelectronics and photonics, are two of the rising themes in the field of metamaterials and metasurfaces. Scalability, stability, and reproducibility are only few of the difficulties that must be overcome in the design and manufacture of metamaterials and metasurfaces.

Communications and computers, energy, medical, military and security, consumer electronics, and environmental monitoring are just some of the potential future uses for metamaterials and metasurfaces.

B. Importance of Active and Passive Metamaterials and Metasurfaces

Different applications require active and passive metamaterials and metasurfaces.

Passive metamaterials and metasurfaces modify electromagnetic radiation without power. Negative index materials, superlenses, perfect absorbers, and electromagnetic cloaks can use them. Passive metamaterials and metasurfaces improve electromagnetic radiation interaction with materials, making it easier to detect subtle environmental changes.

Reconfigurable devices, sensors, and antennas use active metamaterials and metasurfaces, which require an external power source. Frequency conversion and nonlinear optics are nonlinear applications. Active metamaterials and metasurfaces can be camouflaged by changing their appearance.

Active and passive metamaterials and metasurfaces are used in communications, computers, energy, medical, military and security, consumer electronics, and environmental monitoring. They could transform these fields by enabling new technologies and functions.

C. Directions for Future Research and Development

Future research and development in metamaterials and metasurfaces could focus on the following directions:

Multifunctional materials: Developing materials that can exhibit multiple functions simultaneously, such as being both a superlens and a perfect absorber, or having both active and passive functionalities.

Dynamic and adaptive materials: Designing materials that can change their properties in response to external stimuli, such as temperature, light, or electric fields. This could enable materials to have tunable properties, making them useful for a variety of applications.

Miniaturization: Developing metamaterials and metasurfaces that can be made on a small scale, such as at the nanoscale, which would enable the development of miniaturized devices for applications in electronics and photonics.

Integration with other technologies: Integrating metamaterials and metasurfaces with other technologies, such as sensors, actuators, and electronics, could enable the development of new devices with enhanced functionalities.

New fabrication techniques: Developing new fabrication techniques that are faster, cheaper, and more efficient could make metamaterials and metasurfaces more accessible and enable their use in a wider range of applications.

Exploration of new materials: Exploring new materials and their properties, such as 2D materials, could lead to the discovery of new metamaterials and metasurfaces with unique properties and functionalities.

Overall, future research and development in metamaterials and metasurfaces has the potential to lead to the development of new technologies and applications, and could have a significant impact on various fields such as communications, energy, medicine, and defense.

REFERENCES

Bordjiba, T., Mohamedi, M., Dao, L. H., Aissa, B., & El Khakani, M. A. (2007). Enhanced physical and electrochemical properties of nanostructured carbon nanotubes coated microfibrous carbon paper. *Chemical Physics Letters*, *441*(1-3), 88–93. doi:10.1016/j.cplett.2007.04.072

Chen, H. T., O'Hara, J. F., Taylor, A. J., Averitt, R. D., Highstrete, C., Lee, M., & Padilla, W. J. (2007). Complementary planar terahertz metamaterials. *Optics Express*, *15*(3), 1084–1095. doi:10.1364/OE.15.001084 PMID:19532336

Chen, H. T., Padilla, W. J., Cich, M. J., Azad, A. K., Averitt, R. D., & Taylor, A. J. (2009). A metamaterial solid-state terahertz phase modulator. *Nature Photonics*, *3*(3), 148–151. doi:10.1038/nphoton.2009.3

Chen, H. T., Padilla, W. J., Zide, J. M., Gossard, A. C., Taylor, A. J., & Averitt, R. D. (2006). Active terahertz metamaterial devices. *Nature*, *444*(7119), 597–600. doi:10.1038/nature05343 PMID:17136089

Do, T. V., Bui, S. T., Nguyen, T. H., Nguyen, T. T., Nguyen, T. T., Lee, Y., & Vu, D. L. (2012). Design, fabrication and characterization of a perfect absorber using simple cut-wire metamaterials. *Advances in Natural Sciences: Nanoscience and Nanotechnology*, *3*(4), 045014. doi:10.1088/2043-6262/3/4/045014

Fan, K., Strikwerda, A. C., Zhang, X., & Averitt, R. D. (2013). Three-dimensional broadband tunable terahertz metamaterials. *Physical Review B: Condensed Matter and Materials Physics*, *87*(16), 161104. doi:10.1103/PhysRevB.87.161104

Haxha, S., AbdelMalek, F., Ouerghi, F., Charlton, M. D. B., Aggoun, A., & Fang, X. J. S. R. (2018). Metamaterial superlenses operating at visible wavelength for imaging applications. *Scientific Reports*, *8*(1), 16119. doi:10.103841598-018-33572-y PMID:30382113

Kim, M., & Rho, J. (2015). Metamaterials and imaging. *Nano Convergence*, *2*(1), 1–16. doi:10.118640580-015-0053-7 PMID:28191408

Kröling, H., Endresb, A., Nubboc, N., Fleckensteinc, J., Miletzkyb, A., & Schabela, S. (2014). Anisotropy of paper and paper based composites and the modelling thereof. *ECCM16*, 1-8.

Liu, N., Guo, H., Fu, L., Kaiser, S., Schweizer, H., & Giessen, H. (2008). Three-dimensional photonic metamaterials at optical frequencies. *Nature Materials*, *7*(1), 31–37. doi:10.1038/nmat2072 PMID:18059275

O'Hara, J. F., Singh, R., Brener, I., Smirnova, E., Han, J., Taylor, A. J., & Zhang, W. (2008). Thin-film sensing with planar terahertz metamaterials: Sensitivity and limitations. *Optics Express*, *16*(3), 1786–1795. doi:10.1364/OE.16.001786 PMID:18542258

Padilla, W. J., Aronsson, M. T., Highstrete, C., Lee, M., Taylor, A. J., & Averitt, R. D. (2007). Electrically resonant terahertz metamaterials: Theoretical and experimental investigations. *Physical Review B: Condensed Matter and Materials Physics*, *75*(4), 041102. doi:10.1103/PhysRevB.75.041102

Padilla, W. J., Basov, D. N., & Smith, D. R. (2006). Negative refractive index metamaterials. *Materials Today*, *9*(7-8), 28–35. doi:10.1016/S1369-7021(06)71573-5

Padilla, W. J., Taylor, A. J., Highstrete, C., Lee, M., & Averitt, R. D. (2006). Dynamical electric and magnetic metamaterial response at terahertz frequencies. *Physical Review Letters*, *96*(10), 107401. doi:10.1103/PhysRevLett.96.107401 PMID:16605787

Peralta, X. G., Arrington, C. L., Wanke, M. C., Brener, I., Williams, J. D., Smirnova, E., ... Padilla, W. J. (2008, May). Flexible, large-area metamaterials fabricated on thin silicon nitride membranes. In *Conference on Lasers and Electro-Optics* (p. CFY4). Optical Society of America. 10.1109/CLEO.2008.4551092

Peralta, X. G., Smirnova, E. I., Azad, A. K., Chen, H. T., Taylor, A. J., Brener, I., & O'Hara, J. F. (2009). Metamaterials for THz polarimetric devices. *Optics Express*, *17*(2), 773–783. doi:10.1364/OE.17.000773 PMID:19158890

Scheuer, J. (2020). Optical metasurfaces are coming of age: Short-and long-term opportunities for commercial applications. *ACS Photonics*, *7*(6), 1323–1354. doi:10.1021/acsphotonics.9b01719

Schurig, D., Mock, J. J., Justice, B. J., Cummer, S. A., Pendry, J. B., Starr, A. F., & Smith, D. R. (2006). Metamaterial electromagnetic cloak at microwave frequencies. *Science*, *314*(5801), 977–980. doi:10.1126cience.1133628 PMID:17053110

Zhang, X., & Liu, Z. (2008). Superlenses to overcome the diffraction limit. *Nature Materials*, *7*(6), 435–441. doi:10.1038/nmat2141 PMID:18497850

Compilation of References

Abadal, S., Liaskos, C., Tsioliaridou, A., Ioannidis, S., Pitsillides, A., Sole-Pareta, J., Alarcon, E., & Cabellos-Aparicio, A. (2017). Computing and Communications for the Software-Defined Metamaterial Paradigm: A Context Analysis. *IEEE Access : Practical Innovations, Open Solutions*, *5*, 6225–6235. doi:10.1109/ACCESS.2017.2693267

Abdalla, M., Abdelnaby, U., & Mitkees, A. A. (n.d.). Compact and Triple Band Meta- material Antenna for All WiMAX Applications. *Proceedings of ISAP2012.*

Abraham, J. K., Vinoy, K. J., Tellakula, R. A., & Varadan, V. K. (2003). Experimental Results of Planar Meta Materials. *IEEE- APS International Symposium.*

Afacan, E., Lourenço, N., Martins, R., & Dündar, G. (2021). Review: Machine learning techniques in analog/RF integrated circuit design, synthesis, layout, and test. *Integration, 77*, 113-130. https:// doi.org/ doi:10.1016/j.vlsi.2020.11.006

Agarwal, M., Behera, A. K., & Meshram, M. K. (2016). Wide-angle quad-band polarization insensitive metamaterial absorber. *Electronics Letters*, *52*(5), 340–342. doi:10.1049/el.2015.4134

Agrawal, A., & Misra, M. (2021). Angle of Incidence and Polarization Insensitive Ultra-Thin Ultra Wide-Band Metamaterial Absorber based on Novel Non-Linearity in Unit Cell Design. *Journal of Microwaves, Optoelectronics and Electromagnetic Applications*, *20*(2), 556–569. doi:10.1590/2179-10742021v20i31207

Agrawal, A., Misra, M., & Singh, A. (2017). Oblique Incidence and Polarization Insensitive Multiband Metamaterial Absorber with Quad Paired Concentric Continuous Ring Resonators. *Progress in Electromagnetics Research M. Pier M*, *60*, 33–46. doi:10.2528/PIERM17061302

Akinsolu, M. O., Mistry, K. K., Liu, B., Lazaridis, P. I., & Excell, P. (2020). Machine learning-assisted antenna design optimization: A review and the state-of-the-art. *Proc. 14th Eur. Conf. Antennas Propag. (EuCAP),* 1–5. 10.23919/EuCAP48036.2020.9135936

Alavikia, B., Almoneef, T. S., Ramahi, O. M., Alavikia, B., Almoneef, T. S., & Ramahi, O. M. (2014). Electromagnetic energy harvesting using complementary split-ring resonators. *Applied Physics Letters*, *104*(16), 163903-1, 163903–163904. doi:10.1063/1.4873587

Ali, M., Sayem, A., & Kunda, V. (2007, March). A reconfigurable stacked microstrip patch antenna for satellite and terrestrial links. *IEEE Transactions on Vehicular Technology*, *56*(2), 426–435. doi:10.1109/TVT.2007.891412

Almoneef, T. S., & Ramahi, O. M. (2014). A 3-dimensional stacked metamaterial arrays for electromagnetic energy harvesting. *Electromagnetic Waves*, *146*, 109–115. doi:10.2528/PIER14031603

Almoneef, T. S., & Ramahi, O. M. (2015). Metamaterial electromagnetic energy harvester with near unity efficiency. *Applied Physics Letters*, *106*(15), 153902-1, 153902–153904. doi:10.1063/1.4916232

Almpanis, E., Pantazopoulos, P. A., Papanikolaou, N., Yannopapas, V., & Stefanou, N. (2017). A birefringent etalon enhances the Faraday rotation of thin magneto-optical films. *Journal of Optics*, *19*(7), 075102. doi:10.1088/2040-8986/aa7420

Al-Otaibi, S., Cherappa, V., Thangarajan, T., Shanmugam, R., Ananth, P., & Arulswamy, S. (2023). Hybrid K-medoids with energy-efficient sunflower optimization algorithm for Wireless Sensor Networks. *Sustainability (Basel)*, *15*(7), 5759. doi:10.3390u15075759

Alsharari, M. (2023). *Monopole Antenna Modeling and Optimization for 5G Communication System using Machine Learning. In 2023 International Wireless Communications and Mobile Computing*. IWCMC.

Alù, A., & Engheta, N. (2005, July). Achieving transparency with plasmonic and metamaterial coatings. *Physical Review E: Statistical, Nonlinear, and Soft Matter Physics*, *72*(1), 016623. doi:10.1103/PhysRevE.72.016623 PMID:16090123

Alù, A., & Engheta, N. (2017). Magnetoelectric metamaterials. *Nature Reviews. Materials*, *2*(8), 16095.

Alù, A., & Engheta, N. (2017). Nonlinear optics: Breaking linear limits in metamaterials. *Science*, *358*(6362), caan2472.

Alu, A., Engheta, N., Erentok, A., & Ziolkowski, R. W. (2007, February). Single-negative, double-negative, and low-index metamaterials and their electromagnetic applications. *IEEE Antennas & Propagation Magazine*, *49*(1), 23–36. doi:10.1109/MAP.2007.370979

Andryieuski, A., Kuznetsova, S. M., Zhukovsky, S. V., Kivshar, Y. S., & Lavrinenko, A. V. (2015). Water: Promising Opportunities For Tunable All-dielectric Electromagnetic Metamaterials. *Scientific Reports*, *5*(1), 9. doi:10.1038rep13535 PMID:26311410

Anigbogu, W., & Bardaweel, H. (2020). A Metamaterial-Inspired Structure for Simultaneous Vibration Attenuation and Energy Harvesting. *Shock and Vibration*, *2020*, 1–12. doi:10.1155/2020/4063025

Antoniades, M. A., & Edward, S. (2012, December). Multiband Compact Printed Dipole Antennas UsingNRI-TL Metamaterial Loading. *IEEE Transactions on Antennas and Propagation*, *60*(12), 5613–5626. doi:10.1109/TAP.2012.2211324

Antoniades, M. A., Eleftheriades, G. V., & Rogers, E. S. (2009). A broadband dual-mode monopole antenna using NRI-TL metamaterial loading. *IEEE Antennas and Wireless Propagation Letters*, *8*, 258–261. doi:10.1109/LAWP.2009.2014402

Araghi, M. (2022, October). Systematic Design of a Holographic-Based Metasurface Reflector in the Sub-6 GHz Band. *IEEE Antennas and Wireless Propagation Letters*, *21*(10), 1960–1964. doi:10.1109/LAWP.2022.3186922

Arun Kumar, K. (2010). Optimization of Multislotted Rectangular Microstrip Patch Antenna Using ANN Becterial Foraging Optimization. *Asia-Pacific Symposium on Electromagnetic Compatibility (APEMC)*, 449 – 452.

Asili, M., Chen, P., Hood, A., Purser, A., Hulsey, R., Johnson, L., Venkataraman Ganesan, A., Demirci, U., & Topsakal, E. (2015). Flexible microwave antenna applicator for chemo-thermotherapy of the breast. *IEEE Antennas and Wireless Propagation Letters*, *14*(7), 1778–1781. doi:10.1109/LAWP.2015.2423655

Assimonis, S. D., Kollatou, T., Tsiamitros, D., Stimoniaris, D., Samaras, T., & Sahalos, J. N. (2016). High efficiency and triple-band metamaterial electromagnetic energy hervester. *ELECO 2015 - 9th International Conference on Electrical and Electronics Engineering*, 320–323. 10.1109/ELECO.2015.7394527

Augenstein, Y., Repan, T., & Rockstuhl, C. (2023). Neural Operator-Based Surrogate Solver for Free-Form Electromagnetic Inverse Design. *ACS Photonics*, *10*(5), 1547–1557. Advance online publication. doi:10.1021/acsphotonics.3c00156

Bala, B. D., Rahim, M. K. A., & Murad, N. A. (2014, January 30). Small electrical metamaterial antenna based on coupled electric field resonator with enhanced bandwidth. *Electronics Letters*, *50*(3), 138–139. doi:10.1049/el.2013.3884

Balaji, R., & Ramachandran, K. N. (2017). Magnetic resonance imaging of a benign phyllodes tumor of the breast. *Breast Care (Basel, Switzerland)*, *4*(3), 125–137. PMID:20847879

Balanis, C. A. (2012). *Antenna theory: Analysis and design.* John Wiley & Sons. doi:10.1109/LMWC.2004.828009

Bashar & Koziel. (2023). High isolation metamaterial-based dual-band MIMO antenna for 5G millimeter-wave applications. *International Journal of Electronics and Communication, 158*, 154470.

Bashir, D. (2012). Bandwidth Enhanced Microstrip Patch Antenna Using Metamaterials. *IEEE Asia-Pacific Conference on Applied Electromagnetics (APACE),* 280 – 282.

Bassi, M., Caruso, M., Khan, M. S., Bevilacqua, A., Capobianco, A. D., & Neviani, A. (2020). An integrated microwave imaging radar with planar antennas for breast cancer detection. *IEEE Transactions on Microwave Theory and Techniques*, *61*(5), 2108–2118. doi:10.1109/TMTT.2013.2247052

Bellekhiri, A., Chahboun, N., Laaziz, Y., & El, A. (2022). A New Design of 5G Planar Antenna based on metamaterials with a high gain using array antenna. *ITM Web of Conferences, 48.*

Bello, R., Ramírez, J. A., & Pomares, H. (2012). A review on clustering algorithms for fuzzy model-based control. *IEEE Transactions on Fuzzy Systems*, *20*(4), 644–658.

Bezdek, J. C. (1981). *Pattern recognition with fuzzy objective function algorithms*. Plenum Press. doi:10.1007/978-1-4757-0450-1

Bhagat, P. P. (2012). Analysis and Synthesis of Microstrip Patch Antenna using Artificial Neural Networks. *IEEE Asia-Pacific Conference on Antennas and Propagation*. 10.1109/APCAP.2012.6333177

Bhargava, D., & Rattanadecho, P. (2022). Microstrip antenna for radar-based microwave imaging of Breast Cancer: Simulation Analysis. *International Journal on Communications Antenna and Propagation*, *12*(1), 47. doi:10.15866/irecap.v12i1.20702

Bianco, S. (2020). AESA Adaptive Beamforming Using Deep Learning. *Proceedings of 2020 IEEE Radar Conference.*

Bilotti, F., Nucci, L., & Vegni, L. (2006). An SRR based microwave absorber. *Microwave and Optical Technology Letters*, *48*(11), 2171–2175. doi:10.1002/mop.21891

Biswas, A., Zekios, C. L., Ynchausti, C., Howell, L. L., Magleby, S. P., & Georgakopoulos, S. V. (2022). An ultra-wideband origami microwave absorber. *Scientific Reports*, *12*(1), 13449. doi:10.103841598-022-17648-4 PMID:35927331

Bordjiba, T., Mohamedi, M., Dao, L. H., Aissa, B., & El Khakani, M. A. (2007). Enhanced physical and electrochemical properties of nanostructured carbon nanotubes coated microfibrous carbon paper. *Chemical Physics Letters*, *441*(1-3), 88–93. doi:10.1016/j.cplett.2007.04.072

Bourqui, J., Kuhlmann, M., Kurrant, D., Lavoie, B., & Fear, E. (2018). Adaptive monostatic system for measuring microwave reflections from the breast. *Sensors (Basel)*, *18*(5), 1340. doi:10.339018051340 PMID:29701677

Britto, E. C., Danasegaran, S. K., Xavier, S. C., & Lalithakumari, S. (2022). Investigation of electromagnetic wave propagation in a defected photonic Crystal Square lattice structure. *Journal of Electronic Materials*, *52*(2), 1177–1185. doi:10.100711664-022-10058-2

Britto, E. C., Danasegaran, S. K., Xavier, S. C., Sridevi, A., & Batcha, A. R. S. (2022). Study of Various Beamformers and Smart Antenna Adaptive Algorithms for Mobile Communication. In P. K. Malik, J. Lu, B. T. P. Madhav, G. Kalkhambkar, & S. Amit (Eds.), *Smart Antennas. EAI/Springer Innovations in Communication and Computing*. Springer. doi:10.1007/978-3-030-76636-8_10

Brown, W. C. (1984). The History of Power Transmission by Radio Waves. *IEEE Transactions on Microwave Theory and Techniques*, *32*(9), 1230–1242. doi:10.1109/TMTT.1984.1132833

Brunelli, D., Benini, L., Moser, C., & Thiele, L. (2008). An Efficient Solar Energy Harvester for Wireless Sensor Nodes. *Design, Automation and Test in Europe*, 104–109.

Byrne, D., Sarafianou, M., & Craddock, I. J. (2011). Compound radar approach for breast imaging. *IEEE Transactions on Biomedical Engineering*, *64*(1), 40–51. doi:10.1109/TBME.2016.2536703 PMID:26955012

Caloz, C., & Itoh, T. (2006). *Electromagnetic Metamaterials Transmission Line Theory and Microwave Applications*. Wiley.

Campbell, S. D., & Ziolkowski, R. W. (2013). Lightweight, Flexible, Polarization- Insensitive, Highly Absorbing Meta-Films. *IEEE Transactions on Antennas and Propagation*, *61*(March), 1191–1200. doi:10.1109/TAP.2012.2227658

Campione, S., Brener, I., & Marquier, F. (2015). Theory of nonlinear metamaterials: A primer. *Journal of Applied Physics*, *118*(10), 103102.

Chaimool, S., Pinsakul, A., & Akkaraekthalin, P. (2012). Patch Antenna Miniaturization Using Artificial Magneto-Dielectric Metasubstrate. *Proceedings of ISAP2012*, 906-909.

Chandra & Dwivedi. (2020). Simulation analysis of High Gain and Low Loss Antenna with Metallic Ring and spoof SPP Transmission Line. In *7th International Conference on Signal Processing & Integrated Networks*. IEEE Explore.

Chandra & Dwivedi. (2021). Graphene based Radiation Pattern reconfigurable antenna. *International Conference for convergence in Technology*.

Chaudhuri, & Pal. (2022). Design and Analysis of Programmable Cross-Polarization Converter In Terahertz Band Using Photo-excited Split-ring Resonator Based Metasurface. *2022 IEEE International Conference on Emerging Electronics (ICEE)*, 1-4. doi: 10.1109/ICEE56203.2022.10117692

Chen, D., Yang, J., Zhang, J., Huang, J., & Zhang, Z. (2017). Section 1Tunable broadband terahertz absorbers based on multiple layers of graphene ribbons. *Scientific Reports*, *7*(1), 15836. doi:10.103841598-017-16220-9 PMID:29158569

Cheng, Y., Gong, R., & Zhao, J. (2016). A photoexcited switchable perfect metamaterial absorber/reflector with polarization-independent and wide-angle for terahertz waves. *Optical Materials*, *62*, 28–33. doi:10.1016/j.optmat.2016.09.042

Chen, H. (2012). Interference theory of metamaterial perfect absorbers. *Optics Express*, *20*(7), 7165. doi:10.1364/OE.20.007165 PMID:22453398

Chen, H. T., O'Hara, J. F., Taylor, A. J., Averitt, R. D., Highstrete, C., Lee, M., & Padilla, W. J. (2007). Complementary planar terahertz metamaterials. *Optics Express*, *15*(3), 1084–1095. doi:10.1364/OE.15.001084 PMID:19532336

Chen, H. T., Padilla, W. J., Cich, M. J., Azad, A. K., Averitt, R. D., & Taylor, A. J. (2009). A metamaterial solid-state terahertz phase modulator. *Nature Photonics*, *3*(3), 148–151. doi:10.1038/nphoton.2009.3

Chen, H. T., Padilla, W. J., Zide, J. M., Gossard, A. C., Taylor, A. J., & Averitt, R. D. (2006). Active terahertz metamaterial devices. *Nature*, *444*(7119), 597–600. doi:10.1038/nature05343 PMID:17136089

Chen, H., Chen, Z. H., Yang, H., Wen, L. H., Yi, Z., Zhou, Z. G., Dai, B., Zhang, J., Wu, X., & Wu, P. H. (2022). Multi-mode surface plasmon resonance absorber based on dart-type single-layer graphene. *RSC Advances*, *12*(13), 7821–7829. doi:10.1039/D2RA00611A PMID:35424732

Chen, H.-T., Taylor, A. J., & Yu, N. (2016). A review of metasurfaces: Physics and applications. *Reports on Progress in Physics*, *79*(7), 076401. doi:10.1088/0034-4885/79/7/076401 PMID:27308726

Chen, J. M., & Chen, C. M. (2011). Design of fuzzy modelling and control systems: A tutorial review. *International Journal of Fuzzy Systems*, *13*(1), 14–33.

Chen, J., Zhang, S., Zhang, C., & Wang, X. (2015). A novel fuzzy C-means clustering algorithm based on a modified objective function. *Information Sciences*, *293*, 141–158.

Chen, S. M., & Chung, Y. C. (2021). Fuzzy Control System Design and Analysis: A Literature Review and Perspective. *IEEE Transactions on Fuzzy Systems*, *29*(1), 3–15.

Chen, X., Grzegorczyk, T. M., Wu, B., Pacheco, J., & Kong, J. (2004). Robust method to retrieve the constitutive effective parameters of metamaterials. *Physical Review. E*, *70*(1), 016608. Advance online publication. doi:10.1103/PhysRevE.70.016608 PMID:15324190

Chen, Y. (2015). Nanofabrication by electron beam lithography and its applications: A review. *Microelectronic Engineering*, *135*, 57–72. doi:10.1016/j.mee.2015.02.042

Chen, Y. S., Zhu, J. F., Xie, Y. N., Feng, N. X., & Liu, Q. H. (2019). Smart inverse design of graphene-based photonic metamaterials by an adaptive artificial neural network. *Nanoscale*, *11*(19), 9749–9755. doi:10.1039/C9NR01315F PMID:31066432

Chen, Z., Guo, B., Yang, Y., & Cheng, C. (2014). Metamaterials-based enhanced energy harvesting : A review. *Physica B. Physica B, Condensed Matter*, *438*, 1–8. doi:10.1016/j.physb.2013.12.040

Cherappa, V., Thangarajan, T., Meenakshi Sundaram, S. S., Hajjej, F., Munusamy, A. K., & Shanmugam, R. (2023). Energy-efficient clustering and routing using ASFO and a cross-layer-based expedient routing protocol for Wireless Sensor Networks. *Sensors (Basel)*, *23*(5), 2788. doi:10.339023052788 PMID:36904993

Choi, D., Shin, C., Kim, N., & Shin, H. (2005). Design and SAR Analysis of Broadband PIFA with Triple Band. *Progress in Electromagnetics Research Symposium*, *1*(3), 290–293. doi:10.2529/PIERS041212211650

Chung, M., Jeong, H., Kim, Y. K., Lim, S., & Baek, C. W. (2022). Design and Fabrication of Millimeter-Wave Frequency-Tunable Metamaterial Absorber Using MEMS Cantilever Actuators. *Micromachines*, *13*(8), 1354. Advance online publication. doi:10.3390/mi13081354 PMID:36014276

Colaco & Lohan. (2021). Metamaterial Based Multiband Microstrip Patch Antenna for 5G Wireless Technology-enabled IoT Devices and its applications. *J. Phys.: Conf. Ser.*

Collado, A., & Georgiadis, A. (2013). Conformal hybrid solar and electromagnetic (EM) energy harvesting rectenna. *IEEE Transactions on Circuits and Systems. I, Regular Papers*, *60*(8), 2225–2234. doi:10.1109/TCSI.2013.2239154

Costa, F., Monorchio, A., & Manara, G. (2010). Analysis and Design of Ultra Thin Electromagnetic Absorbers Comprising Resistively Loaded High Impedance Surfaces. *IEEE Transactions on Antennas and Propagation*, *58*(5), 1551–1558. doi:10.1109/TAP.2010.2044329

Coupland, S., & Hines, A. (2018). A review of fuzzy rule-based models for pattern classification and their applications. *Information Sciences*, *429*, 211–225.

Cui, T. J., Smith, D. R., & Liu, R. (2014). *Metamaterials: theory, design, and applications*. Springer.

Cummer, S. A., & Schurig, D. (2007, March). One path to acoustic cloaking. *New Journal of Physics*, *9*(3), 45. doi:10.1088/1367-2630/9/3/045

Dadgarpour, A., Zarghooni, B., Virdee, B., & Denidni, T. A. (2014). Beam Tilting Antenna Using Integrated Metamaterial Loading. *IEEE Transactions on Antennas and Propagation*, *62*(May), 2874–2879. doi:10.1109/TAP.2014.2308516

Dai, J. Y., Tang, W., Yang, L. X., Li, X., Chen, M. Z., Ke, J. C., Cheng, Q., Jin, S., & Cui, T. J. (2020). Realization of Multi-Modulation Schemes for Wireless Communication by Time-Domain Digital Coding Metasurface. *IEEE Transactions on Antennas and Propagation*, *68*(3), 1618–1627. doi:10.1109/TAP.2019.2952460

Delaune, D., Guan, N., & Ito, K. (2008). Simple Multiband Antenna for Mobile Phone Application Based on a Dual-arm Monopole Structure. PIERS Online, 4(1).

Delgado, H. J. (2005). A Novel Neural Network for the Synthesis of Antennas andMicrowave Devices. *IEEE Transactions on Antennas and Propagation*, *16*, 1590–1600. PMID:16342499

Deng, R., Li, M., Muneer, B., Zhu, Q., Shi, Z., Song, L., & Zhang, T. (2018). Theoretical Analysis and Design of Ultrathin Broadband Optically Transparent Microwave Metamaterial Absorbers. *Materials (Basel)*, *11*(1), 107. Advance online publication. doi:10.3390/ma11010107 PMID:29324686

Deng, Y., Xu, Z., Zhang, H., & He, Q. (2017). Optimization design of nonlinear metamaterial using an improved hybrid algorithm. *Applied Mathematical Modelling*, *51*, 524–536.

Dhouibi, A., Burokur, S. N., Lustrac, A., & Priou, A. (2012). Compact Metamaterial-Based Substrate-Integrated Luneburg Lens Antenna. *IEEE Transactions On Antennas And Propagation, VOL.*, *11*, 1504–1507.

Dincer, F., Karaaslan, M., Unal, E., & Sabah, C. (2013). Dual-band polarization independent metamaterial absorber based on omega resoanator and octa-starstrip configuration. *Electromagnetic Waves*, *141*, 219–231. doi:10.2528/PIER13061105

Ding, C., Sun, H.-H., Ziolkowski, R. W., & Jay Guo, Y. (2018, December). A dual layered loop array antenna for base stations with enhanced crosspolarization discrimination. *IEEE Transactions on Antennas and Propagation, 66*(12), 6975–6985. doi:10.1109/TAP.2018.2869216

Ding, F., Cui, Y. X., Ge, X. C., Jin, Y., & He, S. L. (2012). Ultra-broadband microwave metamaterial absorber. *Applied Physics Letters, 100*(10), 103506. Advance online publication. doi:10.1063/1.3692178

Dinh, M., Ha-Van, N., Tung, N. T., & Le, M. T. (2021). Dual-Polarized Wide-Angle Energy Harvester for Self-Powered IoT Devices. *IEEE Access : Practical Innovations, Open Solutions, 9*, 103376–103384. doi:10.1109/ACCESS.2021.3098983

Dong, Y., Toyao, H., & Itoh, T. (2011). Compact Circularly-Polarized Patch Antenna Loaded With Metamaterial Structures. *IEEE Transactions on Antennas and Propagation, 59*(11), 4329–4333. doi:10.1109/TAP.2011.2164223

Do, T. V., Bui, S. T., Nguyen, T. H., Nguyen, T. T., Nguyen, T. T., Lee, Y., & Vu, D. L. (2012). Design, fabrication and characterization of a perfect absorber using simple cut-wire metamaterials. *Advances in Natural Sciences: Nanoscience and Nanotechnology, 3*(4), 045014. doi:10.1088/2043-6262/3/4/045014

Drukteinis, J. S., Mooney, B. P., Flowers, C. I., & Gatenby, R. A. (2013). Beyond Mammography: New frontiers in breast cancer screening. *The American Journal of Medicine, 126*(6), 472–479. doi:10.1016/j.amjmed.2012.11.025 PMID:23561631

Dudczyk, J., & ... Applying the neural networks to formation of radiationpattern of microstrip antenna. *IEEE International Radar Symposium*, 1-4.

Dwivedi. (2022). *Design and Analysis of Metamaterial Waveguide Antenna for Broadband Applications.* doi:10.1007/978-981-19-5224-1_42

Dwivedi. (2023). *Antenna Array for reconfiguration.* IGI Global. doi:10.4018/978-1-6684-5955-3

Ejaz, J. W., Anpalagan, A., Imran, M. A., Jo, M., Naeem, M., Bin Qaisar, S., & Wang, W. (2016). *2016 Internet of Things (IoT) in 5G Wireless Communications.* IEEE Access.

Eleftheriades, G. V., Antoniades, M. A., & Qureshi, F. (2007, February). Antenna applications of negativerefractive- index transmission-line (NRI-TL) structures. *IET Microwaves, Antennas & Propagation, 1*(1), 12–22. doi:10.1049/iet-map:20050345

Elnaggar, S. Y., & Milford, G. N. (2017). Description and stability analysis of nonlinear transmission line type metamaterials using nonlinear dynamics theory. *Journal of Applied Physics, 121*(12), 124902. doi:10.1063/1.4979022

Elsaadi, M., & Hamad, R. (2023). Breast cancer detection based on multi-slotted patch antenna at ISM band. *Circuits and Systems, 14*(05), 1–9. doi:10.4236/cs.2023.145001

Elsheakh, D. N., Mohamed, R. A., Fahmy, O. M., Ezzat, K., & Eldamak, A. R. (2023). Complete breast cancer detection and monitoring system by using microwave textile based antenna sensors. *Biosensors (Basel)*, *13*(1), 87. doi:10.3390/bios13010087 PMID:36671922

Engheta, N., & Ziolkowski, R. W. (2006). *Metamaterials: physics and engineering explorations*. Wiley-IEEE Press. doi:10.1002/0471784192

Epstein, N. R., Meaney, P. M., & Paulsen, K. D. (2014). 3D parallel-detection microwave tomography for Clinical Breast Imaging. *The Review of Scientific Instruments*, *85*(12), 124704. Advance online publication. doi:10.1063/1.4901936 PMID:25554311

Erentok, A., & Ziolkowski, R. W. (2007). Efficient MetamaterialInspired ElectricallySmall Electric-Based Antennas: Two- and ThreeDimensional Realizations. *Proc. of Antennas and Propag. Society International Symposium*, 1309-1312.

Esmail, B. A. F., Koziel, S., Golunski, L., Majid, H. B. A., & Barik, R. K. (2022). Overview of Metamaterials-Integrated Antennas for Beam Manipulation Applications: The Two Decades of Progress. *IEEE Access : Practical Innovations, Open Solutions*, *10*, 67096–67116. doi:10.1109/ACCESS.2022.3185260

Eteng, A. A., Goh, H. H., Rahim, S. K. A., & Alomainy, A. (2021). A Review of Metasurfaces for Microwave Energy Transmission and Harvesting in Wireless Powered Networks. *IEEE Access : Practical Innovations, Open Solutions*, *9*, 27518–27539. doi:10.1109/ACCESS.2021.3058151

Fang, N., Lee, H., Sun, C., & Zhang, X. (2005). Sub-diffraction-limited optical imaging with a silver super lens. *Science*, *308*(5721), 534–537. doi:10.1126cience.1108759 PMID:15845849

Fan, J., Sun, G.-Y., & Zhu, W.-R. (2011). Electric and magnetic dipole couplings in split ring resonator metamaterials. *Chinese Physics B*, *20*(11), 114101. doi:10.1088/1674-1056/20/11/114101

Fan, K., Strikwerda, A. C., Zhang, X., & Averitt, R. D. (2013). Three-dimensional broadband tunable terahertz metamaterials. *Physical Review B: Condensed Matter and Materials Physics*, *87*(16), 161104. doi:10.1103/PhysRevB.87.161104

Fedotov, V. A., Rose, M., Prosvirnin, S. L., Papasimakis, N., & Zheludev, N. I. (2010). Sharp trapped-mode resonances in planar metamaterials with a broken structural symmetry. *Physical Review Letters*, *104*(22), 223901. doi:10.1103/PhysRevLett.104.223901 PMID:20867169

Fhafhiem, N., Krachodnok, P., & Wongsan, R. (2012). The Circularly Polarized Resonator Antenna using Double Polarizing Metallic EBG. *IEEE Asia-Pacific Conference on Antennas and Propagation*. 10.1109/APCAP.2012.6333133

Fischer, S. C. L., Hillen, L., & Eberl, C. (2020). Mechanical Metamaterials on the Way from Laboratory Scale to Industrial Applications: Challenges for Characterization and Scalability. *Materials (Basel)*, *13*(16), 3605. doi:10.3390/ma13163605 PMID:32824029

Foroozesh, A., & Shafai, L. (2006). Size reduction of a microstrip antenna with dielectric superstrate using meta-materials: Artificial magnetic conductors versus magneto- dielectrics. *Proc. IEEE Antennas Propag. Soc. Int. Symp.*, 11–14. 10.1109/APS.2006.1710439

Franco-Martínez, F., Grasl, C., Kornfellner, E., Vostatek, M., Cendrero, A. M., Moscato, F., & Lantada, A. D. (2022). Hybrid design and prototyping of metamaterials and metasurfaces. *Virtual and Physical Prototyping*, *17*(4), 1031–1046. doi:10.1080/17452759.2022.2101009

Freitas, G. M. F., Rêgo, S. L., & Vasconcelos, C. F. L. (2011). *Design of Metamaterials Using Artificial Neural Networks. In International Microwave & Optoelectronics Conference*. IMOC.

Fu, X., Zeng, X., Cui, T. J., Lan, C., Guo, Y., Zhang, H. C., & Zhang, Q. (2016). Mode jumping of split-ring resonator metamaterials controlled by high-permittivity BST and incident electric fields. *Scientific Reports*, *6*(31274), 1–9. doi:10.1038rep31274 PMID:27502844

Gao, C., & Wang, Y. (2002). Analysis of EBG Structures Implemented on CPW Components by using EM-ANN Models. *Antennas and Propagation Society International Symposium*, 4, 368 – 371. 10.1109/APS.2002.1016999

Geerthana, S., Sridarshini, T., Balaji, V. R., Sitharthan, R., Madurakavi, K., Thirumurgan, A., & Dhanabalan, S. S. (2023). Ultra compact 2D- PhC based sharp bend splitters for terahertz applications. *Optical and Quantum Electronics*, *55*(778), 778. Advance online publication. doi:10.100711082-023-04956-w

Geetharamani, G., & Aathmanesan, T. (2019). Metamaterial inspired thz antenna for breast cancer detection. *SN Applied Sciences*, *1*(6), 595. Advance online publication. doi:10.100742452-019-0601-6

Gentile, M., Hentschel, M., Taubert, R., Guo, H., Giessen, H., & Fiebig, M. (2011). Investigation of the nonlinear optical properties of metamaterials by second harmonic generation. *Applied Physics. B, Lasers and Optics*, *105*(1), 149–162. doi:10.100700340-011-4766-y

Geyi, W. (2012). Optimum Design of Antennas. *Asia-Pacific Conference on Applied Electromagnetics*, 116-117.

Gianvittorio, J., & Yah, A. (2002, February). Fractal Antennas: A Novel Antenna Miniaturization Technique, and Applications. *IEEE Antennas & Propagation Magazine*, *44*(1), 20–36. doi:10.1109/74.997888

Girija, D. K., Rashmi, M., William, P., & Yogeesh, N. (2023). Framework for Integrating the Synergies of Blockchain with AI and IoT for Secure Distributed Systems. In N. Chaki, N. D. Roy, P. Debnath, & K. Saeed (Eds.), *Proceedings of International Conference on Data Analytics and Insights, ICDAI 2023. ICDAI 2023. Lecture Notes in Networks and Systems, 727*. Springer. https://doi.org/10.1007/978-981-99-3878-0_22.

Gneiding, N., Zhuromskyy, O., Shamonina, E., & Peschel, U. (2014, October). Circuit model optimization of a nano split ring resonator dimer antenna operating in infrared spectral range. *Journal of Applied Physics*, *116*(16), 164311. doi:10.1063/1.4900479

Gogoi, D. J., & Bhattacharyya, N. S. (2017). Embedded dielectric water "atom" array for broadband microwave absorber based on Mie resonance. *Journal of Applied Physics*, *122*(17), 175106. Advance online publication. doi:10.1063/1.4995519

Gogoi, D. J., & Bhattacharyya, N. S. (2018). Metasurface absorber based on water meta "molecule" for X-band microwave absorption. *Journal of Applied Physics, 124*(7), 075106. Advance online publication. doi:10.1063/1.5041450

Gopalakrishnan, R., & Gunasekaran, N. (2005). Design of equilateral triangular microstrip anten using artificial neural networks. *IEEE International Workshop on Antenna Technology: Small Antennas and Novel Metamaterials*, 246 - 249.

Gozel, M. A., Kahriman, M., & Kasar, O. (2019). Design of an efficiency-enhanced Greinacher rectifier operating in the GSM 1800 band by using rat-race coupler for RF energy harvesting applications. *International Journal of RF and Microwave Computer-Aided Engineering, 29*(1), 1–8. doi:10.1002/mmce.21621

Gozhenko, Wigle, Vedral, Hendrix, & Pinchuk. (2012). *Focusing effect of a metamaterial slab on the radiation pattern produced by a patch antenna.* Elsevier B.V.

Griguer, H. (2009). Patch Antenna Bandwidth Enhancement through the Use of Metamaterials. *International Conference on Telecommunications*, 323-327.

GSMA. (2019). *5G the internet of things and wearable devices: Radiofrequency exposure.* GSMA Public Policy.

Guan, Z., Zhao, P., Wang, X., & Wang, G. (2021). Modeling Radio-Frequency Devices Based on Deep Learning Technique. *Electronics (Basel), 10*, 1710. doi:10.3390/electronics10141710

Guo, Y.-X. (2003). Compact internal multiband antennas for mobile handsets. *IEEE Antennas and Wireless Propagation Letters, 2*, 143–146. doi:10.1109/LAWP.2003.815278

Gupta, M. M., & Ojha, A. K. (2010). Design and modelling of fuzzy logic systems. *Studies in Fuzziness and Soft Computing, 259.*

Gupta, A., Rajesh, R., & Manoharan, A. (2020). Comparative analysis of clustering algorithms for unsupervised machine learning. *Soft Computing, 24*(16), 12555–12571.

Gupta, S., Jha, P. C., & Joshi, D. (2019). Fuzzy logic and its applications in bioinformatics: A review. *Genomics, 111*(5), 883–893.

Hamdi & Aguili. (2021). Spectral Floquet analysis devoted to meta-surface applied for 5G and planned 6G antenna designs. *2021 IEEE 19th International Symposium on Antenna Technology and Applied Electromagnetics (ANTEM),* 1-2. doi: 10.1109/ANTEM51107.2021.9518709

Han, Rahim, Masri, & Karim. (2007). Left handed metamaterial design for microstrip antenna application 2007. *APMC 2007. Asia-Pacific, 1-4 Microwave Conference.*

Hao, J., Wang, J., Liu, X., Padilla, W. J., Zhou, L., & Qui, M. (2010). High performance optical absorber based on a plasmonic metamaterial. *Applied Physics Letters, 96*(25), 251104. doi:10.1063/1.3442904

Harutyunyan, H., Volpe, G., Quidant, R., & Novotny, L. (2012). Enhancing the Nonlinear Optical Response Using Multifrequency Gold-Nanowire Antennas. *Physical Review Letters, 108*(21), 217403. doi:10.1103/PhysRevLett.108.217403 PMID:23003302

Hassan Mirzaei & Eleftheriades. (2011). Frequency Reconfigurable Metamaterial Inspired Antenna. IEEE Antennas and Wireless Propagation Letters, 10.

Haxha, S., AbdelMalek, F., Ouerghi, F., Charlton, M. D. B., Aggoun, A., & Fang, X. J. S. R. (2018). Metamaterial superlenses operating at visible wavelength for imaging applications. *Scientific Reports, 8*(1), 16119. doi:10.103841598-018-33572-y PMID:30382113

Hee, T. W., Hall, P. S., & Song, C. T. (2003). Fractal PIFA, Dipole and Monopole Antennas. *IEEE Topical Conference on Wireless Communication Technology,* 275-276. 10.1109/WCT.2003.1321522

Helena, Ramos, Varum, & Matos. (2020). Antenna Design Using Modern Additive Manufacturing Technology: A Review. *IEEE Access, 8*, 177064-177083.

Hidayat, M. V., & Apriono, C. (2019). Simulation of terahertz imaging using microstrip linear array antenna for breast cancer detection. *AIP Conference Proceedings, 2092*(1), 020020. doi:10.1063/1.5096688

Hien Nguyen, T., Tung Bui, S., Tuan Nguyen, T., Tung Nguyen, T., Lee, Y., An Nguyen, M., & Lam Vu, D. (2014). Metamaterial-based perfect absorber: Polarization insensitivity and broadband. *Advances in Natural Sciences: Nanoscience and Nanotechnology, 5*(2), 025013. Advance online publication. doi:10.1088/2043-6262/5/2/025013

Holloway, Dienstfrey, Kuester, O'Hara, Azad, & Taylor. (2009). A Discussion on the Interpretation and Characterization of Metafilms/Metasurfaces: The Two Dimensional Equivalent of Metamaterials. *Metamaterials, 3*(2), 100-112.

Holloway, Kuester, Baker-Jarvis, & Kabos. (2003). A Double Negative (DNG) Composite Medium Composed of Magneto Dielectric Spherical Particles Embedded in a Matrix. *IEEE Transactions on Antennas and Propagation, 51*(10), 2596-2603.

Holloway, C. L., & Kuester, E. F. (2020). Electromagnetic metamaterials and metasurfaces: historical overview, characterization, and the effect of length scales. In *Dielectric Metamaterials* (pp. 1–38). Elsevier. doi:10.1016/B978-0-08-102403-4.00006-2

Howey, D. A., Bansal, A., & Holmes, A. S. (2011). Design and performance of a centimetre-scale shrouded wind turbine for energy harvesting. *Smart Materials and Structures, 20*(8), 1–12. doi:10.1088/0964-1726/20/8/085021

Ho-Yong. (2008). The Design of a Small Monopole Type Antenna using Negative Permittivity Metamaterial Structure. *Antenna Technology: Small Antennas and Novel Metamaterials, 2008. iWAT 2008. International Workshop on, 2008*, 275-278.

Huang, Liu, & Gong. (2015). Multiband Metamaterial-Loaded Monopole Antenna for WLAN/WiMAX Applications. *IEEE Antennas and Wireless Propagation Letters, 14*, 662-665.

Huang, C., Chen, D., Leidich, S., & Gessner, T. (2010). Compact Meta-material Transmission Line Balun Based on Meander Lines Structure and MEMS Technology. *Microelectronic Engineering*, *87*(4), 584–587. doi:10.1016/j.mee.2009.08.018

Huang, D., Poutrina, E., Zheng, H., & Smith, D. R. (2011). Design and experimental characterization of nonlinear metamaterials. *Journal of the Optical Society of America. B, Optical Physics*, *28*(12), 2925. doi:10.1364/JOSAB.28.002925

Huang, H., Li, X., & Liu, Y. (2018, June). A novel vector synthetic dipole antenna and its common aperture array. *IEEE Transactions on Antennas and Propagation*, *66*(6), 3183–3188. doi:10.1109/TAP.2018.2819894

Huang, H., Yang, X.-S., & Wang, B.-Z. (2023). Machine-Learning-Based Generative Optimization Method and Its Application to an Antenna Decoupling Design. *IEEE Transactions on Antennas and Propagation*, *71*(7), 6243–6248.

Huang, Y., Zhou, J., & Li, L. (2022). Wireless Power Transfer and Energy Harvesting Using Metamaterials and Metasurfaces. *2022 International Workshop on Antenna Technology, IWAT 2022*, 280–283. 10.1109/iWAT54881.2022.9811085

Huang, Z., Liu, X., & Zang, J. F. (2019). The inverse design of structural color using machine learning. *Nanoscale*, *11*(45), 21748–21758. doi:10.1039/C9NR06127D PMID:31498348

Hu, Y., Wang, X., Luo, X., Ou, X., Li, L., Chen, Y., Yang, P., Wang, S., & Duan, H. (2020). All-dielectric metasurfaces for polarization manipulation: Principles and emerging applications. *Nanophotonics*, *9*(12), 3755–3780. doi:10.1515/nanoph-2020-0220

Islam, M. R., Islam, M. T., Moniruzzaman, M., Samsuzzaman, M., & Arshad, H. (2021). Penta band single negative meta-atom absorber designed on square enclosed star-shaped modified split ring resonator for S-, C-, X- and Ku- bands microwave applications. *Scientific Reports*, *11*(1), 8784. doi:10.103841598-021-87958-6 PMID:33888759

Jabbar, H., Song, Y. S., & Jeong, T. T. (2010). RF energy harvesting system and circuits for charging of mobile devices. *IEEE Transactions on Consumer Electronics*, *56*(1), 247–253. doi:10.1109/TCE.2010.5439152

Jebril, I., Dhanaraj, P., Abdulsahib, G. M., Palanisamy, S., Prabhu, T., & Khalaf, O. I. (2022). Analysis of electrically couple SRR EBG structure for sub 6 GHz wireless applications. *Advances in Decision Sciences*, *26*(5), 102–123. https://doi.org/10.47654/v26y2022i5p102-123

Jeong, H., Nguyen, T. T., & Lim, S. (2018). Meta-Dome for Broadband Radar Absorbing Structure. *Scientific Reports*, *8*(1), 17893. doi:10.103841598-018-36273-8 PMID:30559365

Jones, A. R., & Wooding, E. R. (1964). Multi-layer Microwave Absorber. *IEEE Transactions on Antennas and Propagation, AP12*(4), 508. doi:10.1109/TAP.1964.1138262

Joshi, J. G., Pattnaik, S. S., & Devi, S. (2011). *Metamaterial loaded dual band microstrip patch antenna. In Antenna Week.* IAW.

Joshi, J. G., Pattnaik, S. S., Devi, S., Lohokare, M. R., & Vidyasagar, C. (2009). Offset fed diamond shaped split ring (DSSR) planar metamaterial antenna. *IEEE Conference on Applied Electromagnetics*, 1-4. 10.1109/AEMC.2009.5430686

Jouvaud, C., de Rosny, J., & Ourir, A. (2013, April 11). Adaptive metamaterial antenna using coupled tunable split-ring resonators. *Electronics Letters*, *49*(8), 518–519. doi:10.1049/el.2013.0398

Jumaat, H., Hong Ping, K., Abd Rahman, N. H., Yon, H., Mohd Redzwan, F. N., & Awang, R. A. (2021). A compact modified wideband antenna with CBCPW, Stubline and notch-staircase for Breast Cancer Microwave Imaging Application. *AEÜ. International Journal of Electronics and Communications*, *129*, 153492. doi:10.1016/j.aeue.2020.153492

Jung, J., Park, H., Park, J., Chang, T., & Shin, J. (2020). Broadband metamaterials and metasurfaces: A review from the perspectives of materials and devices. *Nanophotonics*, *9*(10), 3165–3196. doi:10.1515/nanoph-2020-0111

Karakaya, E., Bagci, F., Yilmaz, A. E., & Akaoglu, B. (2019). Metamaterial-Based Four-Band Electromagnetic Energy Harvesting at Commonly Used GSM and Wi-Fi Frequencies. *Journal of Electronic Materials*, *48*(4), 2307–2316. doi:10.100711664-019-06962-9

Karamanos, T. D., Dimitriadis, A. I., Kantartzis, N. V., & Member, S. (2012). Compact Double-Negative Metamaterials Based on Electric and Magnetic Resonators. *IEEE Antennas and Wireless Propagation Letters*, *11*(2), 480–483. doi:10.1109/LAWP.2012.2197170

Karimi Habil, M., Ghahremani, M., & Zapata-Rodriguez, C. J. (2022). Multi-octave metasurface-based refractory superabsorber enhanced by a tapered unit-cell structure. *Scientific Reports*, *12*(1), 17066. doi:10.103841598-022-21740-0 PMID:36224251

Kataja, J., & Nikoskinen, K. (2011). The Parametric Optimization of Wire Dipole Antenna. *IEEE Transactions on Antennas and Propagation*, *59*(2), 350–356. doi:10.1109/TAP.2010.2096181

Kaur, G., & Kaur, A. (2020). Monostatic radar-based microwave imaging of breast tumor detection using a compact cubical dielectric resonator antenna. *Microwave and Optical Technology Letters*, *63*(1), 196–204. doi:10.1002/mop.32557

Kavitha, V., Balaji, V. R., Dhanabalan, S. S., Sridarshini, T., Robinson, S., Radhoune, M., Hedge, G., & Sugesh, R. G. (2023). Design and performance analysis of eight channel demultiplexer using 2D photonic crystal with trapezium cavity. *Journal of Optics*, *25*(6), 065102. Advance online publication. doi:10.1088/2040-8986/acceb3

Keiser, G. R., Fan, K., Zhang, X., & Averitt, R. D. (2013). Towards Dynamic, Tunable, and Nonlinear Metamaterials via Near Field Interactions: A Review. *Journal of Infrared, Millimeter and Terahertz Waves*, *34*(11), 709–723. doi:10.100710762-013-9993-3

Kern, D. J., Werner, D. H., Wilhelm, M. J., Church, K. H., & Mittra, R. (2002). Multi-band high impedance frequency selective surfaces. *Proc IEEE AP-S/URSI Symp Dig*, 264.

Khan, S., & Eibert, T. F. (2017). A multi-resonant meta-absorber as an electromagnetic energy harvester. *2017 IEEE International Symposium on Antennas and Propagation & USNC/URSI National Radio Science Meeting*, 1091–1092. 10.1109/APUSNCURSINRSM.2017.8072588

Kim, D., & Kim, M. (2008, January 31). Narrow-beamwidth T-shaped monopole antenna fabricated from metamaterial wires. *Electronics Letters*, *44*(3), 179–180. doi:10.1049/el:20082854

Kim, H. K., Lee, D., & Lim, S. (2016). A Fluidically Tunable Metasurface Absorber for Flexible Large-Scale Wireless Ethanol Sensor Applications. *Sensors (Basel)*, *16*(8), 1–10. doi:10.339016081246 PMID:27509498

Kim, H. K., Lee, D., & Lim, S. (2016). Wideband-Switchable Metamaterial Absorber Using Injected Liquid Metal. *Scientific Reports*, *6*(1), 31823. doi:10.1038rep31823 PMID:27546310

Kim, M., & Rho, J. (2015). Metamaterials and imaging. *Nano Convergence*, *2*(1), 1–16. doi:10.118640580-015-0053-7 PMID:28191408

Klir, G. J., & Yuan, B. (1995). *Fuzzy sets and fuzzy logic: theory and applications*. Prentice Hall.

Koziel, S., & Pietrenko-Dabrowska, A. (2023). Improved-Efficacy EM-Driven Optimization of Antenna Structures Using Adaptive Design Specifications and Variable-Resolution Models. *IEEE Transactions on Antennas and Propagation*, *71*(2), 1863–1874.

Kraus, J. D., & Marhefka, R. J. (2003). *Antennas for all applications* (3rd ed.). McGraw-Hill.

Kröling, H., Endresb, A., Nubboc, N., Fleckensteinc, J., Miletzkyb, A., & Schabela, S. (2014). Anisotropy of paper and paper based composites and the modelling thereof. *ECCM16*, 1-8.

Kshetrimayum, R. S. (2004). A brief intro to metamaterials. *IEEE Potentials*, *23*(50), 44–46. doi:10.1109/MP.2005.1368916

Kumar, Fusco, & Yurduseven. (2022). Preliminary Study of Breast Cancer Detection Using A Computational Microwave Imaging System. *2022 52nd European Microwave Conference (EuMC),* 195-198. doi: 10.23919/EuMC54642.2022.9924492

Kumar, P. S., Boopathy, S., Dhanasekaran, S., & Anand, K. R. G. (2021 October 8-9). *Optimization of Multi-Band Antenna for Wireless Communication Systems using Genetic Algorithm.* 2021 International Conference on Advancements in Electrical, Electronics, Communication, Computing and Automation (ICAECA), Coimbatore, India. 10.1109/ICAECA52838.2021.9675686.

Kumar, D. S., Caroline, B. E., Preethi, N. S., & Lokeshwar, M. (2021, July 30-31). Design and Analysis of High Pace Terahertz Antenna. *2021 International Conference on System, Computation, Automation and Networking (ICSCAN)*, Puducherry, India. 10.1109/ICSCAN53069.2021.9526543

Kumar, D. S., Caroline, B. E., & Thilagavathi, S. (2020, July 03-04). Investigation of Equilateral Triangular Microstrip Patch Antenna using Photonic Crystal. *2020 International Conference on System, Computation, Automation and Networking (ICSCAN)*, Pondicherry, India. 10.1109/ICSCAN49426.2020.9262319

Labadi, M., Jamel, B. T., & Choubani, F. (2011). New design of antenna array using Left Handed Meta-Material "LHM" based on circular Split Ring Resonator "SRR". *Mediterranean Microwave Symposium (MMS),* 52-56.

Landy, N. I., Sajuyigbe, S., Mock, J. J., Smith, D. J., & Padilla, W. J. (2008). Perfect Metamaterial Absorber. *Physical Review Letters, 100*(20), 207402. Advance online publication. doi:10.1103/PhysRevLett.100.207402 PMID:18518577

Lapine, M. (2014). Magnetic response of fano resonant metamaterials: Theory and experiments. *Progress in Quantum Electronics, 38*(1), 1–74.

Lapine, M., Gorkunov, M., Shadrivov, I. V., & Kivshar, Y. S. (2014). Colloquium: Nonlinear metamaterials. *Reviews of Modern Physics, 86*(3), 1093–1123. doi:10.1103/RevModPhys.86.1093

Lapine, M., & Tretyakov, S. (2007). Contemporary notes on metamaterials. *IET Microwaves, Antennas & Propagation, 1*(1), 3–11. doi:10.1049/iet-map:20050307

Lee, Y. P., Rhee, J. Y., Yoo, Y. J., Kim, K. W. (2016). Metamaterials for Perfect Absorption. *Springer Series in Materials Science,* 236.

Lee, Y., Ga, D., & Choi, J. (2011). Design of a MIMO Antenna with Improved Isolation Using MNG Metamaterial. *International workshop on Antenna Technology,* 231 – 234.

Lee, C. H. L., & Itoh, T. (2006). Compact dual-band antenna using an anisotropic metamaterial. *Proceedings of the 36th European Microwave Conference*, 1044-1047. 10.1109/EUMC.2006.281112

Lee, D., Jeong, H., & Lim, S. (2017). Electronically Switchable Broadband Metamaterial Absorber. *Scientific Reports*, *7*(1), 4891. doi:10.103841598-017-05330-z PMID:28687811

Lei, A., Xu, R., Thyssen, A., Stoot, A. C., Christiansen, T. L., Hansen, K., Lou-Møller, R., Thomsen, E. V., & Birkelund, K. (2011). MEMS-Based Thick Film PZT Vibrational Energy Harvester. *2011 IEEE 24th International Conference on Micro Electro Mechanical Systems,* 125–128. 10.1109/MEMSYS.2011.5734377

Li. (2013). Design of Electrically Small Metamaterial Antenna With ELC and EBG Loading. *IEEE Antennas and Wireless Propagation Letters, 12*, 678-681.

Li, H., Yuan, L., Zhou, B., Shen, X., Cheng, Q., & Cui, T. J. (2011). Ultrathin multiband gigahertz metamaterial absorbers. *Journal of Applied Physics, 110*(1), 014909. doi:10.1063/1.3608246

Li, J., & Chan, C. T. (2004). Double-negative acoustic metamaterial. *Physical Review E: Statistical Physics, Plasmas, Fluids, and Related Interdisciplinary Topics, 70*(5), 4. doi:10.1103/PhysRevE.70.055602 PMID:15600684

Li, J., Mei, J., Wu, Y., & Zhang, L. (2018). Advances in 3D-printed metamaterials. *Materials Today*, *21*(6), 697–698.

Li, K., He, X., Wang, X., & Jiang, S. (2019). A Nonlinear Electromagnetic Energy Harvesting System for Self-Powered Wireless Sensor Nodes. *Journal of Sensor and Actuator Networks*, *8*(1), 18. doi:10.3390/jsan8010018

Li, L.-W. (2008). Design of a Novel Rectangular Patch Antenna with Planar Metamaterial Patterned Substrat. *International Workshop on Antenna Technology: Small Antennas and Novel Metamaterials*, 124-126.

Li, L., Wen, J., Wang, Y., Jin, Y., Wen, Y., Sun, J., Zhao, Q., Li, B., & Zhou, J. (2023). A Transparent broadband all-dielectric water-based metamaterial absorber based on laser cutting. *Physica Scripta*, *98*(5), 055516. Advance online publication. doi:10.1088/1402-4896/accc15

Li, P., Chen, S., Dai, H., Yang, Z., Chen, Z., Wang, Y., Chen, Y., Peng, W., Shan, W., & Duan, H. (2021). Recent advances in focused ion beam nanofabrication for nanostructures and devices: Fundamentals and applications. *Nanoscale*, *13*(3), 1529–1565. doi:10.1039/D0NR07539F PMID:33432962

Liu, Y., & Zhao, X. (2014). Perfect Absorber Metamaterial for Designing Low-RCS Patch Antenna. IEEE Antennas and Wireless Propagation Letters, 13.

Liu, H., Wang, Z. H., Li, L., Fan, Y. X., & Tao, Z. Y. (2019). Vanadium dioxide-assisted broadband tunable terahertz metamaterial absorber. *Scientific Reports*, *9*(1), 5751. doi:10.103841598-019-42293-9 PMID:30962484

Liu, N., Guo, H., Fu, L., Kaiser, S., Schweizer, H., & Giessen, H. (2008). Three-dimensional photonic metamaterials at optical frequencies. *Nature Materials*, *7*(1), 31–37. doi:10.1038/nmat2072 PMID:18059275

Liu, N., Mesch, M., Weiss, T., Hentschel, M., & Giessen, H. (2010). Infrared perfect absorber and it's application as plasmonic sensor. *Nano Letters*, *10*(7), 2342–2348. doi:10.1021/nl9041033 PMID:20560590

Liu, W., Chen, Z., & Qing, X. (2014, September). 60-GHz Thin Broadband High-Gain LTCC Metamaterial- Mushroom Antenna Array. *IEEE Transactions on Antennas and Propagation*, *62*(9), 4592–4601. doi:10.1109/TAP.2014.2333052

Liu, X. M., Lan, C. W., Bi, K., Li, B., Zhao, Q., & Zhou, J. (2016). Dual band metamaterial perfect absorber based on Mie resonances. *Applied Physics Letters*, *109*(6), 062902. Advance online publication. doi:10.1063/1.4960802

Liu, X., Starr, T., Starr, A. F., & Padilla, W. J. (2010). Infrared spatial and frequency selective metamaterial with near-unity absorbance. *Physical Review Letters*, *104*(20), 207403. doi:10.1103/PhysRevLett.104.207403 PMID:20867064

Liu, Y., & ... . (2017). Nonlinear Fano-resonant metamaterials: A review. *Advanced Optical Materials*, *5*(8), 1600844.

Liu, Z. C., Zhu, D. Y., Rodrigues, S. P., Lee, K. T., & Cai, W. S. (2018). Generative Model for the Inverse Design of Metasurfaces. *Nano Letters*, *18*(10), 6570–6576. doi:10.1021/acs.nanolett.8b03171 PMID:30207735

Liu, Z. D., Hall, P., & Wake, D. (1997, October). Dual-frequency planar inverted-F antenna. *IEEE Transactions on Antennas and Propagation*, *45*(10), 1451–1458. doi:10.1109/8.633849

Liu, Z., Lee, H., Xiong, Y., Sun, C., & Zhang, X. (2007). Far-field optical hyper-lens magnifying sub-diffraction limited objects. *Science*, *315*(5819), 1686. doi:10.1126cience.1137368 PMID:17379801

Li, X., Liu, H., Sun, Q., & Huang, N. (2015). Ultra-broadband and polarization-insensitive wide-angle terahertz metamaterial absorber. *Photonics and Nanostructures*, *15*, 81–88. doi:10.1016/j.photonics.2015.04.002

Li, X., Zhang, H., & Luo, C. (2020). A fuzzy neural network based on the improved fuzzy min-max clustering algorithm and its application in nonlinear metamaterials. *Journal of Intelligent & Fuzzy Systems*, *39*(5), 7657–7667.

Li, Y. Z., Wang, Y. P., Qi, S. T., Ren, Q., Kang, L., Campbell, S. D., Werner, P. L., & Werner, D. H. (2020). Predicting Scattering From Complex Nano-Structures via Deep Learning. *IEEE Access : Practical Innovations, Open Solutions*, *8*, 139983–139993. doi:10.1109/ACCESS.2020.3012132

Li, Y., Baker, E., Reissman, T., Sun, C., & Liu, W. K. (2017). Design of mechanical metamaterials for simultaneous vibration isolation and energy harvesting. *Applied Physics Letters*, *111*(25), 251903-1, 251903–251905. doi:10.1063/1.5008674

Li, Z. (2000). PBG, PMC and PEC Ground Planes: A Case Study of Dipole Antennas. *IEEE Antennas and Propagation Society International Symposium (APSURSI)*, 2, 674 - 677.

Li, Z., Li, B., Zhao, Q., & Zhou, J. (2020). A metasurface absorber based on the slow-wave effect. *AIP Advances*, *10*(4), 045311. Advance online publication. doi:10.1063/1.5143408

Louertani & Chio. (2014). Low-profile broadband spiral antenna with meta materials. *IEEE Antennas and Propagation Society International Symposium (APSURSI)*, 1131-1332.

Luna, D. R., Vasconcelos, C. F. L., & Cruz, R. M. S. (2013). *Using Natural Optimization Algorithms and Artificial Neural Networks in the Design of Effective Permittivity of Metamaterials*. IMOC. doi:10.1109/IMOC.2013.6646572

Luo, Y., Han, Z., & Toshiyoshi, H. (2014). A Novel J-Shape Active MEMS Meta-material Antenna. *IEEE Antennas and Propagation Society International Symposium (APSURSI), 569-570*. 10.1109/APS.2014.6904615

Lu, Z. Q., Zhao, M., Yang, Z. Y., Wu, L., & Zhang, P. (2013, August 15). Helical Metamaterial Absorbers: Broadband and Polarization-Independent in Optical Region. *Journal of Lightwave Technology*, *31*(16), 2762–2768. doi:10.1109/JLT.2013.2272632

Ma, K., Tan, T., Yan, Z., Liu, F., Liao, W. H., & Zhang, W. (2021). Metamaterial and Helmholtz coupled resonator for high-density acoustic energy harvesting. *Nano Energy, 82*, 105693. doi:10.1016/j.nanoen.2020.105693

Mahmoud & Hu. (2012). A Compact Dual Band Meta-material Antenna for Wireless Applications. *Loughborough Antennas & Propagation Conference.*

Mahmoud & Ibrahim. (2013). Compact and Closely Spaced Metamaterial MIMO Antenna With High Isolation for Wireless Applications. IEEE Antennas and Wireless Propagation Letters, 12.

Mahmud, M. Z., Islam, M. T., Rahman, M. N., Alam, T., & Samsuzzaman, M. (2017). A miniaturized directional antenna for Microwave Breast Imaging Applications. *International Journal of Microwave and Wireless Technologies, 9*(10), 2013–2018. doi:10.1017/S1759078717000927

Mahmud, M. Z., Islam, M. T., Samsuzzaman, M., Kibria, S., & Misran, N. (2017). Design and parametric investigation of directional antenna for Microwave Imaging Application. IET Microwaves, Antennas &amp. *IET Microwaves, Antennas & Propagation, 11*(6), 770–778. doi:10.1049/iet-map.2016.0774

Majid, H.A., & Rahim, M.K.A. (2007). Investigation of LH-meta material in microstrip antenna. *Asia-Pacific Conference on Applied Electromagnetics*, 1-4.

Manoj Prasath, & Hari Hara Suthan, Srivastava, Rohith, Kaur, & Balaram. (2023). Novel Approach of Breast Cancer Detection Based on Wireless Communication System. *Journal of Pharmaceutical Negative Results, 14*(2), 987–992. doi:10.47750/pnr.2023.14.S02.119

Markley, L., & Eleftheriades, G. V. (2011, March). Meta-screens and near-field antenna-arrays: A new perspective on subwavelength focusing and imaging. *Metamaterials (Amsterdam), 5*(2-3), 97–106. doi:10.1016/j.metmat.2011.03.004

Martín, F. (2012). Metamaterials for Wireless Communications, Radiofrequency Identification, and Sensors. *ISRN Electronics, 2012*, 1–29. doi:10.5402/2012/780232

Ma, W., Cheng, F., & Liu, Y. M. (2018). Deep-Learning-Enabled On-Demand Design of Chiral Metamaterials. *ACS Nano, 12*(6), 6326–6334. doi:10.1021/acsnano.8b03569 PMID:29856595

Ma, W., Cheng, F., Xu, Y. H., Wen, Q. L., & Liu, Y. M. (2019). Probabilistic Representation and Inverse Design of Metamaterials Based on a Deep Generative Model with Semi-Supervised Learning Strategy. *Advanced Materials, 31*(35), 1901111. Advance online publication. doi:10.1002/adma.201901111 PMID:31259443

Ma, X., Huang, C., Pan, W., Zhao, B., Cui, J., & Luo, X. (2014). A Dual Circularly Polarized Horn Antenna in Ku-Band Based on Chiral Metamaterial. *IEEE Transactions on Antennas and Propagation, 62*(4), 2307–2311. doi:10.1109/TAP.2014.2301841

Ma, X., Tian, F., Li, X., Guo, L., & Huang, X. (2020). Broadband with enhanced oblique incidence metamaterial absorber. *Materials Research Express, 7*(9), 095803. Advance online publication. doi:10.1088/2053-1591/abba9e

McClintock, H. D., Doshi, N., Iniguez-Rabago, A., Weaver, J. C., Jafferis, N. T., Jayaram, K., Wood, R. J., & Overvelde, J. T. B. (2021). A Fabrication Strategy for Reconfigurable Millimeter-Scale Metamaterials. *Advanced Functional Materials*, *31*(46), 2103428. Advance online publication. doi:10.1002/adfm.202103428

Mendel, J. M. (2001). Fuzzy logic systems for engineering: A tutorial. *Proceedings of the IEEE*, *89*(3), 345–377. doi:10.1109/5.364485

Mendel, J. M. (2020). Fuzzy Logic Systems for Engineering: A Tutorial. *IEEE Transactions on Fuzzy Systems*, *28*(6), 1207–1232.

Mendhe, S. E., & Kosta, Y. P. (2013). Ultra Wide Band Three Layer Microstrip Patch Antenna Using Single Layer Helical Resonating Metamaterial Cover. *International Conference on Wireless and Optical Communications Networks (WOCN)*, 1-5. 10.1109/WOCN.2013.6616250

Mercier, L. (2005). Steerable and tunable "EBG resonator antennas" using smart metamaterials. *IEEE Antennas and Propagation Society International Symposium*, *4B*, 406-409.

Milius, C., Andersen, R. B., Lazaridis, P. I., Zaharis, Z. D., Muhammad, B., & Jes, T. B. (2021). Metamaterial-Inspired Antennas: A Review of the State of the Art and Future Design Challenges. *IEEE Access : Practical Innovations, Open Solutions*, *9*, 89846–89865. doi:10.1109/ACCESS.2021.3091479

Minovich, A. E., Miroshnichenko, A. E., Bykov, A. Y., Murzina, T. V., Neshev, D. N., & Kivshar, Y. S. (2015). Functional and nonlinear optical metasurfaces. *Laser & Photonics Reviews*, *9*(2), 195–213. doi:10.1002/lpor.201400402

Mishra, Elbir, & Zaghloul. (2022). *Machine Learning for Metasurfaces Design and Their Applications*. Academic Press.

Mishra, R. K., & Patnaik, A. (1998). Neural network-based CAD model for the design of square patch antennas. *IEEE Transactions on Antennas and Propagation*, *46*(12), 1890–1891. doi:10.1109/8.743842

Mishra, R. K., & Patnaik, A. (2000). Neurospectral Analysis of Coaxial Fed Rectangular Patch Antenna. *IEEE Antennas and Propagation Society International Symposium*, 2, 1062 – 1065. 10.1109/APS.2000.875405

Mishra, S., Choubey, S., Choubey, A., Yogeesh, N., Rao, J. D. P., & William, P. (2022). Data extraction approach using natural language processing for sentiment analysis. In *2022 International Conference on Automation, Computing and Renewable Systems (ICACRS)* (pp. 970-972). IEEE. 10.1109/ICACRS55517.2022.10029216

Mittra, R. (2008). To Use or Not to Use the Effective Medium Approach for Designing Performance-Enhanced Small Antennas-That is the Question. *IEEE International Workshop on Antenna Technology: Small Antennas and Novel Metamaterials*, 55-58. 10.1109/IWAT.2008.4511289

Mookiah, P., & Dandekar, K. R. (2009). Metamaterial-Substrate Antenna Array for MIMO Communication System. *IEEE Transactions on Antennas and Propagation, 27*(October), 3283–3292. doi:10.1109/TAP.2009.2028638

Moussa, M. N., Madi, M. A., & Kabalan, K. Y. (2022). Breast Tumor Detection, Sizing and Localization Using a 24-Element Antenna Array. *IEEE Journal of Biomedical and Health Informatics, 26*(10), 5109–5121. doi:10.1109/JBHI.2022.3189640 PMID:35816522

Mujawar, M. (2021). Compact Microstrip Patch Antenna Design with Three I-, Two L-, One E- and One F-Shaped Patch for Wireless Applications. CRC Press eBooks. doi:10.1201/9781003093558-7

Mujawar, M., & Gunasekaran, T. (2022). Multiband Slot Microstrip Antenna for Wireless Applications. In EAI/Springer Innovations in Communication and Computing (pp. 23–34). Springer International Publishing. doi:10.1007/978-3-030-76636-8_3

Mujawar, M., Saradhi, D. V., Naz, M. Y., & Muduli, A. (2022). Design and performance comparison of arrays of circular, square and hexagonal meta-material structures for wearable applications. *Journal of Magnetism and Magnetic Materials, 553*, 169235. doi:10.1016/j.jmmm.2022.169235

Munk, B. A. (2000). *Frequency Selective Surfaces: Theory and Design.* Wiley. doi:10.1002/0471723770

Mussetta, M. (2010). Characterization of microstrip reflectarray square ring elements by means of an Artificial Neural Network. *Proceedings of the Fourth European Conference on Antennas and Propagation (EuCAP),* 1-4.

Nakano, H. (2014). Natural and Metamaterial-based Spiral and Helical Antennas. *Loughborough Antennas and Propagation Conference (LAPC).* 10.1109/LAPC.2014.6996405

Nguyen, T. T., & Lim, S. (2017). Wide Incidence Angle-Insensitive Metamaterial Absorber for Both TE and TM Polarization using Eight-Circular-Sector. *Scientific Reports, 7*(1), 1–11. doi:10.103841598-017-03591-2 PMID:28600523

Nisha, U. N., Manikandan, A., Venkatramanan, C., & Dhanapal, R. (2023). A score based link delay Aware Routing Protocol to improve energy optimization in wireless sensor network. *Journal of Engineering Research, 100115.* Advance online publication. doi:10.1016/j.jer.2023.100115

Norouzi, M., Jarchi, S., Ghaffari-Miab, M., Esfandiari, M., Lalbakhsh, A., Koziel, S., Reisenfeld, S., & Moloudian, G. (2023). 3D metamaterial ultra-wideband absorber for curved surface. *Scientific Reports, 13*(1), 1043. doi:10.103841598-023-28021-4 PMID:36658245

O'Hara, J. F., Singh, R., Brener, I., Smirnova, E., Han, J., Taylor, A. J., & Zhang, W. (2008). Thin-film sensing with planar terahertz metamaterials: Sensitivity and limitations. *Optics Express, 16*(3), 1786–1795. doi:10.1364/OE.16.001786 PMID:18542258

Odabasi, H., & Fernando, L. (2013). Complementary Electric-Field-Coupled (CELC) Based Resonator Antennas. *Antennas and Propagation Society International Symposium (APSURSI),* 772-773.

Oliaei, M., & Abrishamian, M. (2014). Corrugated SIW K band horn antenna. *AEÜ. International Journal of Electronics and Communications, 68*(12), 1199–1204. doi:10.1016/j.aeue.2014.06.009

Ouerghi, K., Fadlallah, N., Smida, A., Ghayoula, R. F., & Boulejfen, N. (2017, September 12-14). *Circular antenna array design for breast cancer detection.* Sensors Networks Smart and Emerging Technologies (SENSET), Beirut, Lebanon. 10.1109/SENSET.2017.8125016

Ourir, A., Burokur, S. N., & de Lustrac, A. (2007). Phase-Varying Metamaterial for Compact Steerable Directive Antennas. *Electronics Letters, 43*(9), 493–494. doi:10.1049/el:20070298

Padilla, W. J., Aronsson, M. T., Highstrete, C., Lee, M., Taylor, A. J., & Averitt, R. D. (2007). Electrically resonant terahertz metamaterials: Theoretical and experimental investigations. *Physical Review B: Condensed Matter and Materials Physics, 75*(4), 041102. doi:10.1103/PhysRevB.75.041102

Padilla, W. J., Basov, D. N., & Smith, D. R. (2006). Negative refractive index metamaterials. *Materials Today, 9*(7-8), 28–35. doi:10.1016/S1369-7021(06)71573-5

Padilla, W. J., Taylor, A. J., Highstrete, C., Lee, M., & Averitt, R. D. (2006). Dynamical electric and magnetic metamaterial response at terahertz frequencies. *Physical Review Letters, 96*(10), 107401. doi:10.1103/PhysRevLett.96.107401 PMID:16605787

Palandoken, M., Gocen, C., Kaya, A., Gunes, F., Baytore, C., & Can, F. C. (2018). A Novel Split-Ring Resonator and Voltage Multiplier Based Rectenna Design for 900 MHz Energy Harvesting Applications. *Radioengineering, 27*(3), 711–717. doi:10.13164/re.2018.0711

Pandeeswari, R. (2012). Artificial Neural Network Model for MNG-Metamaterial Spiral Resonator. *Electromagnetics Research Symposium Proceedings*, 29-33.

Pandey, G. K., Singh, H. S., Bharti, P. K., & Meshram, M. K. (2014, August 28). Metamaterial-based UWB antenna. *Electronics Letters, 50*(18), 1266–1268. doi:10.1049/el.2014.2366

Pan, Z. Y. (2012). Terahertz Metamaterials Fabrication from Narrowband to Broadband. *IEEE Asia-Pacific Conference on Antennas & Propagation*, 68 - 69. 10.1109/APCAP.2012.6333151

Park & Lee. (n.d.). Omnidirectional circularly polarized antenna utilizing zeroth- order resonance of epsilon negative transmission line. *IEEE Trans. on Antennas and Propag., 59*, 2717-2720.

Patnaik, A. (1997). An Artificial Neural Network Model for Effective Dielectric Constant of Microstrip Line. IEEE Trans. Antennas Propag., 45.

Patnaik, A. (2005). A Frequency Reconfigurable Antenna Design Using Neural Netwoks. *IEEE Antennas and Propagation Society International Symposium*, 409-412. 10.1109/APS.2005.1551830

Paul, P.M. (2013). Miniaturization of Square Patch Antenna using Complementary Split Ring Resonators. *Advances in Computing and Communications (ICACC),* 122 – 125.

Pelosi, G. (2004). Inter-Element Coupling In Large Arrays by an ANN Approach. *IEEE Antennas and Propagation Society International Symposium,* 1, 523 – 526. 10.1109/APS.2004.1329711

Pendry, J. B. (2000). Negative refraction makes a perfect lens. *Physical Review Letters*, *85*(18), 3966–3969. doi:10.1103/PhysRevLett.85.3966 PMID:11041972

Pendry, J. B., Holden, A. J., Robbins, D. J., & Stewart, W. J. (1999). Magnetism from conductors and enhanced nonlinear phenomena. *IEEE Transactions on Microwave Theory and Techniques*, *47*(11), 2075–2084. doi:10.1109/22.798002

Pendry, J. B., Holden, A. J., Stewart, W. J., & Youngs, I. (1996). Extremely low frequency plasmons in metallic microstructures. *Physical Review Letters*, *76*(25), 4773–4776. doi:10.1103/PhysRevLett.76.4773 PMID:10061377

Pendry, J.-B., Schurig, D., & Smith, D. R. (2006, June). Controlling Electromagnetic Fields. *Science*, *312*(5781), 1780–1782. doi:10.1126cience.1125907 PMID:16728597

Peng, Y. (2010). Ultra thin low profile U band folded meta-material wideband dipole antenna for multi-Gb/s data transmission using 65 nm CMOS technology. *Antennas and Propag Society Inter Symposium (APSURSI)*, 1-4.

Peralta, X. G., Arrington, C. L., Wanke, M. C., Brener, I., Williams, J. D., Smirnova, E., ... Padilla, W. J. (2008, May). Flexible, large-area metamaterials fabricated on thin silicon nitride membranes. In *Conference on Lasers and Electro-Optics* (p. CFY4). Optical Society of America. 10.1109/CLEO.2008.4551092

Peralta, X. G., Smirnova, E. I., Azad, A. K., Chen, H. T., Taylor, A. J., Brener, I., & O'Hara, J. F. (2009). Metamaterials for THz polarimetric devices. *Optics Express*, *17*(2), 773–783. doi:10.1364/OE.17.000773 PMID:19158890

Pestourie, R., Mroueh, Y., Nguyen, T. V., Das, P., & Johnson, S. G. (2020). Active learning of deep surrogates for PDEs: Application to metasurface design. *npj Computational Materials*, *6*(1), 164. Advance online publication. doi:10.103841524-020-00431-2

Pintos, J. F., Louzir, A., Minard, P., Perraudeau, J., Mattei, J. L., Souriou, D., & Queffelec, P. (2010). *Ultra Miniature UHF Antenna Using Magneto-Dielectric Material*. ANTEM. doi:10.1109/ANTEM.2010.5552480

Piñuela, M., Mitcheson, P. D., & Lucyszyn, S. (2013). Ambient RF energy harvesting in urban and semi-urban environments. *IEEE Transactions on Microwave Theory and Techniques*, *61*(7), 2715–2726. doi:10.1109/TMTT.2013.2262687

Polemi, A., Maci, S., & Kildal, P.-S. (2011). Dispersion Characteristics of a Metamaterial-Based Parallel-Plate Ridge Gap Waveguide Realized by Bed of Nails. *IEEE Transactions on Antennas and Propagation*, *59*(March), 904–1003. doi:10.1109/TAP.2010.2103006

Ponnupillai, A., Arunakiri, S., Danasegaran, S. K., Somasundaram, P., & Rajkumar, I. (2023). Investigation of the antenna characteristics with the impact of hexagonal metamaterial. *International Journal of Communication Systems*, *36*(7), e5454. Advance online publication. doi:10.1002/dac.5454

Porter, E., Coates, M., & Popovic, M. (2016). An early clinical study of time-domain microwave radar for Breast Health Monitoring. *IEEE Transactions on Biomedical Engineering, 63*(3), 530–539. doi:10.1109/TBME.2015.2465867 PMID:26259214

Porter, E., Kirshin, E., Santorelli, A., Coates, M., & Popovic, M. (2013). Time-domain multistatic radar system for microwave breast screening. *IEEE Antennas and Wireless Propagation Letters, 12*(4), 229–232. doi:10.1109/LAWP.2013.2247374

Pranav, U. S., Sudheesh, S., Stanly, P., Sankar, S., Devika, R., & Pradeep, A. (2016). Metamaterial Based Energy Harvester. *Procedia Computer Science, 93*(September), 74–80. doi:10.1016/j.procs.2016.07.184

Pyo, Baik, Cho, & Kim. (2009). Metamaterial-based antenna with triangular slotted ground for efficiency improvement. *IEEE Electronics Letters, 45*(3).

Qi, S. T., Wang, Y. P., Li, Y. Z., Wu, X., Ren, Q., & Ren, Y. (2020). Two-Dimensional Electromagnetic Solver Based on Deep Learning Technique. *IEEE Journal on Multiscale and Multiphysics Computational Techniques, 5*, 83–88. doi:10.1109/JMMCT.2020.2995811

Qi, S., Oudich, M., Li, Y., & Assouar, B. (2016). Acoustic energy harvesting based on a planar acoustic metamaterial. *Applied Physics Letters, 108*(26), 263501-1, 263501–263504. doi:10.1063/1.4954987

Quevedo-Teruel, O., & Chen, Q. (2021). Classification and opportunities of metasurfaces for antenna designs. *2021 International Symposium on Antennas and Propagation (ISAP)*, 1–2. 10.23919/ISAP47258.2021.9614437

Quevedo-Teruel, O., Rajo-Iglesias, E., Inclan-Sanchez, L., & Vazquez-Roy, J. (2010). Reconfigurable loaded planar inverted F-antenna by making use of varactor diodes. *Proc. IEEE APSURSI*, 1–4. 10.1109/APS.2010.5560948

Qureshi, F., Antoniades, M. A., & George, V. (2005). A Compact and Low-Profile Metamaterial Ring Antenna With Vertical Polarization. IEEE Antennas and Wireless Propagation Letters, 4.

Rahmatinia, S., & Fahimi, B. (2016, November 13-16). Magneto-thermal modeling of biological tissues: A step toward breast cancer detection. *2016 IEEE Conference on Electromagnetic Field Computation*, Miami, FL, USA. 10.1109/CEFC.2016.7816017

Rahm, M., Schurig, D., Roberts, D. A., Cummer, S. A., Smith, D. R., & Pendry, J. B. (2008). Design of electromagnetic cloaks and concentrators using form-invariant coordinate transformations of Maxwells equations. *Photonics and Nanostructures, 6*(1), 87–95. doi:10.1016/j.photonics.2007.07.013

Ramaccia, D., Toscano, A., & Bilotti, F. (2021). Time-varying metamaterials and metasurfaces for antennas and propagation applications. *Journal of Physics: Conference Series, 2015*(1), 012121. doi:10.1088/1742-6596/2015/1/012121

Ramahi, O. M., Almoneef, T. S., Alshareef, M., & Boybay, M. S. (2012). Metamaterial particles for electromagnetic energy harvesting. *Applied Physics Letters*, *173903*(17), 173903-1, 173903–173905. doi:10.1063/1.4764054

Ranjan, K., & Singh, G. (2013). Effect of low dielectric permittivity on microstrip antenna at terahertz frequency. *Optik (Stuttgart)*, *124*(22), 5777–5780. doi:10.1016/j.ijleo.2013.04.061

Rao, P. K., & Mishra, R. (2020). Elliptical shape flexible MIMO antenna with high isolation for breast cancer detection application. *Journal of the Institution of Electronics and Telecommunication Engineers*, *69*(1), 325–333. doi:10.1080/03772063.2020.1819887

Rashmi, M., William, P., Yogeesh, N., & Girija, D. K. (2023). Blockchain-Based Cloud Storage Using Secure and Decentralised Solution. In N. Chaki, N. D. Roy, P. Debnath, & K. Saeed (Eds.), *Proceedings of International Conference on Data Analytics and Insights, ICDAI 2023. ICDAI 2023. Lecture Notes in Networks and Systems, 727*. Springer. https://doi.org/10.1007/978-981-99-3878-0_23.

Ripin, N., Awang, R., Sulaiman, A., Baba, N., & Subahir, S. (2012). Rectangular microstrip patch antenna with EBG structure. *SCOReD 2012 - 2012 IEEE Student Conference on Research and Development*, 266-271. . doi:10.1109/SCOReD.2012.6518651

Robustillo, P. (2011). ANN Element Characterization for Reflectarray Antenna Optimization. *Proceedings of the 5th European Conference onAntennas and Propagation*, 957 – 960.

Rockstuhl, C., & Lederer, F. (2012). Metamaterials: A nonlinear perspective. *Applied Physics Reviews*, *4*(2), 021302.

Rotman, W. (1962, January). Plasma simulation by artificial dielectrics and parallelplate media. *IRE Transactions on Antennas and Propagation*, *10*(1), 82–95. doi:10.1109/TAP.1962.1137809

Rozanov, K. N. (2000). Ultimate Thickness to Bandwidth Ratio of Radar Absorbers. *IEEE Transactions on Antennas and Propagation*, *48*(8), 1230–1234. doi:10.1109/8.884491

Sabban, A. (2013). New Printed Meta Material Antenna. *Antennas and Propagation Society International Symposium (APSURSI)*, 1790 – 1791.

Sabban, A. (2014). New microstrip metamaterial antennas. In *Antennas and Propagation Society International Symposium (APSURSI)*. IEEE.

Saeed Majedi. (2012). *Compact MNG zeroth order resonator antenna*. Elsevier B.V.

Saha, C., & Siddiqui, J. Y. (2012, July). Theoretical model for estimation of resonance frequency of rotational circular split-ring resonators. *Electromagnetics*, *32*(6), 345–355. doi:10.1080/02726343.2012.701540

Salhi, R., Laabidi, M., & Choubani, F. (2016). Modeling and simulation of practical metamaterial structures: SRR and HIS structures. *2016 IEEE International Conference on Computational Electromagnetics (ICCEM)*, 259–261. 10.1109/COMPEM.2016.7588663

Salisbury, W. W. (1954). *Absorbent body for electromagnetic waves*. United States Patent 2599944.

Saraswat, R. K., & Kumar, M. (2020, October). A quad band metamaterial miniaturized antenna for wireless applications with gain enhancement. *Wireless Personal Communications*, *114*(4), 3595–3612. doi:10.100711277-020-07548-z

Sarmah, K., & Sarma, K. K. (n.d.). Design Optimization of Loop Antenna Using Competitive Leakning ANN. *National Conference on Emerging Trends and Applications in Computer.*

Sathish Kumar, D., Elizabeth Caroline, B., Sagadevan, K., Sakthiganesh, G., & Saravanan, R. (2020, July 23-24). Investigation of high directional gain pentagonal shaped patch antenna. In *7th International Conference on Smart Structures and Systems (ICSSS)*. IEEE. 10.1109/ICSSS49621.2020.9202223

Scheuer, J. (2020). Optical metasurfaces are coming of age: Short-and long-term opportunities for commercial applications. *ACS Photonics*, *7*(6), 1323–1354. doi:10.1021/acsphotonics.9b01719

Schurig, D., Mock, J. J., Justice, B. J., Cummer, S. A., Pendry, J. B., Starr, A. F., & Smith, D. R. (2006). Metamaterial electromagnetic cloak at microwave frequencies. *Science*, *314*(5801), 977–980. doi:10.1126cience.1133628 PMID:17053110

Schurig, D., Mock, J. J., & Smith, D. J. (2006). Electric-field-coupled resonators for negative permittivity metamaterials. *Applied Physics Letters*, *88*(4), 041109. doi:10.1063/1.2166681

Sehrai, D. A., Muhammad, F., Kiani, S. H., & Kim, S. (2020). Gain-Enhanced Metamaterial Based Antenna for 5G Communication Standards. *Computers, Materials & Continua*, *64*(3), 1587–1599. doi:10.32604/cmc.2020.011057

Sendrea, R. E., Zekios, C. L., & Georgakopoulos, S. V. (2023). Multifidelity Surrogate Modeling Based on Analytical Eigenfunction Expansions. *IEEE Transactions on Antennas and Propagation*, *71*(2), 1673–1683.

Shadrivov, I. V., Lapine, M., & Kivshar, Y. S. (Eds.). (2015). *Nonlinear, Tunable and Active Metamaterials* (Vol. 200). Springer International Publishing. doi:10.1007/978-3-319-08386-5

Shamonina & Solymar. (2007). Metamaterials: How the Subject Started. *Metamaterials*, *1*(1), 12-18.

Shang. (2010). An Effective Method for RFID Tag Antenna Optimization Based on Artificial Neural Network. *IEEE Youth Conference on Information Computing and Telecommunications*, 383 – 386.

Shang, S., Yang, S., Shan, M., Liu, J., & Cao, H. (2017). High performance metamaterial device with enhanced electromagnetic energy harvesting efficiency. *AIP Advances*, *7*(August), 105204-1, 105204–105207. doi:10.1063/1.5002165

Sharma, Zhang, & Xin. (2020). Machine Learning Technique for Optimizing Design of Double T-Shaped Monopole Antenna. *IEEE Trans. on Ant. and Prop.*, *68*(7).

Shena, Y., & Gongc, L. (2014). Investigation of gain effect of multi-band patch antenna based oncomposite rectangular SRRs. *Optik (Stuttgart)*, *124*(22), 933–935.

Shen, B., Zhai, W., & Zheng, W. (2014). Ultrathin Flexible Graphene Film: An Excellent Thermal Conducting Material with Efficient EMI Shielding. *Advanced Functional Materials*, *24*(28), 4542–4548. doi:10.1002/adfm.201400079

Sievenpiper, D., Zhang, L., Broas, R. F. J., Alex'opoulos, N. G., & Yablonovitch, E. (1999). Artificial Magnetic conductor Surfaces with a forbidden frequency band. *IEEE Transactions on Microwave Theory and Techniques*, *47*(11), 2059–2074. doi:10.1109/22.798001

Sihvola, A. (2007). Metamaterials in electromagnetics. *Metamaterials (Amsterdam)*, *1*(1), 2–11. doi:10.1016/j.metmat.2007.02.003

Silin, R. A., & Chepurnykh, I. P. (2001). On Media with Negative Dispersion. *Commun. Technol. Electron.*, *46*, 1121–1125.

Sim, M. S., You, K. Y., & Esa, F. (2020). Electromagnetic metamaterials in microwave regime. In Handbook of Research on Recent Developments in Electrical and Mechanical Engineering (pp. 64–86). doi:10.4018/978-1-7998-0117-7.ch002

Sim, M. S., You, K. Y., Esa, F. B., & Chan, Y. L. (2017). Broadband metamaterial microwave absorber for X-Ku band using planar split ring-slot structures. *Progress in Electromagnetics Research Symposium-Fall (PIERS-FALL)*, 215–221. 10.1109/PIERS-FALL.2017.8293138

Simion, S., Marcelli, R., & Sajin, G. (2007, August 16). Small-size CPW silicon resonating antenna based on transmission-line meta-material approach. *Electronics Letters*, *43*(17), 908–909. doi:10.1049/el:20071639

Sim, M. S., You, K. Y., Esa, F., & Chan, Y. L. (2021). Nanostructured electromagnetic metamaterials for sensing applications. In *Applications of Nanomaterials in Agriculture* (pp. 141–164). Food Science, and Medicine. doi:10.4018/978-1-7998-5563-7.ch009

Sim, M. S., You, K. Y., Esa, F., Dimon, M. N., & Khamis, N. H. (2018). Multiband metamaterial microwave absorbers using split ring and multiwidth slot structure. *International Journal of RF and Microwave Computer-Aided Engineering*, *28*(7), 1–13. doi:10.1002/mmce.21473

Singh, A. K., Pandurang, M., Shiban, A., & Koul, K. (2019). A Triple Band Polarization Insensitive Ultrathin Metamaterial Absorber for s- C- and X-Bands. *Progress in Electromagnetics Research M. Pier M*, *77*, 187–194. doi:10.2528/PIERM18110601

Singh, N., & Singh, S. (2010). Effect of Photonic Band Gap Structure on Planar Antenna Configuration. *Mediterranean Microwave Symposium (MMS)*, 81 – 85. 10.1109/MMW.2010.5605140

Singh, N., Singh, S., & Kumar, H. (2011). A study on Applications of Meta-Material based Antennas. *Conference on Electronics Computer Technology (ICECT)*, 192 - 196. 10.1109/ICECTECH.2011.5941588

Sivasangari, A., Deepa, D., Ajitha, P., Gomathi, R. M., Vignesh, R., Danasegaran, S., & Poonguzhali, S. (2023). Performance analysis of metamaterial patch antenna characteristics for advanced high-speed wireless system. *Journal of Electronic Materials*, *52*(7), 4785–4792. doi:10.100711664-023-10420-y

Sivukhin, D. V. (1957). The Energy of Electromagnetic Fields in Dispersive Media. *Opt. Spektrosk.*, *3*, 308–312.

Slimi, M., Jmai, B., Dinis, H., & Gharsallah, A. (2022). Microwave imaging for breast tumor detection using a CPW antenna. *Indian Journal of Science and Technology*, *15*(13), 554–560. doi:10.17485/IJST/v15i13.1974

Slimi, M., Jmai, B., Dinis, H., Gharsallah, A., & Mendes, P. M. (2022). Metamaterial Vivaldi antenna array for breast cancer detection. *Sensors (Basel)*, *22*(10), 3945. doi:10.339022103945 PMID:35632355

Smith, D. J., Vier, D. C., Koschny, T., & Soukoulis, C. M. (2005). Electromagnetic parameter retrieval from inhomogeneous metamaterials. *Physical Review. E*, *71*(3), 036617. Advance online publication. doi:10.1103/PhysRevE.71.036617 PMID:15903615

Smith, D. R., Padilla, W. J., Vier, D. C., Nemat-Nasser, S. C., & Schultz, S. (2000). Composite Medium with Simultaneously Negative Permeability and Permittivity. *Physical Review Letters*, *84*(18), 4184–4186. doi:10.1103/PhysRevLett.84.4184 PMID:10990641

Smith, D. R., Pendry, J. B., & Wiltshire, M. C. (2004). Metamaterials and negative refractive index. *Science*, *305*(5685), 788–792. doi:10.1126cience.1096796 PMID:15297655

Smith, D. R., Schultz, S., Markos, P., & Soukoulis, C. M. (2002). Determination of effective permittivity and permeability of metamaterials from reflection and transmission coefficients. *Physical Review B: Condensed Matter*, *65*(19), 5. doi:10.1103/PhysRevB.65.195104

Soheilifar, M. R., & Sadeghzadeh, R. A. (2014). Design, fabrication and measurement of two-layered quadruple-band microwave metamaterial absorber. *Journal of Communication Engineering*, *3*(1), 13–22.

Song, C., Zhu, X., Cao, C., Wang, L., & He, Q. (2017). A robust fuzzy C-means clustering algorithm based on new kernel-induced distance measure. *Information Sciences*, *379*, 237–257.

Song, H., Kono, H., Seo, Y., Azhari, A., Somei, J., Suematsu, E., Watarai, Y., Ota, T., Watanabe, H., Hiramatsu, Y., Toya, A., Xiao, X., & Kikkawa, T. (2015). A radar-based breast cancer detection system using CMOS integrated circuits. *IEEE Access : Practical Innovations, Open Solutions*, *3*, 2111–2121. doi:10.1109/ACCESS.2015.2496101

Song, X.-H., Wu, W.-Y., Shen, T.-G., & Zhou, Y.-Q. (2011). Investigation of A Patch Antenna Based on I-shaped Left-handed Material. *Optik (Stuttgart)*, *122*(16), 1426–1429. doi:10.1016/j.ijleo.2010.09.021

Suh, Y.-H., & Chang, K. (2002). A high-efficiency dual-frequency rectenna for 2.45- and 5.8-GHz wireless power transmission. *IEEE Transactions on Microwave Theory and Techniques*, *50*(7), 1784–1789. doi:10.1109/TMTT.2002.800430

Sun, K. H., Kim, J. E., Kim, J., & Song, K. (2017). Sound energy harvesting using a doubly coiled-up acoustic metamaterial cavity. *Smart Materials and Structures*, *26*(7), 075011. doi:10.1088/1361-665X/aa724e

Sun, W., Zhong, K., Liu, Y., Xiao, H., Zhao, D., Yan, Z., & Tan, T. (2022). Enhanced metamaterial vibration for high-performance acoustic piezoelectric energy harvesting. *Composites Communications*, *35*(July), 101342. doi:10.1016/j.coco.2022.101342

Sun, X., Fu, Q., Fan, Y., Wu, H., Qiu, K., Yang, R., Cai, W., Zhang, N., & Zhang, F. (2019). Thermally controllable Mie resonances in a water-based metamaterial. *Scientific Reports*, *9*(1), 5417. doi:10.103841598-019-41681-5 PMID:30931963

Tadesse, A. D., Acharya, O. P., & Sahu, S. (2020). Application of metamaterials for performance enhancement of planar antennas: A review. *International Journal of RF and Microwave Computer-Aided Engineering*, *30*(5). Advance online publication. doi:10.1002/mmce.22154

Tamayama, Y., Yasui, K., Nakanishi, T., & Kitano, M. (2014). A linear-to-circular polarization converter with half transmission and half reflection using a single-layered metamaterial. *Applied Physics Letters*, *105*(2), 021110. doi:10.1063/1.4890623

Tamilarasan, A. K., Krishnadhas, S. K., Sabapathy, S., & Thankamony Sarasam, A. S. (2021). A novel design of Rogers RT/duroid 5880 material based two turn antenna for intracranial pressure monitoring. *Microsystem Technologies*, *27*(9), 3579–3588. doi:10.100700542-020-05122-y

Tang, Z., Liu, J., Cai, Y.-M., Wang, J., & Yin, Y. (2018, April). A wideband differentially fed dual-polarized stacked patch antenna with tuned slot excitations. *IEEE Transactions on Antennas and Propagation*, *66*(4), 2055–2060. doi:10.1109/TAP.2018.2800764

Tan, T., Yan, Z., Zou, H., Ma, K., Liu, F., Zhao, L., Peng, Z., & Zhang, W. (2019). Renewable energy harvesting and absorbing via multi-scale metamaterial systems for Internet of things. *Applied Energy*, *254*(August), 113717-1, 113717–15. doi:10.1016/j.apenergy.2019.113717

Tan, Y. K., & Panda, S. K. (2011). Light and Thermal Energy Sources for Enhanced Performance of Wireless Sensor Nodes. *IEEE Transactions on Industrial Electronics*, *58*(9), 4424–4435. doi:10.1109/TIE.2010.2102321

Tao, H., Landy, N. I., Bingham, C. M., Zhang, X., Averitt, R. D., & Padilla, W. J. (2008). A metamaterial absorber for the terahertz regime: Design, fabrication and characterization. *Optics Express*, *16*(10), 7181–7188. doi:10.1364/OE.16.007181 PMID:18545422

Tatartschuk, E., Gneiding, N., Hesmer, F., Radkovskaya, A., & Shamonina, E. (2012, May). Mapping inter-element coupling in metamaterials: Scaling down to infrared. *Journal of Applied Physics*, *111*(9), 094904. doi:10.1063/1.4711092

Thakur, A., Thakur, P., & Hsu, J.-H. (2011). Novel magnetodielectric nano-materials with matching permeability and permittivity for the very high frequency applications. *Scripta Materialia*, *642*(2), 205–208. doi:10.1016/j.scriptamat.2010.09.045

Thippeswamy, M. C., Kuchibhatla, S. A. R., & Rajagopal, P. (2021). Concentric shell gradient index metamaterials for focusing ultrasound in bulk media. *Ultrasonics*, *114*, 106424. doi:10.1016/j.ultras.2021.106424 PMID:33819870

Troudi, Z., Machac, J., & Osman, L. (2019, October). Miniaturised planar band-pass filter based on interdigital arm SRR. *IET Microwaves, Antennas & Propagation*, *13*(12), 2081–2086. doi:10.1049/iet-map.2018.5708

Truong, B. C. Q., Tuan, H. D., Fitzgerald, A. J., Wallace, V. P., & Nguyen, H. T. (2015). A dielectric model of human breast tissue in terahertz regime. *IEEE Transactions on Biomedical Engineering*, *62*(2), 699–707. doi:10.1109/TBME.2014.2364025 PMID:25347869

Tung, B. S., Khuyen, B. X., Kim, Y. J., Lam, V. D., Kim, K. W., & Lee, Y. (2017). Polarization-independent, wide-incident-angle and dual-band perfect absorption, based on near-field coupling in a symmetric metamaterial. *Scientific Reports*, *7*(1), 11507. doi:10.103841598-017-11824-7 PMID:28912553

Turpin, J. P. (2011). Anisotropic Metamaterial Realization of a Flat Gain-enhancing Lens for Antenna Applications. *Symposium on Antennas and Propagation (APSURSI), 2011 IEEE International*, 2886-2889. 10.1109/APS.2011.5997130

Turpin. (2012). Low Cost and Broadband Dual-Polarization Metamaterial Lens for Directivity Enhancement. *IEEE Transactions on Antennas and Propagation, 12.*

Van Blarigan, L., Danzl, P., Moehlis, J., Van Blarigan, L., Danzl, P., & Moehlis, J. (2012). A broadband vibrational energy harvester. *Applied Physics Letters*, *100*(25), 253904-1, 253904–4. doi:10.1063/1.4729875

Vellucci, S., Monti, A., Barbuto, M., Longhi, M., Toscano, A., & Bilotti, F. (2022). Metasurface Coatings Enabling Scattering, Frequency, and Radiation Tunability for Next-Generation Antenna Systems. *2022 Microwave Mediterranean Symposium (MMS)*, 1–5. 10.1109/MMS55062.2022.9825540

Venkatrami Reddy, S., Saurav, K., Sarkar, D., Singh, A., & Srivastava, K. V. (2014). Design of a Novel Dual Band Patch Antenna Loaded with CELC Resonator for WLAN/WiMAX Applications. *Twentieth National Conference on Communications (NCC),* 1-6.

Veselago, V. G. (1968). The Electrodynamics of Substances with Simultaneously Negative Values of ε and μ. *Physics Uspekhi*, *10*, 509–514. doi:10.1070/PU1968v010n04ABEH003699

Walia, S., Shah, C. M., Gutruf, P., Nili, H., Chowdhury, D. R., Withayachumnankul, W., Bhaskaran, M., & Sriram, S. (2015). Flexible metasurfaces and metamaterials: A review of materials and fabrication processes at micro- and nano-scales. *Applied Physics Reviews*, *2*(1), 011303. doi:10.1063/1.4913751

Walser, R. M. (2001). Electromagnetic Metamaterials. *Proceeding SPIE Complex Mediums II beyond Linear Isotropic Dielectrics, 4467*, 1-15. https://doi.org/10.1117/12.432921

Wang, C., Chen, Y., & Yang, S. (2018, November). Dual-band dual-polarized antenna array with flat-top and sharp cutoff radiation patterns for 2G/3G/LTE cellular bands. *IEEE Transactions on Antennas and Propagation, 66*(11), 5907–5917. doi:10.1109/TAP.2018.2866596

Wang, J., Zhu, Z., Sun, Y., Shen, T., & Gong, L. (2013). Study on left-handed effect of composite helices and its application in square frame patch antennas. *Optik (Stuttgart), 124*(21), 5189–5192. doi:10.1016/j.ijleo.2013.03.052

Wang, S. (2009). Design of broadband array elements based on neuralnetwork and genetic algorithm. *IET Journals & Magazines, Volume, 45*, 1011–1012.

Wang, S.-Y., & Chen, F.-C. (2012). The Multi-Band EBG Characteristic of The Mushroom Structure with The Suspending Microstrip line. *Proceedings of the 42nd European Microwave Conference*, 663-666. 10.23919/EuMC.2012.6459242

Wartak, M. S., Tsakmakidis, K., & Hess, O. (2011). Introduction to Metamaterials. *Physics In Canada, 67*(1), 30–34.

Welch, H. G., Prorok, P. C., O'Malley, A. J., & Kramer, B. S. (2016). Breast-cancer tumor size, overdiagnosis, and Mammography Screening Effectiveness. *The New England Journal of Medicine, 375*(15), 1438–1447. doi:10.1056/NEJMoa1600249 PMID:27732805

Wen, J., Ren, Q., Peng, R., & Zhao, Q. (2021). An all-dielectric metasurface absorber based on surface wave conversion effect. *Applied Physics Letters, 119*(6), 062902. Advance online publication. doi:10.1063/5.0057529

Wen, J., Ren, Q., Peng, R., & Zhao, Q. (2022a). Multi-functional tunable ultra-broadband water-based metasurface absorber with high reconfigurability. *Journal of Physics. D, Applied Physics, 55*(28), 285103. Advance online publication. doi:10.1088/1361-6463/ac683e

Wen, J., Ren, Q., Peng, R., & Zhao, Q. (2022c). Ultrabroadband Saline-Based Metamaterial Absorber With Near Theoretical Absorption Bandwidth Limit. *IEEE Antennas and Wireless Propagation Letters, 21*(7), 1388–1392. doi:10.1109/LAWP.2022.3169467

Wen, J., Zhao, Q., Peng, R., Yao, H., Qing, Y., Yin, J., & Ren, Q. (2022). Progress in water-based metamaterial absorbers: A review. *Optical Materials Express, 12*(4), 1461. Advance online publication. doi:10.1364/OME.455723

Wen, L.-H., Gao, S., Mao, C.-X., Luo, Q., Hu, W., Yin, Y., & Yang, X. (2018). A wideband dual-polarized antenna using shorted dipoles. *IEEE Access : Practical Innovations, Open Solutions, 6*, 39725–39733. doi:10.1109/ACCESS.2018.2855425

Werner, D. H., Haupt, R. L., & Werner, P. L. (1999, October). Fractal Antenna Engineering "The Theory and Design of Fractal Antenna Arrays". *IEEE Antennas & Propagation Magazine, 41*(5), 37–59. doi:10.1109/74.801513

William, P., Yogeesh, N., Vimala, S., Gite, P., & SKS. (2022). Blockchain technology for data privacy using contract mechanism for 5G networks. In *2022 3rd International Conference on Intelligent Engineering and Management (ICIEM)* (pp. 461-465). IEEE. doi:10.1109/ICIEM54221.2022.9853118

Wnuk, M. (2013). The Utilization of Neural Networks to Radiation Pattern of Microstrip Antenna Formation. *International Conference on Signal Processing and Communication Systems*, 1-5.

Wong, K.-L., Lee, G.-Y., & Chiou, T.-W. (2003, January). A Low-Profile Planar Monopole Antenna for Multiband Operation of Mobile Handset. *IEEE Transactions on Antennas and Propagation*, *51*(1), 121–125. doi:10.1109/TAP.2003.809044

Wu, C.-Y., Yeh, S. H., & Lu, T.-H. (2007). Planar High Gain Antenna for 5.8-GHz WiMAX Operation. *Antennas and Propagation Society International Symposium*, 3488 – 3491.

Wu, M. F., Meng, F. Y., Wu, Q., Wu, J., & Li, L. W. (2005). A compact equivalent circuit model for the SRR structure in metamaterials. *Asia-Pacific Microwave Conference Proceedings, APMC*, *1*, 5–8. 10.1109/APMC.2005.1606214

Xie, M. T., & Guo, Q. G. (2010). Design of a Novel Artificial Magnetic Conductor Plane and Its Application for Low-Profile Dipole. *International Conference on Microwave and Millimeter Wave Technology (ICMMT),* 2085 – 2087. 10.1109/ICMMT.2010.5525228

Xue, X., Makota, C., Khalaf, O. I., Jayabalan, J., Samui, P., & Abdulsahib, G. M. (2023). Machine Learning Approach for Prediction of Lateral Confinement Coefficient of CFRP-Wrapped RC Columns. *Symmetry*, *15*, 545. https://doi.org/10.3390/sym15020545

Xue, X., Marappan, R., Raju, S. K., Raghavan, R., Rajan, R., Khalaf, O. I., & Abdulsahib, G. M. (2023). Modelling and Analysis of Hybrid Transformation for Lossless Big Medical Image Compression. *Bioengineering (Basel, Switzerland)*, *10*, 333. https://doi.org/10.3390/bioengineering10030333

Xue, X., Poonia, M., Abdulsahib, G. M., Bajaj, R. K., Khalaf, O. I., Dhumras, H., & Shukla, V. (2023). On Cohesive Fuzzy Sets, Operations and Properties with Applications in Electromagnetic Signals and Solar Activities. *Symmetry*, *15*, 595. https://doi.org/10.3390/sym15030595

Xue, X., Shanmugam, R., Palanisamy, S., Khalaf, O. I., Selvaraj, D., & Abdulsahib, G. M. (2023). A hybrid cross layer with Harris-hawk-optimization-based efficient routing for wireless sensor networks. *Symmetry*, *15*(2), 438. doi:10.3390ym15020438

Xue, X., Shanmugam, R., Palanisamy, S., Khalaf, O. I., Selvaraj, D., & Abdulsahib, G. M. (2023). A Hybrid Cross Layer with Harris-Hawk-Optimization-Based Efficient Routing for Wireless Sensor Networks. *Symmetry*, *15*, 438. https://doi.org/10.3390/sym15020438

Yan, T. W., Ma, M. J., Li, C., & Wang, L. (2020). A Low Profile Quadruple-Band Polarization Insensitive Metamaterial Absorber. *Progress In Electromagnetics Research M, 90*, 69-79. http://www.jpier.org/PIERM/pier.php?paper=20010505

Yang, D., Liu, S., & Zhao, Z. (2017, May). A broadband dual-polarized printed dipole antenna with low cross-polarization and high isolation for base station applications. *Microwave and Optical Technology Letters, 59*(5), 1107–1111. doi:10.1002/mop.30467

Yang, J. K., Zhou, Z., & Yang, H. Y. (2016). Metamaterials for perfect absorption: A review. *Advanced Optical Materials, 4*(5), 524–535.

Yang, S.-L., Kishk, A., & Lee, K.-F. (2008). Frequency reconfigurable U-slot microstrip patch antenna. *IEEE Antennas and Wireless Propagation Letters, 7*, 127–129. doi:10.1109/LAWP.2008.921330

Yang, X., Chen, Y., Zhao, Y., Pan, J., Guo, J., & Yang, D. (2023). Application of KNN for Linear Array Pattern Prediction Based on the Active Element Pattern Method. *IEEE Antennas and Wireless Propagation Letters, 22*(5), 1134–1138.

Yang, Y., Li, Q., & Wang, G. P. (2008). Design and fabrication of diverse metamaterial structures by holographic lithography. *Optics Express, 16*(15), 11275. doi:10.1364/OE.16.011275 PMID:18648445

Yan, W., Shen, X., & Liu, X. (2017). Modelling of nonlinear metamaterials: Progress, challenges, and perspectives. *IEEE Transactions on Antennas and Propagation, 65*(1), 70–87.

Yasuda, H., & Yang, J. (2015). Reentrant origami-based metamaterials with negative poisson's ratio and bistability. *Physical Review Letters, 114*(18), 185502-1, 185502–185506. doi:10.1103/PhysRevLett.114.185502 PMID:26001009

Yogeesh, N. (2021). Mathematical approach to representation of locations using K-means clustering algorithm. *International Journal of Mathematics And its Applications (IJMAA), 9*(1), 127-136.

Yogeesh, N. (2021). Mathematical Approach to Representation of Locations Using K-Means Clustering Algorithm. *International Journal of Mathematics and Its Applications, 9*(1), 127-136.

Yogeesh, N. (2023). Fuzzy Logic-Based Beat Tracking in Music Signals. *Musik In Bayern, 88*(9), 145-157. https://doi.org/10.15463/gfbm-mib-2023-343

Yogeesh, N., Girija, D. K., Rashmi, M., & Divyashree, J. (2023). Exploring the Potential of Fuzzy Domination Graphs in Aquatic Animal Health and Survival Studies. *Journal of Survey in Fisheries Sciences (SFS), 10*(4S), 3133-3147.

Yogeesh, N. (2012). A conceptual discussion about an intuitionistic fuzzy-sets and its applications. *International Journal of Advanced Research in IT and Engineering, 1*(6), 45–55.

Yogeesh, N. (2012a). A conceptual discussion about an intuitionistic fuzzy-sets and its applications. *International Journal of Advanced Research in IT and Engineering, 1*(6), 45–55.

Yogeesh, N. (2012b). Operations on intuitionistic fuzzy directed graphs. *Journal of Advances and Scholarly Researches in Allied Education, 3*(6), 1–4.

Yogeesh, N. (2014). Graphical representation of Solutions to Initial and boundary value problems Of Second Order Linear Differential Equation Using FOOS (Free & Open Source Software)-Maxima. *International Research Journal of Management Science and Technology*, *5*(7), 168–176.

Yogeesh, N. (2015). Solving Linear System of Equations with Various Examples by using Gauss method. *International Journal of Research and Analytical Reviews*, *2*(4), 338–350.

Yogeesh, N. (2019). Graphical Representation of Mathematical Equations Using Open Source Software. *Journal of Advances and Scholarly Researches in Allied Education*, *16*(5), 2204–2209.

Yogeesh, N. (2020). Study on clustering method based on K-means algorithm. *Journal of Advances and Scholarly Researches in Allied Education*, *17*(1), 2230–7540.

Yoon, G., Kim, I., & Rho, J. (2016). Challenges in fabrication towards realization of practical metamaterials. *Microelectronic Engineering*, *163*, 7–20. doi:10.1016/j.mee.2016.05.005

Yoo, Y. J., Ju, S., Park, S. Y., Ju Kim, Y., Bong, J., Lim, T., Kim, K. W., Rhee, J. Y., & Lee, Y. (2015). Metamaterial Absorber for Electromagnetic Waves in Periodic Water Droplets. *Scientific Reports*, *5*(1), 14018. doi:10.1038rep14018 PMID:26354891

Yuan, B., Liu, Y., Zhang, C., Li, G., & Fang, D. (2018). Nonlinear behaviour of metamaterials: Modelling, simulation and experiment. *Composite Structures*, *189*, 404–413.

Yu, C., Fan, S., & Sun, Y., & Pickwell-MacPherson, E. (2011). The potential of terahertz imaging for cancer diagnosis: A review of investigations to date. *Quantitative Image in Medicine and Surgery*, *2*(1), 33–45. doi:10.3978/j.issn.2223-4292.2012.01.04 PMID:23256057

Yu, Z., & Qi Wang, A. (2007). Multi-Band Small Antenna Based on Deformed Split Ring Resonators of Left-Handed Meta-Materials. *International Symposium on Microwave, Antenna, Propagation and EMC,* 531-534. 10.1109/MAPE.2007.4393672

Yu, Z., Veronis, G., & Fan, S. (2011, August). Resonant electromagnetic fields and local density of states in disordered metamaterials. *Physical Review B: Condensed Matter and Materials Physics*, *84*(7), 075119. doi:10.1103/PhysRevB.84.075119

Zadeh, L. A. (1965). Fuzzy sets. *Information and Control*, *8*(3), 338–353. doi:10.1016/S0019-9958(65)90241-X

Zani, M. Z. M., Jusoh, M. H., Sulaiman, A. A., Baba, N. H., Awang, R. A., & Ain, M. F. (2010). Circular Patch Antenna On Metamaterial. *IEEE International Conference on Electronic Devices, Systems and Applications (ICEDSA),* 313-316. 10.1109/ICEDSA.2010.5503053

Zeng, Z., Li, X., Bermak, A., Tsui, C. Y., & Ki, W. H. (2016). A WLAN 2.4-GHz RF energy harvesting system with reconfigurable rectifier for wireless sensor network. *Proceedings - IEEE International Symposium on Circuits and Systems,* 2362–2365. 10.1109/ISCAS.2016.7539059

Zerrad, F., Taouzari, M., Makroum, E. M., Aoufi, J. E., Qanadli, S. D., Karaaslan, M., Al-Gburi, A. J., & Zakaria, Z. (2023). Microwave Imaging Approach for breast cancer detection using a tapered slot antenna loaded with parasitic components. *Materials (Basel)*, *16*(4), 1496. doi:10.3390/ma16041496 PMID:36837126

Zhang, H., El-Rayis, A. O., Haridas, N., Noordin, N. H., Erdogan, A. T., & Arslan, T. (2011, November 14-15). *A smart antenna array for brain cancer detection.* Loughborough Antennas & Propagation Conference, Loughborough, UK. 10.1109/LAPC.2011.6114045

Zhang, J., Feng, F., & Na, W. (2022). Machine Learning Technologies for RF/Microwave CAD: Past, Present and Future Perspectives. *2022 52nd European Microwave Conference (EuMC),* 16-16. doi: 10.23919/EuMC54642.2022.9991872

Zhang, Y., Hong, W., Yu, C., Kuai, Z. Q., Don, Y. D., & Zhou, J. Y. (2008). Planar ultrawideband antennas with multiple notched bands based on etched slots on the patch and or split ring resonators on the feed line. *IEEE Trans. Antennas Propag., 56*(9), 3063–3068. doi:.928815 doi:10.1109/tap.2008

Zhang, G., Lan, C., Gao, R., & Zhou, J. (2017). Permanent magnetic ferrite based power-tunable metamaterials. *Journal of Magnetism and Magnetic Materials*, *436*, 57–60. doi:10.1016/j.jmmm.2017.04.027

Zhang, H., Li, X., Li, Z., & Luo, C. (2019). Adaptive fuzzy modelling for predicting the nonlinear behaviour of metamaterials. *IEEE Access : Practical Innovations, Open Solutions*, *7*, 10319–10326.

Zhang, P., Zhang, X., & Li, L. (2022). An Optically Transparent Metantenna for RF Wireless Energy Harvesting. *IEEE Transactions on Antennas and Propagation*, *70*(4), 2550–2560. doi:10.1109/TAP.2021.3137166

Zhang, Q., Ma, Q., Yan, S., Wu, F., He, X., & Jiang, J. (2015). Tunable terahertz absorption in graphene-based metamaterial. *Optics Communications*, *353*, 70–75. doi:10.1016/j.optcom.2015.05.017

Zhang, S., Chen, X., & Pedersen, G. F. (2019). Mutual coupling suppression with decoupling ground for massive MIMO antenna arrays. *IEEE Transactions on Vehicular Technology*, *68*(8), 7273–7282. doi:10.1109/TVT.2019.2923338

Zhang, X., Liu, H., & Li, L. (2017). Tri-band miniaturized wide-angle and polarization-insensitive metasurface for ambient energy harvesting. *Applied Physics Letters*, *111*(7), 071902. Advance online publication. doi:10.1063/1.4999327

Zhang, X., & Liu, Z. (2008). Superlenses to overcome the diffraction limit. *Nature Materials*, *7*(6), 435–441. doi:10.1038/nmat2141 PMID:18497850

Zhang, X., Yan, F., Du, X., Wang, W., & Zhang, M. (2020). Broadband water-based metamaterial absorber with wide angle and thermal stability. *AIP Advances*, *10*(5), 055211. Advance online publication. doi:10.1063/5.0006166

Zhang, Y., Wei, K., Zhang, Z., Li, Y., & Feng, Z. (2015). A Compact Dual-Mode Metamaterial-Based Loop Antenna for Pattern Diversity. *IEEE Antennas and Wireless Propagation Letters*, *14*, PP394–PP397. doi:10.1109/LAWP.2014.2363847

Zhang, Z., Xie, H., & Zheng, L. (2008). V- Shape wired structure H.F bowtie antenna. *International Symposium on Antennas, Propagation and EM Theory*, 207–210.

Zhao, J., Cheng, Q., Chen, J., Qi, M. Q., Jiang, W. X., & Cui, T. J. (2013). A tunable metamaterial absorber using varactor diodes. *New Journal of Physics*, *15*(4), 043049. Advance online publication. doi:10.1088/1367-2630/15/4/043049

Zhao, J., & Cheng, Y. (2016). Ultrabroadband Microwave Metamaterial Absorber Based on Electric SRR Loaded with Lumped Resistors. *Journal of Electronic Materials*, *45*(10), 5033–5039. doi:10.100711664-016-4693-0

Zhao, X., Lee, Y., Choi, J. H., & Jung, K. Y. (2012, August 21). Design of a metamaterial-inspired size-reduced wideband loop antenna with frequency scanning characteristic with frequency scanning characteristic. *IET Microwaves, Antennas & Propagation*, *6*(11), 1227–1235. doi:10.1049/iet-map.2012.0006

Zhi. (2014). A Low-Profile High-Gain Substrate-Integrated Waveguide Slot Antenna Enabled by an Ultrathin Anisotropic Zero-Index Metamaterial Coating. *IEEE Transactions On Antennas And Propagation, 62*(March), 1173–1184.

Zhong, H. T., Yang, X. X., Tan, C., & Yu, K. (2016). Triple-band polarization-insensitive and wide-angle metamaterial array for electromagnetic energy harvesting. *Applied Physics Letters*, *109*(25), 1–5. doi:10.1063/1.4973282

Zhong, M. (2014). Influence of dielectric layer on negative refractive index and transmission of metal-dielectric-metal sandwiched metamaterials. *Chinese Optics Letters*, *12*(4), 10–12. doi:10.3788/COL201412.041601.Metamaterials

Zhou, Y., Qin, Z., Liang, Z., Meng, D., Xu, H., Smith, D. R., & Liu, Y. (2021). Ultra-broadband metamaterial absorbers from long to very long infrared regime. *Light, Science & Applications*, *10*(1), 138. doi:10.103841377-021-00577-8 PMID:34226489

Zhu, H. L., Cheung, S. W., Chung, K. L., & Yuk, T. I. (2013). Linear-to-circular polarization conversion using metasurface. *IEEE Transactions on Antennas and Propagation*, *61*(9), 4615–4623. doi:10.1109/TAP.2013.2267712

Zhu, J., & Eleftheriades, G. V. (2009). A Compact Transmission-Line Metamaterial Antenna With Extended Bandwidth. *IEEE Antennas and Wireless Propagation Letters*, *8*, 295–298. doi:10.1109/LAWP.2008.2010722

About the Contributors

Shilpa Mehta has completed her PhD degree from Auckland University of Technology, New Zealand. She has five years of teaching experience from different reputed colleges and universities of India. Currently, she is working as a Teaching Assistant at Auckland University of Technology (AUT), Auckland, New Zealand. She was awarded Summer Doctoral Research Scholarship for her PhD research. She has worked on various interdisciplinary research projects and her research interests include Radio Frequency Integrated Circuits, RF front ends, Optimization, Internet of Things, Wireless Communication, Artificial Intelligence, Healthcare, Radars, and Software-defined Radios, Smart cities. She is working as a Series Editor. As an editor, she has already edited two books out of which one has been published by IGI Global and the other one will be published by Nova Science (under publication phase). Additionally, she is currently working on four edited books to be published by IGI Global, Apple Academic Press, Scrivener-Wiley, and Springer Nature, respectively.

Arij Naser Abougreen received the M.Sc degree in Communication engineering from the University of Tripoli, Libya, in 2018. Her research interests include image processing, machine learning, Internet of Things and wireless communication. She served as a technical Program Committee member and a reviewer for several conferences such as 7th IEEE World Forum on the Internet of Things (WFIoT2021), New Orleans, USA; 11th series of IEEE Symposium on Computer Applications and Industrial Electronics (ISCAIE 2021); 44th International Conference on Telecommunications and Signal Processing (TSP), 2021; 30th International Conference on Computing and Communication Networks (ICCCN-2021), organized by Manchester Metropolitan University, United Kingdom. She also served as a reviewer for for journal of Multimedia Tools and Applications. She has authored/ co-authored four book chapters and two papers.

* * *

Alkesh Agrawal is an Associate Professor in the Department of Electronics and Communication Engineering, Shri Ramswaroop Memorial University, Lucknow-Deva Road, Barabanki, Uttar Pradesh, India. He received B.Tech. Degree in Electronics and Telecommunication from J. K. Institute of Applied Physics and Technology, University of Allahabad, India in the year 2002. He received M.Tech. degree in Digital Communication from UPTU, India in 2008. He received Ph.D degree in 2019 for his research work on "Broad-band EM Absorber based on Metamaterial". He has a total academic experience of 21 years with paper presentations and publications in reputed conferences and journals indexed in Web of Science. He has been an active reviewer of papers of many reputed journals- IEEE Access, Journal of Physics Communication (IOP), Physica Scripta (IOP), Journal of Physics D: Applied Physics (IOP), Microwave and Optical Technology Letters (Wiley). He is professional member of IEEE. He is the recipient of many awards and scholarships in his academic and professional career- Suhrota Roy Memorial Scholarship, Malvea Memorial Scholarship, Smt. Chandramani Agrawal Memorial Scholarship, Mitra Memorial Scholarship, Ewing Christian College Gold Medal (for highest marks in B.Sc. 1995-98 among PCM/PSM/PEM group), Best Faculty Award (session: 2002-03 and 2009-10), Best paper awards for his research work presented in international conferences, ICMARS-2014 and IEEE-UPCON-2016. His area of research include design, simulation, fabrication, and testing of Metamaterial Absorbers. Proficient in: CST Microwave Studio, LabVIEW, Matlab, Scilab, Microsoft Visio, HTML, CSS, JS.

Elizabeth Caroline Britto was awarded her PhD by Anna University, Chennai, Tamil Nadu, India, in 2010. She has concurrent teaching, research, and industrial experience of more than two decades. Currently, she is working as a professor in the Department of ECE at IFET College of Engineering, Villupuram. Her areas of interests include signal processing, radio frequency, antenna, and optical computing. She is a lifetime member of ISTE and Broadcasting Engineers Society. She has received grants from AICTE and Council of Scientific and Industrial Research (CSIR). She has published more than 30 international journals and 30 international conference proceedings.

Sathish Kumar Danasegaran received B.E. in Electronics and Communication Engineering degree in 2008 from Kanchi Pallavan Engg. College, Anna University, Kanchipuram. He obtained his master's degree in communication systems in the year 2011 from Easwari Engineering College (Anna University), Chennai. He was awarded his Ph.D. by Anna University, Chennai, Tamil Nadu, India in 2022. He is having concurrent Teaching and Research experience of more than ten years. Currently, he is working as an Associate Professor in the ECE Department at IFET College of Engineering, Villupuram. His areas of interest include Wireless Communication, Photonic Devices, Antenna and Optical Computing. He has published more than 30 International Journals, Book Chapters and IEEE conferences including Elsevier, Optical Engineering - SPIE, Springer, Wiley, Taylor & Francis, etc.,

Fahmiruddin Esa graduated from Universiti Putra Malaysia (UPM) with a Bachelor's degree in Physics in 2006. Two years later, he pursued his studies at Universiti Teknologi Malaysia (UTM) and obtained a Master of Science (Physics) degree through coursework in 2010. In 2011, he began his Ph.D. studies in Microwave at the Faculty of Science, UPM, and completed it in 2015. Currently, he holds a position as a lecturer at the Department of Physics and Chemistry, Faculty of Applied Sciences and Technology, Pagoh Higher Education Hub, Universiti Tun Hussein Onn Malaysia (UTHM), Johor, Malaysia. His research interest lies in experimental and simulation techniques for the characterization of microwave-absorbing materials.

Fandi Hamid is working as a military Signal Officer in the Malaysian Armed Forces. He was born in Sarawak, Malaysia, in 1981. He received his Bachelor of Engineering (Electrical – Electronics) from the Universiti Teknologi Malaysia, Johor, Malaysia, in 2003 and Master of Engineering (Telecommunications) from the Universiti Malaya, Kuala Lumpur, Malaysia, in 2016. He is currently pursuing his Doctor of Philosophy (PhD) at the Faculty of Electrical Engineering, Universiti Teknologi Malaysia, Johor, Malaysia. His research interest includes electromagnetic pulse (EMP), metamaterials, microwave antennas, and antenna theory.

Shanthalakshmi Revathy J. completed B.E Information Technology from Mohamed Sathak Engineering College, Kilakkarai and M.E. Computer Science and Engineering from PSG College of Technology, Coimbatore.

Sagadevan K. has received his BE degree in ECE from Kamban Engineering college in Tiruvannamalai (Anna University, Chennai, India) in 2009 and completed his ME degree in Arunai college of Engineering in Tiruvannamalai (Anna University, Chennai, India). Currently, he is working as a Associate Professor in the department of ECE at IFET college of Engineering, Villupuram. He has published papers in more than 3 international journals and 3 international conference proceedings.

Rupesh Kumar was born in Bokaro Steel City, India, in 1983. He received B.Tech degree in Electronics & Communication engineering from SMIT, India, in 2008, and MS and PhD degrees in Electronics & Communication from Institut Mines-Telecom, Paris, France, in 2011 and 2014. Currently, he is working as a Professor at SSRM AP University, India. His research is focused on indoor localization system, FMCW-CW Radars, and Reconfigurable Array Antenna, RFID Technology, Wireless Sensing, Machine Learning.

Yogeesh N., Government First Grade College, Tumkur, is one of the leading government institutes in Karnataka state among the 412 government colleges across the state. At Government First Grade College, Tumkur, Dr. Yogeesh N, has been working as an assistant professor and head of the department of mathematics since 2013–14. He started his teaching career in the academic year 2005-06 as a lecturer in the College of Diploma in Education, where he taught methodology in mathematics teaching and educational psychology. After that, he worked as a principal and educational officer at the Sacred Heart Group of Institutions from 2007 to 2009. He has now worked in many government-first colleges since 2009, and he has served on the Board of Studies and the Board of Examination at Tumkur University, as well as in many autonomous colleges. He hosted state-level and national-level seminars as organizing secretary, and he also worked in various online and offline journals as a chief editor and editorial member. He wrote several manuals for mathematics labs using FOSS and some academic books, along with paper publications in various journals.

Subuh Pramono was born in Indonesia. He received the B. Eng degree in Electrical Engineering from Telkom University, Indonesia in 2003, and the M. Eng degree in Electrical Engineering from Institut Teknologi Bandung (ITB), Indonesia, in 2009. He has published three books and three Indonesia patents.Now, he is currently pursuing his Ph.D at Graduate School of Science and Engineering, Chiba University, Japan. He has published 92 papers in the national/international journals and conferences, 2 books, and 3 book chapters. He is currently a lecturer at Department of Electrical Engineering, Faculty of Engineering, Universitas Sebelas Maret, Surakarta-Indonesia. He is a member of IEEE, IEEE Comsoc, and IEEE Antennas & Propagation Society. His main areas of research interests are wireless and mobile communications, antenna and propagation, wireless sensor networks, and synthetic aperture radar.

Anuj Kumar Sharma Department of Electronics & Communication Engineering, Faculty of Engineering & Technology, Gurukul Kangri Deemed to be University, Haridwar, India. He is currently working as an Assistant Professor in the Department of Electronics & Communication Engineering at Faculty of Engineering & Technology, Gurukula Kangri, Haridwar. He is pursuing Ph.D in Metamaterial Antenna from Gurukula Kangri, Haridwar. He did M.Tech in Electronics and Communication Engineering from National Institute of Technology, Jalandhar. His Research interests includes Wireless Communication, Embedded System Design, Antenna Fabrication, Cognitive radios, MIMO Communication. He has published 2 books and 5 research papers in the field of wireless communication. He received Summer Faculty Research Fellowship in 2020 from IIT Delhi.

Vipul Sharma works in the Department of Electronics & Communication Engineering, Faculty of Engineering & Technology, Gurukul Kangri Deemed to be University, Haridwar, India. He belongs to Haridwar, Uttarakhand, India. He completed M.Tech. in electronics and communication engineering from IIT, Roorkee, Uttarakhand, India and Ph.D. degree in electronics and communication engineering from PTU, Jalandhar, Punjab, India. He has more than 20 years of teaching experience. He is working as an Professor and Head in the Electronics and Communication Engineering Department, GK (DU), Uttarakhand, India. He is an antenna researcher and he has got many patents, research articles and books in this particular area. His research interest involves in FDTD, Soft Computing Applications in Electromagnetics. He is a Member of Editorial Board of Journal of Natural and Physical Sciences and reviewer in many national and international Scopus and SCI journals.

Man Seng Sim was born in Johor, Malaysia, in 1993. He obtained his Bachelor of Science in Applied Physics with Honors from Universiti Tun Hussein Onn Malaysia (UTHM) in 2017. In 2019, he completed his Master of Philosophy in Electrical Engineering from Universiti Teknologi Malaysia (UTM). He is currently a Ph.D. research student in the Faculty of Electrical Engineering at UTM. His research interests include microwave metamaterials, microwave absorbers, microwave sensors, and material characterization.

Bhagwant Singh is presently a research scholar in Shri Ramswaroop Memorial University, Barabanki under the supervision of Dr. Alkesh Agrawal. He is working in the field of Microwave Absorbers.

Devendra Kumar Somwanshi did B.E. in 2007 from Govt. Eng. College, Bikaner and M.E. in Electronics Instrumentation & Control from Thapar University, Patiala in 2009 and submitted his PHD in Artificial Intelligence and Deep learning. He started his teaching career as Assistant Professor, in JNIT, Jaipur (RTU) in 2009. He authored 3 technical books. He joined Poornima Group, as M. Tech. Coordinator and worked in Advanced Studies and Research Center for more than 8 years. During this time he coordinated and supported to design and execute the curriculum for M. Tech and Ph. D. course work. He guided more than 40 M. Tech. research scholars till now. He has expertise in Artificial Intelligence, Data Analysis, Image Processing, Instrumentation and Process Control. He has published more than 80 research papers in National and International conferences and journals including IEEE, ACM, Springer, Elsevier etc. He has also expertise to organize International Conferences as publication chair including IEEE, Springer, and Elsevier conferences. He worked as IEEE Student Branch chair for more than 4 years. Currently he is working as Registrar, Poornima College of Engineering since 2019.

Yinpeng Wang received the B.S. degree in Electronic and Information Engineering and the M.S. degree in Electronic Science and Technology both at Beihang University in 2020 and 2023, respectively. From 2017 to 2018, he was a researcher at the Physical Experiment Center, Beihang University. In 2018, he worked as a research assistant at the Spintronics Interdisciplinary Center. Since 2018, he has been a member of the Institute of EMC Technology. Since 2018, Mr. Wang has published one academic monograph and more than 20 peer-reviewed technical papers in international journals and conferences. He serves as a reviewer for Springer, IOP, Elsevier, and IEEE journals.

Yuhang Wang was an exchange student in Germanistik of Martin-Luther-Universität, Halle, Germany, in 2019 and received the B.S. degree in German from Mianyang Normal University, Mianyang, China, in 2020. He is currently working toward the M.S. degree in Comparative Institutionlism with the School of International Relations of Sichuan International Studies University, Chongqing, China. His research interests include German politics and Sino-European Relations. He completed part of the data acquisition and translation of "Advanced Design of Metamaterial Absorber."

Jingda Wen is now a PhD student at Tsinghua University in 2023. Before joining the PhD program, he received the Master degree in Beihang University in 2022, where he also got a Bachelor degree in 2020. His research interest including metamaterial and metasurface. He has authored more than 8 academic articles including SCI journals such as Applied Physics Letters, IEEE AWPL, and so on. Moreover, he serves as a reviewer and editorial board member for SCI journals. In addition, he recevied more than 15 kinds of awards and scholarships from Beihang University and the Chinese Ministry of Education.

Kok Yeow You was born in 1977. He obtained his B.Sc. Physics (Honours) degree from Universiti Kebangsaan Malaysia (UKM) in 2001. He pursued his M.Sc. in Microwave at the Faculty of Science in 2003 and his Ph.D. in Wave Propagation at the Institute for Mathematical Research in 2006 at Universiti Putra Malaysia (UPM), Serdang, Selangor, Malaysia. He is currently working as a senior lecturer at the Faculty of Electrical Engineering, Faculty of Engineering, Universiti Teknologi Malaysia (UTM), Skudai, Johor, Malaysia. His main research interest lies in the theory, simulation, and instrumentation of electromagnetic wave propagation at microwave frequencies, with a focus on the development of microwave passive devices and sensors for medical and agricultural applications.

Index

V

W

www.ingramcontent.com/pod-product-compliance
Lightning Source LLC
LaVergne TN
LVHW081350300926
849161LV00008B/2120